FIRST ECOLOGY

FIRST ECOLOGY

Alan Beeby and Anne-Maria Brennan
School of Applied Science
University of the South Bank
London, UK

CHAPMAN & HALL
London • Weinheim • New York • Tokyo • Melbourne • Madras

Published by Chapman & Hall, 2–6 Boundary Row, London SE1 8HN, UK

Chapman & Hall, 2–6 Boundary Row, London SE1 8HN, UK

Chapman & Hall GmbH, Pappelallee 3, 69469 Weinheim, Germany

Chapman & Hall USA, 115 Fifth Avenue, New York, NY 10003, USA

Chapman & Hall Japan, ITP-Japan, Kyowa Building, 3F, 2-2-1 Hirakawacho, Chiyoda-ku, Tokyo 102, Japan

Chapman & Hall Australia, 102 Dodds Street, South Melbourne, Victoria 3205, Australia

Chapman & Hall India, R. Seshadri, 32 Second Main Road, CIT East, Madras 600 035, India

First edition 1997

Typeset in 10/11 pt Times by Saxon Graphics Ltd, Derby

Printed in Italy by Vincenzo Bona, Turin

ISBN 0 412 63060 5

A catalogue record for this book is available from the British Library

It is true: Man is the microcosm: I am my world.

Ludwig Wittgenstein

Contents

Guest contributors

Anthony Bradshaw is Emeritus Professor of Botany at the University of Liverpool. He graduated from Cambridge in Natural Sciences and later received a PhD from the University of Wales. He then taught at the University of Wales at Bangor followed by the University of Liverpool, where he became known for his work on land reclamation. He is considered to be one of the pioneers of restoration ecology. A fellow of the Royal Society, he has acted as adviser to many organizations and institutions involved in ecology and conservation, including the Natural Environment Research Council, Nature Conservancy Council, British Ecological Society and the Institute of Ecology and Environmental Management and is currently President of the UK Society for Ecological Restoration. A tireless advocate for the professional recognition of ecologists, he has done much to develop and maintain the professional standards by which ecologists operate.

Gary Haynes is Professor of Anthropology at the University of Nevada at Reno. He took an MA and PhD in Anthropology at the Catholic University of America, specializing in human hunting during the Pleistocene. Professor Haynes's studies on the demise of the mammoths and mastodonts of the ice ages are detailed in his book *Mammoths, Mastodonts and Elephants: Biology, Behaviour and the Fossil Record*, written after extensive study of North American and European collections of remains, including those at the Russian Academy of Sciences in Krakow. His interest in the behaviour and ecology of extant large mammals has meant 13 years study on the free-roaming elephant of Zimbabwe. He co-directs the Hwange Research Trust in Zimbabwe, a non-profit organization dedicated to research into the wildlife of the Park.

Lynn Margulis is Distinguished University Professor in the Department of Biology at the University of Massachusetts. She graduated with an AB in Liberal Arts from the University of Chicago before taking an MS in Genetics and Zoology at the University of Wisconsin and a PhD in Genetics from the University of California, Berkeley. Professor Margulis has held a Sherman Fairchild Fellowship and a Guggenheim Fellowship at the California Institute of Technology and was elected to membership of the National Academy of Sciences in 1983. She chaired the National Academy Space Science Committee that helped NASA formulate research strategies into Planetary Biology and Chemical Evolution and continues to work on NASA's planetary biology programme. Her principal research interests are in cell biology and microbial evolution. Professor Margulis collaborates with James E. Lovelock on the Gaia hypothesis and continues her research on symbiogenesis, 30 years after she originally proposed the theory to explain the origins of the main phyla.

Sir Robert May is a Royal Society Research Professor in the Zoology Department of Oxford University and Imperial College, University of London. He graduated from Sydney University with a BSc and later a PhD in Theoretical Physics and remained there as Senior Lecturer, Reader and finally Professor. He then made the transition from physical to biological sciences first as Professor of Biology and later Zoology at Princeton University. Over the years Professor May has undertaken numerous advisory roles and appointments within universities and research institutes throughout the world; these include the California Institute of Technology, United Kingdom Atomic Energy Authority, World Wide Fund for Nature, British Ecological Society, British Museum (Natural History) and Royal Botanic Gardens, Kew. Most recently he was appointed the UK Government's Chief Scientist and is able to advise decision-makers on the importance of biological diversity and sustainable development in the future of the planet.

Mike Newman is senior research scientist at the University of Georgia, Savannah River Ecology Laboratory. He took a BA (Biological Sciences) and MA (Zoology) at the University of Connecticut before joining Rutgers University to complete an MS and a PhD in Environmental Sciences.

Thereafter he was a postdoctoral fellow at the University of California, San Diego, and joined the University of Georgia in 1983. Dr Newman has research interests in the uptake and effects of pollutants in freshwater ecosystems, and has looked particularly at quantitative methods for ecological risk assessment and toxicology. He was founder and first president of the Carolinas Chapter of the Society of Environmental Toxicology and has written a number of books on ecotoxicology.

First words

Our primary aim has been to produce a first-year undergraduate textbook for the twenty-first century. This has two implications: that the book should draw upon common experience and also that it should take a global perspective.

In the first instance, we have tried to put some of the basic ideas in ecology into a human context. There is a great need to see humankind as part of the ecological processes of the planet, a species constrained by its ecology just as plants and animals are. We have done this in various ways – by opening with a review of our own origins, by approaching topics from a human dimension and by reviewing the ecological implications of our endeavours throughout the book.

The present century has been marked by an increased global awareness amongst the peoples of the planet. It has not simply been the technological, social, economic and political changes over the last decades that have prompted this, but also a growing appreciation of the scale of ecological change. Ecology, the science, now embraces these larger spatial and time-scales and we have sought to show how widely ecological thought now ranges.

Our guest contributors help us to do this. In the Prologue, Mike Newman describes the changed perspective that followed the advent of space exploration, whilst Lynn Margulis sets out the theory that planetary ecology is tightly regulated. Gary Haynes shows how our history is itself intertwined with species extinctions and the insight this gives us on the predicament of protected animals today. Robert May counts the lost species and shows that the scale of our disruption ranks with mass extinctions in the geological past. However, ecologists are not powerless and Tony Bradshaw describes how our attempts to restore degraded habitats represent a useful test of our ecological understanding.

Books are not just written by their authors. A collection of people around them help to make it happen. We wish to thank our respective families for their unstinting love, support and patience as they endured a book in the making. Thanks are also due to the numerous friends and colleagues who cheered and encouraged us along the way. We acknowledge the kind help and cooperation of individuals and institutions who provided material in the form of the written word, images and ideas. We therefore thank Ambio, Martin Angel, Jackie Beeby, Blackwell Scientific Publications, Tony Bradshaw, the Broads Authority, Bob Carling (Chapman & Hall), Cambridge University Press, Luca Cavalli-Sforza, Paddy Coker, D.J. Currie, John Dodd, John Feltwell, French Ministry of Culture, Ken Giller, Elliot Gingold, Lee Hannah and Conservation International, Mike Harding, Gary Haynes, Alan Hopkins, John Hopkins, T. Jaffre, B. Jedrzejewska, Susan Jenks, Joint Nature Conservation Committee, Tim Johns and John Mitchell of the Hadley Centre, Hefin Jones, Nancy Laurenson, Rachel Leech, George Lees, the Linnean Society of London, Joe Lopez-Real, Lynn Margulis, Sir Robert May, Chris Mead, Patrick Morgan, NASA, Natural History Museum, Kate Neale, Mike Newman, Mike Nicholls, Simon Parfitt, Panos Institute, Arthur Penhally, Val Porter, George Potts, Dominic Recaldin (C & H), Roger Reeves, John Rodwell, the Royal Entomological Society of London, Peter Savill, Candy d'Sa, Helen Sharples (C & H), Liz Sestito, Roger Tidman, Martin Tribe (C & H), Will Wadell, Robert Wayne and all the others who helped in any way. Finally, our grateful thanks to Jonathan Mitchley, who read and commented on drafts of the manuscript and whose wit was as invaluable as his wisdom.

Acknowledgements

Whilst every effort has been made to trace the owners of copyright material, we offer our apologies to any copyright holder whose rights we may unwittingly have infringed. We are grateful to all of the following for kind permission to reproduce illustrations, tables and data.

Ardea: Figure 2.2
Broads Authority: Figure 6.10
Paddy Croker (University of Greenwich): Plates 46 and 47
John C. Dodd (International Institute of Biotechnology): Figure 6.6
John Feltwell: Frontispiece, Chapters 3, 9; Plates 3a, 3d, 18, 19, 27, 30, 35, 36, 39, 43, 44, 45; Figures 7.8, 8.9.
French Ministry of Culture: Frontispiece, Chapter 1 (image on Netscape screen)
Ken Giller (Wye College, University of London): Figure 6.8a
Mike Harding: Figure 9.12
Hefin Jones, Centre for Population Biology, Imperial College: Figure 9.3
Nancy Laurenson: Frontispiece, Chapter 4; Plates 10, 11, 40; Figure 3.9
Rachel Leech (University of York): Figure 5.2
George Lees: Plate 21
The Linnean Society of London: Figure 2.1
Joe Lopez-Real (Wye College, University of London): Figure 6.8b
Morgan/A. Pengally: Frontispiece, Chapter 6
NASA: Prologue Figure 1; Plates 2, 22, 48
Natural History Museum: Plate 4
Kate Neale: Plate 29
Michael Newman: Plates 1, 12
Panos Pictures: Plate 37
Candy d'Sa: Plate 42
Liz Sestito: Plate 49; Figure 3.8
Roger Tidman: Plate 3(c)

Other photographs

Plate 23: provided by Jonathan Mitchley, Wye College, University of London; photographed by Mike Griggs for TML (copyright).
Plate 24: provided by Roger Reeves and reproduced by permission of T. Jaffre.
Plate 33: provided by John Mitchell and Tim Johns of the Hadley Centre for Climate Prediction and Research, The Meteorological Office, London Road, Bracknell, Berkshire RG12 2SY, UK. Both figures are Crown copyright and are used with their kind permission. Figure (a) is a derivation of Figure 2a in Mitchell *et al.*, *Nature*, **376**, 501–504, but using different running means; Figure (b) is previously unpublished.
Plate 34: provided by Lee Hannah, Conservation International, 1015 18th Street NW, Suite 1000, Washington DC 20036, USA, from an article in *Ambio* (and with the permission of the publisher), **23**, 246–250 (1994).

Authors' photographs

Alan Beeby: Frontispiece, Chapters 7, 8; Plates 8, 26, 28, 32, 38.
Anne-Maria Brennan: Frontispiece, Chapters 2, 5; Plates 3b, 5, 6, 7, 9, 13, 14, 15, 16, 17, 20, 25, 31, 41; Figures 2.2b, 2.5, 2.6, 2.12, 2.13, 4.5, 4.8, 4.12, 5.10, 7.7.

Line drawings

Prologue Figure 2: Modified from Machita (1973) in *Carbon and the Biosphere, United States*, NTIS, Washington, DC.
Figure 1.8: With kind permission of Professor Lynn Margulis, University of Massachusetts, Amherst, Massachusetts, MA 01003 USA.
Figure 2.11: Based on findings of Schluter, D. (1994) Experimental evidence that competition promotes divergence in adaptive radiation. *Science*, **266**, 798–801.
Figure 2.14: Redrawn from Ammerman, A. and Cavalli-Sforza, L.L. (1971) Measuring the rate of spread of early farming in Europe. *Man*, **6**, 674–688.
Figure 3.6: Data redrawn with kind permission from Jedrzejewska, B., Okarma, H., Jedrzejewska, W. and Milkowski, L. (1994) Effects of exploitation and protection on forest structure, ungulate density and wolf predation in

Bialowieza Primeval Forest, Poland. *Journal of Applied Ecology*, **31**, 664–676.

Figure 3.10: Redrawn from Ashley, M.V., Melnick, D.J. and Western, D. (1990) Conservation genetics of the black rhinoceros (*Diceros bicornis*). 1. Evidence from the mitochondrial DNA of three populations. *Conservation Biology*, **4**, 71–77.

Figure 4.6: Redrawn from Crombie, A.C. (1946) Further experiments on insect competition. *Proceedings of the Royal Society of London, Series B*, **133**, 76–109; and Tilman, D., Mattson, M. and Langer, S. (1981) Competition and nutrient kinetics along a temporal gradient: an experimental test of a mechanistic approach to niche theory. *Limnology and Oceanography*, **26**, 1020–1033.

Figure 4.7: Redrawn from Grime, J.P. (1974) Vegetation classification by reference to strategies. *Nature*, **250**, 26–31.

Figure 4.10: Used with kind permission from Potts, G.R. and Aebischer, N.J. (1989) Control of size in birds: the grey partridge as a case study. in *Toward a More Exact Ecology* (eds P.J. Grubb and J.B. Whittaker), Blackwell, Oxford.

Figure 4.11: Redrawn from LaPage, G. (1963) *Animals Parasitic in Man*, Dover, New York; and Pierkarski, G. (1962) *Medical Parasitology*, Bayer, Leverkusen.

Figure 5.3: Redrawn from Zscheile, F.P. and Comar, C.L. (1941) Influence of preparative procedure on the purity of chlorophyll components as shown by absorption spectra. *Botanical Gazette*, **102**, 463–481.

Figure 5.5: Derived from data from Schultz, E.D. (1970) Der CO_2-Gaswechsel de Buche (*Fagus sylvatica* L.) in Abhangigkeit von den Klimafaktoren in Feiland. *Flora Jena*, **159**, 177–232; Schultz, E.D., Fuchs, M. and Fuchs, M.I. (1977a) Spatial distribution of photosynthetic capacity and performance in a mountain spruce forest in northern Germany. I. Biomass distribution and daily CO_2 uptake in different crown layers. *Oecologia*, **29**, 43–61; and Schultz, E.D., Fuchs, M. and Fuchs, M.I. (1977b) Spatial distribution of photosynthetic capacity and performance in a mountain spruce forest in northern Germany. III. The significance of the evergreen habit. *Oecologia*, **30**, 239–248.

Figures 5.11 and 5.12: Drawn from data from Golley, F.B. (1960) Energy dynamics of an old-field community. *Ecological Monographs*, **30**, 187–200.

Figure 5.13(a,b): Data from Whittaker, R.H. (1975) Communities and Ecosystems (2nd edn), Macmillan, New York.

Figures 5.13(c) and 5.14: Data from Varley, G.C. (1970) The concept of energy flow applied to a woodland community, in *Animal Populations in Relation to their Food Resource* (ed. A. Watson), Blackwell, Oxford.

Figure 5.15: Drawn using data from Andrews, S.M., Johnson, M.S. and Cooke, J.A. (1989) Distribution of trace element pollutants in a contaminated grassland ecosystem established on metalliferous fluorspar tailings. 1: Lead. *Environmental Pollution* **58**, 73–85; and Andrews, S.M., Johnson, M.S. and Cooke, J.A. (1989) Distribution of trace element pollutants in a contaminated grassland ecosystem established on metalliferous fluorspar tailings. 2: Zinc. *Environmental Pollution*, **59**, 241–252.

Figure 5.16: Redrawn with modifications from Duckham, A.N. (1976) Environmental constraints, in *Food Production and Nutrient Cycles* (eds A.N. Duckham, J.G.W. Jones and E.H. Roberts), North Holland Publishing Company, Amsterdam.

Figure 5.18: After Balch, C.C. and Reid, J.T. (1976) The efficiency of conversion of feed energy and protein into animal products, in *Food Production and Nutrient Cycles* (eds A.N. Duckham, J.G.W. Jones and E.H. Roberts), North Holland Publishing Company, Amsterdam.

Figure 5.19: After Tivy, J. (1990) *Agricultural Ecology*, Longman, Harlow.

Figure 5.20: Based on data from Slesser, M. (1975) Energy requirements of agriculture, in *Food, Agriculture and the Environment* (eds J. Lenihan and W.W. Fletcher), Blackie, Glasgow and London; and Tivy, J. (1990) *Agricultural Ecology*, Longman, Harlow.

Figure 5.21: Derived from Smith, D.F. and Hill, D.M. (1975) Natural agricultural ecosystems. *Journal of Environmental Quality*, **4** , 143–145.

Figures 6.2 and 9.8: Redrawn from Ricklefs, R.E. (1990) *Ecology* (3rd edn), Freeman, New York.

Figure 6.5: Redrawn from Odum, E.P. (1989) *Ecology and Our Endangered Life-support System*, Sinauer Associates, Sunderland, Massachusetts.

Figure 6.11: After Bradshaw, A.D. (1984) Ecological principles and land reclamation practice. *Landscape Planning*, **11**, 35–48.

Figure 6.14: From Pan, J.-X. (1591) *A review of river flooding control* (in Chinese).

Figures 7.2 and 7.3: Modified after Trabaud, L. (1981) Man and fire: impacts on mediterranean vegetation, in *Mediterranean-type Shrublands* (eds F. Di Castri *et al.*), Elsevier, Amsterdam.

Figure 7.4: Modified after Di Castri, F. (1981) Mediterranean-type shrublands of the world, in *Mediterranean-type Shrublands* (eds F. Di Castri *et al.*), Elsevier, Amsterdam.

Figure 7.5: Drawn using data from Cody, M.L. and Mooney, H.A. (1978) Convergence versus non-convergence in mediterranean-climate ecosystems. *Annual Review of Ecology and Systematics*, **9**, 265–321.

Figure 7.9: After MacArthur, R.H. and Wilson, E.O. (1967) *The Theory of Island Biogeography*, Princeton University Press, Princeton.

Figure 7.11: Reprinted with kind permission from Rodwell, J.S. (1991) *British Plant Communities. Volume 2: Mires and Heaths*, Cambridge University Press, Cambridge.

Figure 7.12: Reprinted with kind permission from Rodwell, J.S. (1992) *British Plant Communities. Volume 3: Grassland and Montane Communities*, Cambridge University Press, Cambridge.

Figure 8.4: After Whittaker, R.H. (1975) *Communities and Ecosystems* (2nd edn), MacMillan, New York.

Figures 8.13 and 8.15: Redrawn using data from Houghton, J.T., Jenkins, G.J. and Ephraums, J.J. (eds) (1990) *Climate Change: The IPCC Scientific Assesment*, Cambridge University Press, Cambridge.

Figure 8.14: Redrawn using data principally from UK Natural Environment Research Council (1989) *Our Future World: Global Environmental Research.*

Figure 8.16: Data from Whittaker, R.H. and Likens, G.E. (1973) cited by Paul Colinvaux in *Ecology 2*, John Wiley, New York

Figure 9.5: Redrawn from Tilman, D. and Downing, J.A. (1994) Biodiversity and stability in grasslands. *Nature*, **367**, 3633–3635.

Figure 9.6: Used with the kind permission of Dr M.V. Angel of the Southampton Oceanographic Centre, Empress Dock, Southampton SO14 3ZH, UK.

Figure 9.7: Used with the kind permission of Dr D.J. Currie and the Editor of the *American Naturalist*, University of Chicago Press.

Figure 9.9: Used with the kind permission the Editor of the *American Naturalist*, University of Chicago Press.

Figure 9.10: Redrawn from Tuckwell, H.C. and Koziol, J.A. (1992) World population. *Nature*, **359**, 200.

Tables

Prologue Table 1: Compiled from Kelly, K.W. (ed.) (1988) *The Home Planet*, Mir Publishers, Moscow, and Allen, J. (1991) *Biosphere 2. The Human Experiment*, Viking Press, New York.

Prologue Table 2: Modified from Lovelock, J.E. (1987) *Gaia. A New Look at Life on Earth*, Oxford University Press, Oxford.

Table 4.2: Adapted from Harborne, J.B. (1982) *Introduction to Ecological Biochemistry* (2nd edn), Academic Press, London.

Table 5.1: Adapted from Whittaker, R.H. (1975) *Communities and Ecosystems* (2nd edn), Macmillan, New York.

Table 5.2: After Cooper, J.P. (ed.) (1975) *Photosynthesis and Productivity in Different Environments*, Cambridge University Press, Cambridge.

Table 5.3: After Humphreys, W.F. (1979) Production and respiration in animal populations. *Journal of Animal Ecology*, **48**, 427–474.

Table 5.4: After Price, P.W. (1975) *Insect Ecology*, Wiley, New York.

Table 5.5: After Spedding, C.R.W. (1975) *Biology and Agricultural Systems*, Academic Press, London.

Table 5.6: After Slesser, M. (1975) Energy requirements of agriculture, in *Food, Agriculture and the Environment* (eds J. Lenihan and W.W. Fletcher), Blackie, Glasgow and London.

Table 6.3: After Bradshaw, A.D. (1983) The reconstruction of ecosystems. *Journal of Applied Ecology*, **20**, 1–17.

Table 8.1: Data from Houghton, J.T., Jenkins, G.J. and Ephraums J.J. (eds) (1990) *Climate Change: The IPCC Scientific Assessment*, Cambridge University Press, Cambridge.

Prologue

Michael C. Newman
Savannah River Ecology Laboratory, University of Georgia, USA

Figure PR.1 The Earth rising above the lunar horizon as seen by Michael Collins in the *Apollo 11* Command Module. Below Collins, as he took this photograph, Neil Armstrong and Edwin Aldrin were about to leave their Lunar Excursion Module, the *Eagle*, to become the first humans to walk on the Moon.

A NEW PERSPECTIVE

As this prologue is written, we celebrate the 25th anniversary of *Apollo 11* landing on the Moon. Back in 1969 we left the Earth and, for the first time, set foot upon another world. Humanity made a giant leap outward as the lunar excursion module settled on the plains of the *Mare Tranquillitatis*.

Impressive as these technological and scientific achievements were, the Apollo missions brought another, more subtle advance. A profound shift in perspective took place when we looked up collectively from Neil Armstrong's fresh footprint to gaze back at our planet (Figure 1). As Rusty Schweickart said after his spacewalk outside *Apollo 9*:

On that small blue and white planet below is everything that means anything to you; ... all on that little spot in the cosmos. National boundaries and human artifacts no longer seem real. Only the biosphere, whole and home of life.

This new perspective is evident from the comments of many astronauts and cosmonauts (Table 1). For them it was clear that the survival of humanity depended on the limited resources of this 'one spot in the cosmos'.

Table Pr.1 The 'Overview Effect' evident in quotes from cosmonauts and astronauts of many nations and cultural backgrounds

When we look into the sky it seems to us endless ... then you sit aboard a spacecraft, and you tear away from Earth ... the "boundless" blue sky, the ocean which gives us breath and protects us from the endless black and death, is [then seen as] but an infinitesimally thin film. How dangerous it is to threaten even the smallest part of this gossamer covering, this conserver of life. **Vladimir Shatalov (USSR, Soyuz 4, 8, 10: 1969, 1969, 1971)**

On that small blue and white planet below is everything that means anything to you; all history and music and poetry and art, death and birth, love, joy, games ... all on that little spot in the cosmos. National boundaries and human artifacts no longer seem real. Only the biosphere, whole and home of life. **Rusty Schweickhart (USA, Apollo 9: 1969)**

Now I know why I'm here. Not for a closer look at the Moon, but to look back at our home, the Earth. **Alfred Worden (USA, Apollo 15: 1971)**

... only when I saw it from space, in all its ineffable beauty and fragility, did I realize that humankind's most urgent task is to cherish and preserve it for future generations. **Sigmund Jähn (GDR, Soyuz 31: 1978)**

During the eight days I spent in space, I realized that mankind needs height primarily to better know our long-suffering Earth, to see what cannot be seen close up. Not just to love her beauty, but to ensure that we do not bring even the slightest harm to the natural world. **Pham Tuam (Vietnam, Soyuz 37: 1980)**

For the first time in my life I saw the horizon as a curved line. It was accentuated by a thin seam of dark blue light – our atmosphere. Obviously this was not the ocean of air I had been told it was so many times in my life. I was terrified by its fragile appearance. **Ulf Merhold (FRG, Shuttle Columbia 6: 1983)**

A Chinese tale tells of some men sent to harm a young girl who, upon seeing her beauty, become her protectors rather than her violators. That's how I felt seeing the Earth for the first time. I could not help but love and cherish her. **Taylor Wang (China/USA, Shuttle Challenger 7: 1985)**

But while Rusty Schweickart looked down in wonder at the signs of life below, earthbound scientists were beginning to see signs that all was not well in the biosphere. In 1962, Rachel Carson published *Silent Spring*, a book detailing the consequences of widespread pesticide release into the environment. Her pioneering work showed that the chemical control of agricultural pests was having an impact far beyond the fields being sprayed. Elsewhere our wastes incurred costs in both environmental and human health, from toxic metals poisoning water, soils and people, to the tainted air of city smogs. Evidence of a long-term rise in atmospheric carbon dioxide levels led to early speculations about change in the global climate. Our long-held assumption that the Earth and its living biosphere could absorb these insults was challenged. The convenient belief that 'the solution to pollution is dilution' was slowly replaced by a recognition of the banana-skin syndrome: 'what you release into the environment can come back and hurt you'.

As we continue to probe our solar system and build space stations, we also expend considerable effort assessing the health of the biosphere. Our satellites show the scale of the damage we have inflicted. Today we can trace the path of oil slicks, the burning of tropical forests, and take the temperature of the atmosphere. Along with its new perspective, space technology has given us the means to follow the threats to our planet. We are learning more about the biosphere itself, information essential to our efforts to protect these living systems.

BIOSPHERE 2: LIFE IN A BOTTLE

Another research tool was recently forged in the hot deserts of Arizona. Biosphere 2 has been built to study the processes of the biosphere and as a prototype for future space colonies (Plate 1). It consists of a glass and steel enclosure of 1.8 hectares complete with a stainless steel liner below ground. In it are representative plants and animals of five natural communities (tropical rainforest, savanna, desert, marsh and an ocean), comprising more than 3000 species, along with an agricultural area and living quarters for a human crew.

Unlike a spacecraft, Biosphere 2 is designed to regenerate consumables and eliminate the wastes of living systems using the biological, geological and chemical processes that happen on Earth. The Soviet Bios 1-3 project had a similar goal, but

Biosphere 2 mimics the Earth's biosphere at a scale never before attempted. Not only will this provide new insights; it will also test our present understanding of the real biosphere.

Biosphere 2 has just completed its two-year maiden voyage that began when its airlock was sealed behind five American, two British and one Belgian 'biospherians'. Several telling lessons were learnt from this first trip. Despite careful planning, life in Biosphere 2 quickly modified its atmosphere. Levels of carbon dioxide rose soon after closure. All the natural processes for removing the gas were functioning inside the microcosm, but their capacity could not match the rate of production. On Earth a molecule of carbon dioxide will pass through the atmosphere in eight to ten years, whereas in the enclosure it was cycled in just two to four days. This was a consequence of the ratio of carbon in the biomass (the weight of living material) to that in the atmosphere: on Earth it is 1:1, yet in Biosphere 2 it is 9:1. The daily fluctuations in carbon dioxide became extraordinary because of the small volume of atmosphere relative to that of the soil and plants in the microcosm.

Reluctantly, a chemical scrubber system was installed to alleviate the imbalance, incorporating carbon dixoide in calcium carbonate. This played an equivalent role to geological processes on Earth. The trapped carbon could later be released by heating in a furnace, much as carbon dioxide is released from carbonate rocks by volcanic activity.

Oxygen concentrations in Biosphere 2's atmosphere decreased to 14.5% from the normal level of 20.9%. Sixteen months after closure, 14 000 kg of oxygen were pumped into the microcosm to alleviate the symptoms of altitude sickness suffered by several biospherians. Still, oxygen levels continued to fall by 0.25% each month until the experiment ended.

Our ecological and technological acumen could not balance the system over the two years. It cost nearly 20 million dollars per crew member to maintain the 18 000 m^3 of atmosphere for the duration of the experiment. In contrast, the balance of the Earth's atmosphere has been maintained, *gratis*, for many hundreds of millions of years.

EARTH'S BIOSPHERE: COMING ALIVE

Reflecting on these brief sojourns away from the Earth's life support system, it seems remarkable that the biosphere is so efficient at maintaining itself with such apparent ease. How did this come to be and will it remain so forever? We can start to answer both questions by comparing our 'spot in the cosmos' with its lifeless neighbours.

Unlike Venus and Mars, Earth has supported life for almost as long as it has existed. Our planet was formed 4.6 billion years ago and life appeared around 3.85 billion years ago. Since this time the intensity of light from the Sun has increased by 30%; yet, paradoxically, the Earth has maintained a temperature range that fosters life. The temperature of a planet is determined by the solar luminosity, its ability to reflect light (albedo), and also the composition of its atmosphere. The proportions of carbon dioxide and water vapour are particularly important since these gases absorb heat reflected back off the surface, a process called the **greenhouse effect**.

Earth, Venus and Mars began with a similar collection of atmospheric gases. However, our planetary neighbours differ in two crucial respects – their size and their distance from the Sun. The cascade of events which follow from these differences have determined their local conditions and the presence or absence of life.

The Sun's luminosity defines a zone in which life might be expected to survive, where it is neither too hot nor too cold. Although the orbit of Mars is within the habitable zone, it remains cold and lifeless. Mars is smaller than Earth and this has allowed its interior to cool quickly so that the period of venting of carbon dioxide from volcanic activity was much shorter than Earth's. Levels of atmospheric carbon dioxide and water vapour necessary to establish temperatures conducive to life only existed on Mars during its first billion years. Mars has slowly lost its atmospheric carbon dioxide to its surface and to space. The more massive Earth is taking much longer to cool and consequently escaped the fate of Mars.

At the other extreme, Venus has a similar mass to Earth but orbits too close to the Sun, outside the habitable zone. Intense solar luminosity produced surface temperatures that heated its early oceans to such an extent that a runaway greenhouse effect occurred: large amounts of water vapour entered the atmosphere and trapped heat reflected back from the planet's surface. Ocean waters boiled away and were lost to space as temperatures increased beyond those that could support life.

The early success of the Earth on its way to producing life resulted largely from its position within the zone of habitability and its mass. Caught between the too hot Venus and too cold Mars, conditions on Earth were – like Goldilocks' porridge – 'just right'. Its size allowed volcanic activity to persist so carbon locked in carbonate rocks could be recycled back to the atmosphere through volcanic venting. The carbon dioxide in the atmosphere helped to maintain appropriate temperatures through a greenhouse effect. If the Earth had orbited more than 5% closer to the Sun, it would have succumbed to the hot and lifeless fate of Venus. Given the present rate of increase in the Sun's luminosity, and assuming all else were to remain equal, the Earth's ability to support life will continue for another billion years. At that time, it will experience the runaway greenhouse effect of Venus. Of course, this assumes that human activity does not perturb the system in the meantime.

Life appeared shortly after stable conditions developed on the Earth's surface. It arose because of the primordial atmospheric, geochemical and solar conditions. In turn, it quickly came to dominate the cycling of elements on the Earth. Life also grew to fill a key role in the flow of solar energy.

Atmospheric oxygen levels began to rise two billion years ago as photosynthetic organisms evolved to produce molecular oxygen (O_2). Six to eight hundred million years ago, life's signature was clearly present in the atmosphere. Oxygen levels went from less than 1% to the modern level of 20.9%. This amount of oxygen is remarkable because, in the absence of life (Table 2), the two major gases in our atmosphere, molecular oxygen and nitrogen (N_2), could not exist together. They would slowly combine to form nitrate (NO_3). Life has created and sustains the composition of the Earth's atmosphere.

An abundance of molecular oxygen also enabled a layer of ozone (O_3) gas to form in the upper atmosphere. Ozone absorbs lethal ultraviolet radiation arriving from the Sun, and its presence in the atmosphere allowed life to emerge from the protective waters, to cover the land about 450 million years ago.

THE GAIA HYPOTHESIS

Since then, the distribution and cycling of elements between the atmospheric, hydrological, geological and biological components of the Earth have remained remarkably constant. Molecular oxygen levels in the atmosphere have stayed between 15% and 25% and carbon dioxide has remained at extremely low levels, below 0.1% – approximately 30 times lower than those expected for a lifeless Earth. As we have seen, the regulation of carbon dioxide is extremely critical. The chemist, James Lovelock, suggests that at levels above 1% the Earth's atmosphere would rapidly increase in temperature, boiling off the oceans. The Earth would become like Venus.

These stable conditions are maintained despite changes in the particular species making up the biosphere at different times in its history. This constancy also survived major impacts of meteors and asteroids, occurring approximately once every 70 million years. Each collision releases 10 000 times the energy of the entire nuclear arsenal on Earth, creating a fireball that would ignite fires over much of the planet. One collision 66 million years ago (the Cretaceous–Tertiary boundary) is estimated to have eliminated 70% of all the species, including the dinosaurs. Poisonous gases and dusts released into the atmosphere would have choked both aquatic and terrestrial ecosystems, blocked out sunlight, and lowered temperatures and photosynthesis for months.

Table Pr.2 The atmospheric composition of Earth, Venus and Mars compared with Earth before life appeared (3.8 billion years ago) and a hypothetical Earth without life

Atmospheric quality	*Venus*	*Mars*	*Earth*		
			Lifeless	*Pre-life*	*With life*
Carbon dioxide (CO_2)	98%	95%	98%	50%	0.03%
Nitrogen (N_2)	1.9%	2.7%	1.9%	50%	78%
Oxygen (O_2)	trace	0.13%	trace	trace	20.9%
Temperature (°C)	477	–53	290	57	13
Pressure (bars[a])	90	0.0064	60	2	1

[a] One Bar is Earth's present atmospheric pressure at sea level.

Given such cataclysmic events in the past, how is it that conditions on Earth have apparently remained so stable? James Lovelock and microbiologist Lynn Margulis argue that this consistency is evidence of some self-regulating or 'cybernetic' control of the biosphere. They suggest that the biosphere acts as a homeostatic system, one that can resist large-scale changes which otherwise would destroy it. This idea is called the **Gaia hypothesis** after the Greek Earth goddess named Gaia or Ge. Lovelock and Margulis suggest that the temperature and composition of the Earth's atmosphere, waters and solid surface are regulated by life (Plate 2). This regulation resists change and maintains favourable conditions in the biosphere through the key role that living systems play in the cycling of the major elements.

TESTING GAIA'S PATIENCE

Humanity has now emerged as another major factor in material cycling and energy flow in the biosphere. The widespread burning of fossil fuels, deforestation and other human activities have increased atmospheric concentrations of carbon dioxide (Figure 2). It may be that we face similar problems to the biospherians in Biosphere 2.

Other greenhouse gases have risen with human activity – including methane from intensive agriculture and chlorofluorocarbons from refrigerants and propellants. A rise in average atmospheric temperatures will increase the water vapour in the atmosphere, itself a greenhouse gas, leading to further warming. The thermal balance of the atmosphere appears to have been modified and that global warming is now underway.

In addition, the protective ozone layer that allowed life to emerge on land is also being affected by our wastes. This layer is maintained by the balance between the destruction and production of ozone from molecular oxygen in the stratosphere. An estimated 350 million kilograms of ozone are generated and destroyed each day. Yet the balance has now shifted to a more rapid destruction of ozone. Chlorofluorocarbons, first manufactured in the 1930s and accumulating in the stratosphere ever since, are tipping this balance. The thinning of the layer has raised questions about increased incidence

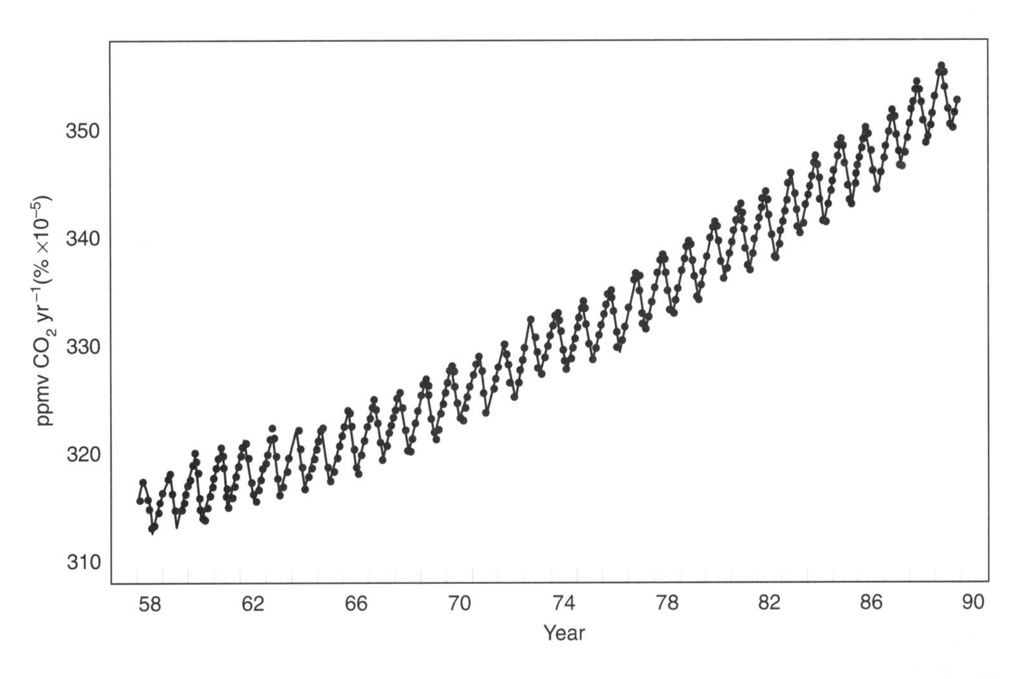

Figure PR.2 Recent increases in atmospheric carbon dioxide attributed to human activities measured at the Muana Loa Observatory (Hawai'i) and the South Pole. Annual oscillations reflect seasonal changes in photosynthesis and respiration. The seasonal changes are much shallower in the less productive polar region of the Earth than toward the Equator. The steady increase in atmospheric concentrations at Muana Loa Observatory shown here has continued its slow rise. By 1990, levels increased beyond 0.0355%.

Box PR.1

TESTING GAIA

Professor Lynn Margulis
University of Massachusetts – Amherst

The Gaia hypothesis states that the lower atmosphere and surface of the Earth, together with the biota (the sum of all organisms) form a physiological system: they are regulated with respect to their chemical composition, their acidity and alkalinity (pH), and their temperature. This environmental regulation is affected by many interacting processes to achieve an extraordinary, constantly changing balance. The most striking evidence for gaian environmental control comes from the persistence of the reactive gas oxygen (O_2) in the presence of many other gases which would tend to remove it: nitrogen (which would make nitrogen oxides, and nitrate in the presence of water), methane (which would form carbon dioxide), ammonia (which should form nitrogen or nitrogen oxides) and some 40 other gases of biological origin. Other evidence for planetary-wide gaian control comes from the observation that the mean mid-latitude temperature of the Earth hovers between 18 and 22°C; that is, around room temperature. Even in the midst of the last ice age, the mean temperature did not fall more than eight degrees below its present value. Furthermore, while the atmospheres of the neigbouring planets, Mars and Venus, are carbon dioxide-rich (more than 95% carbon dioxide), the Earth's atmosphere is depleted in CO_2 (less than 0.04%). Mars and Venus are acidic whereas as the Earth is alkaline: the pH of the oceans is 8.1 to 8.2 pH units. The Gaia hypothesis states that our planetary anomalies are due to a biological control system and are entirely dependent on the persistence of life on the planet.

Although the details of gaian regulation are under investigation by researchers, the broad outlines of the control systems are clear. First, changes in temperature, chemical composition including oxidation state, alkalinity, etc. are sensed. The sensors include heat and light (eyespots and eyes), vibration (touch sensors), acidity (taste and smell receptors) and all other sensors of the biota (plants, animals and the microbiota: fungi, protoctists and bacteria). The sensors communicate with the amplifiers in the entire system: the main amplifier is the tendency toward exponential growth of all populations of organisms of the estimated 30 million species that inhabit the soils, waters and air of the Earth. Both negative and positive feedback systems accelerate and inhibit this amplifier. Another name for the negative feedback system of Gaia is 'Darwinian natural selection', the inhibition of population growth coupled to the metabolic activity of some organisms at the expense of others. Biodiversity is utterly required for the gaian system to persist.

The recognition that the Earth's surface is not simply run by chemistry and physics but that it is a physiological system is attributable to Dr James E. Lovelock, an English chemist who worked extensively with the international space programme. Dr Lovelock, an inventor, gas chromatographer and student of atmospheric chemistry, is a Fellow of the Royal Society of London. He developed the very sensitive electron capture detector that permitted exceedingly tiny quantities of chlorinated and fluorinated gases to be measured in the atmosphere at very great distances from where they had been released. His studies on board ships and in his home-rigged laboratory in the English countryside, coupled with much reading of the scientific literature and consultations with colleagues, led him to frame the Gaia hypothesis in the 1970s. Now his mathematical models and collaborative research efforts permit us to test the hypothesis – promoting it to the status of a theory. The theory, dubbed Gaia by novelist William Golding (author of *Lord of the Flies*), was named after the terrifying but productive ancient Greek goddess of the Earth: Gaia (or Ge, or Gaea). Gaia has the same root as 'geo' which we use in the words geology, geometry, geography, Geos (the satellite) and geosynchronous.

of human skin cancers and damage to other species. Scientists were shocked to find large holes in or thinning of the ozone layer in the early 1980s, first over the Antarctic and then the Arctic regions. Unfortunately chlorofluorocarbons will remain a long time in the stratosphere, and their effects will linger for many years after their emissions are brought under control.

Whether these and our other impacts can be compensated by the cybernetic control of the biosphere is a major question facing humankind today. Unlike astronauts and biospherians, we have nowhere to go if our knowledge and resources are inadequate to answer this question.

TAKING RESPONSIBILITY FOR THE BIOSPHERE

Within Biosphere 2, the biospherians quickly had to adjust their operations to ensure their survival. They compromised by applying technological remedies to adjust carbon dioxide and oxygen to life-sustaining levels. All of the biospherians lost weight during their two-year journey. They spent several hours each day coaxing food from the small agricultural unit of their microcosm. Much time was also spent fighting agricultural pests, without the aid of pesticides that would have quickly poisoned their small world. Sacrifices were made to leave the regenerative capacities of Biosphere 2 intact.

As the number of humans increases on Earth, our impact on the material cycling and energy flow grows to unprecedented levels. We are still ignorant of the processes necessary to stay the course – sustaining life. Only now are we learning how the planet functions, perhaps as its capacity to maintain itself is finally being overwhelmed by our activity. Now the responsibility for restoring the system must be shared by us.

The stakes involved in this human endeavour are high. If our understanding of the biosphere grows fast enough, we will enhance our quality of life and, perhaps, open the way to colonization of other planets. If our understanding fails, we could cheat ourselves of a billion more years of life and prematurely join Venus as a hot, lifeless planet.

But how do we take responsibility for something we do not understand? The most effective means of understanding our living world has been through scientific analysis. Science builds upon our collective knowledge using methods of objective observation and by testing our ideas through experimentation. Its goal is to understand the world by organizing and classifying knowledge based on explanatory principles. This book uses this approach to show how environmental systems are organized, and how we can better understand the one biosphere we all share. The science that attempts to explain how organisms interact with their environment is called ecology.

FURTHER READING

Allen, J. (1991) *Biosphere 2. The human experiment*, Viking Press, New York.

Kasting, J.F., Toon, O.B. and Pollack, J.B. (1988) How the climate evolved on the terrestrial planets. *Scientific American* **258**(2): 90–97.

Kelly, K.W. (ed.) (1988) *The Home Planet*, Mir Publishers, Moscow.

Lovelock, J.E. (1987) *Gaia. A new look at life on Earth*, Oxford University Press, Oxford.

Sagan, D. (1990) *Biospheres. Metamorphosis of Planet Earth*, McGraw-Hill Publishing Company, New York.

White, F. (1987) *The Overview Effect, Space Exploration and Human Evolution*, Houghton Mifflin, Boston.

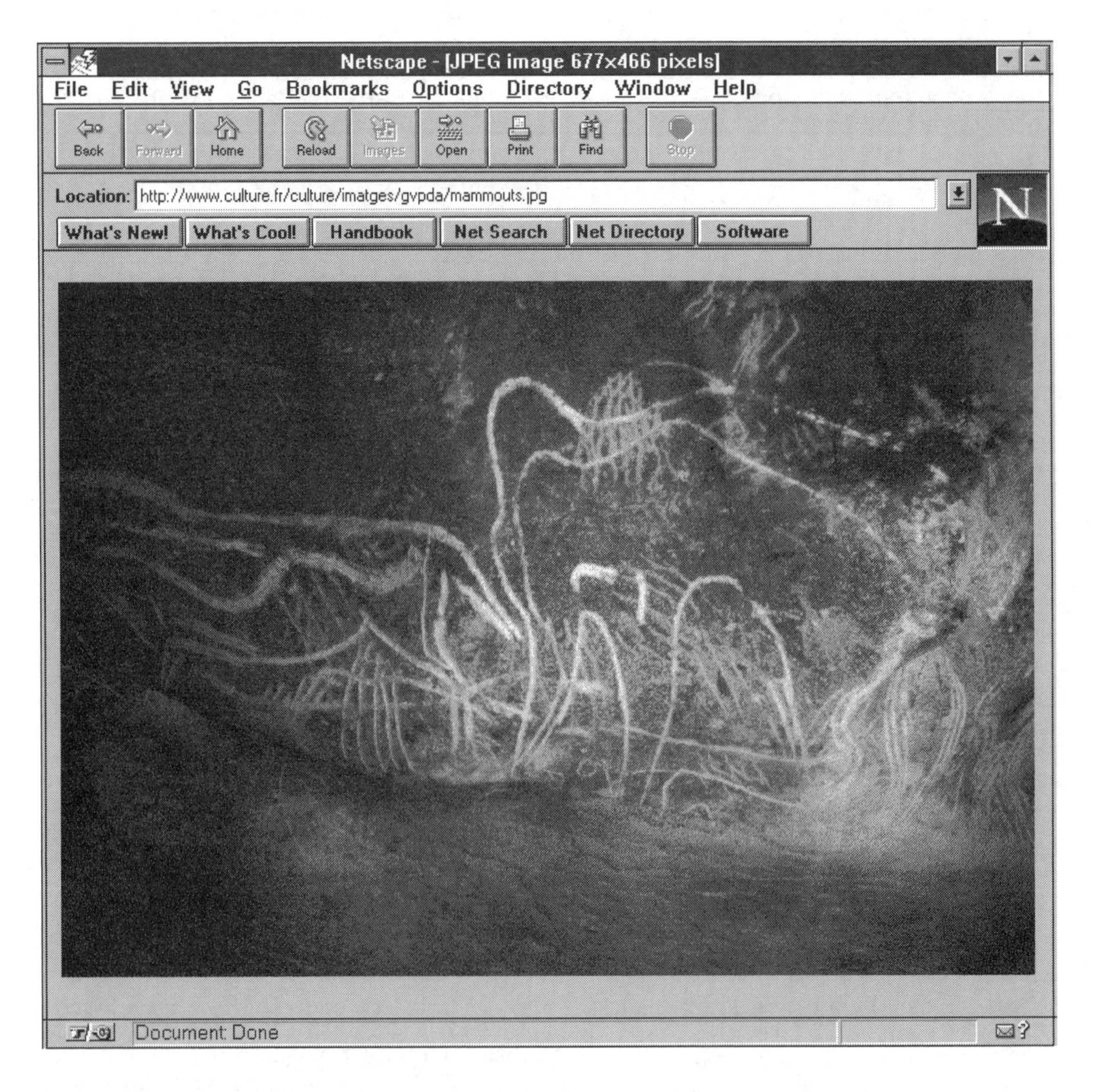

The evolution of humans – evolution by natural selection – adaptation – the selective advantages of sexual reproduction – variation and the genetic code – chemical evolution and the origins of life – the evolution of the major groups of living organisms.

1 Origins

We are products of our environment. Like that of all living creatures, our biology has been shaped by the world in which we live. Human beings are terrestrial animals adapted to warm climates. But we are conscious, thinking warm animals and we talk, to ourselves and to each other. Being a gregarious species we form social groups and through our language inherit cultures of collected wisdom and beliefs.

Human societies define themselves by their patterns of social behaviour, language and customs. Their values are passed from one generation to the next; some ideas change in the process, but others are preserved because their usefulness is undiminished by time. Traditions and conventions are often symbolic, retained because they give a society its identity, form bonds amongst its members and elicit their allegiance.

As individuals, we are shaped by the cultures into which we are born. Our values and ethics are moulded by our teachers and peers, and by our forefathers. For many of us, social pressure is the most important selective pressure and our social organization is the key factor in much human ecology.

The science of **ecology** attempts to explain the relationship between living organisms and their surroundings – where they live, what they live on, how their numbers increase, how they interact with each other. As a science it is a little over a hundred years old, but some of its ideas are found in the earliest writings and embedded in many traditions. Today its terminology is part of a new global vocabulary.

Ecology developed at the same time as the most fundamental of all biological theories, the theory of evolution by natural selection. This is now widely accepted as the most plausible explanation of how different species have arisen. It is primarily an ecological theory, attributing the diversity of life to the evolution of adaptations for survival in a changing world.

The human species is a product of this evolutionary process. However, human beings distinguish themselves from other animals by their enjoyment in talking about themselves and so we start our exploration of ecology by looking at our own origins. Later we look in detail at the theory of evolution and the mechanism of natural selection. We then go on to look at the very origins of life on Earth.

1.1 ORIGINS OF HUMANITY

The human body is very obviously adapted to its physical environment, for life in relatively warm terrestrial habitats where fresh water is available. We may walk upon the moon or dive down into ice-cold seas, but this is not due to any great flexibility in human physiology. Instead it is our mental agility, expressed as technology, that enables us to survive these harsh environments. As a species, our greatest asset is the mind, with its creativity, ingenuity and problem-solving ability.

Compared with our ape-like ancestors, humans have a very large brain (though not the largest). The dramatic increase in brain size in later fossil skulls suggest that greater mental skills were the key to the evolution of the human species (Figure 1.1). But why should this ability become so important in this particular group of animals?

The earliest hominid (human-like) fossils are found in eastern and southern Africa in deposits

Box 1.1

A FEW DEFINITIONS

The formidable collection of names in biology can be a major barrier to understanding. A glossary is provided at the end of the book, but a few important definitions are needed to help clarify your study from the outset.

A **species** is any group of individuals that can actually or potentially breed with each other to produce viable and fertile offspring. Because there are degrees of genetic difference between all individuals, the demarcation of where one species ends and another begins can be very difficult to draw. **Hybrids** derived from two different (if closely related) species are possible, but these are often infertile and therefore a genetic dead-end. This definition of a species is fine for sexually reproducing organisms, but becomes problematical in others. We look at the concept of the species in greater detail in Chapter 2.

Within a species there may be several **populations** – individuals of a single species occupying a particular location at a particular time. Notice that both place and time need to be defined carefully (Chapter 3).

Populations of different species are collected together into **communities** (Chapters 4–7). These interact, feeding on each other, competing for resources or cooperating through special relationships. Various types of community consisting of regular associations of plant species – **biomes** – are found in different parts of the globe (Chapter 8).

A community together with its physical environment is an **ecosystem**. Very often physical features, such as light, moisture and so on, define the sort of community which can be supported. Within the system, nutrients and energy move between the living ('**biotic**') components and the non-living environment (its '**abiotic**' components) (Chapters 5 and 6).

We have just described a hierarchy moving from the species to the ecosystem. This is the sequence we use in this book to explore the science of ecology.

dating from about five million years ago, a period of considerable environmental change. The climate had become cooler and drier, causing a retreat of the forests. Our early ancestors were apes probably forced to forage in the open grasslands. Here an ability to walk upright, freeing the hands to carry food and infants, would have been a distinct advantage. Around three and a half million years ago these early hominids had a brain the size of a chimpanzee's, but stood upright (Figure 1.2).

Walking on two feet (**bipedalism**) marks the start of the evolutionary path that gave rise to humans. Some suggest that holding the body upright was a necessary adaptation to life in the open savanna, since it reduced the area directly exposed to the sun and increased the body's capacity to lose heat. Similarly, an increase in brain size may have followed when the cooling system for the brain, the blood vessels draining it, became more efficient. Certainly the fossil record suggests that bipedalism pre-dated any increase in brain size by over a million years.

In fact these fossils indicate that the earliest hominids had not entirely relinquished the protection of the trees. Features of their anatomy show they retained adaptations for climbing, perhaps returning to the trees for shelter and protection. Nevertheless, being able to stand up and look around would have been useful in deep grass harbouring large predators and wily serpents. It may have helped them to find food more easily. We know from their teeth that these animals were herbivores, but later fossils indicate a shift to flesh-eating. Whether these hominids were primarily hunters or scavengers remains a matter of some debate, though like many other predators they prob-

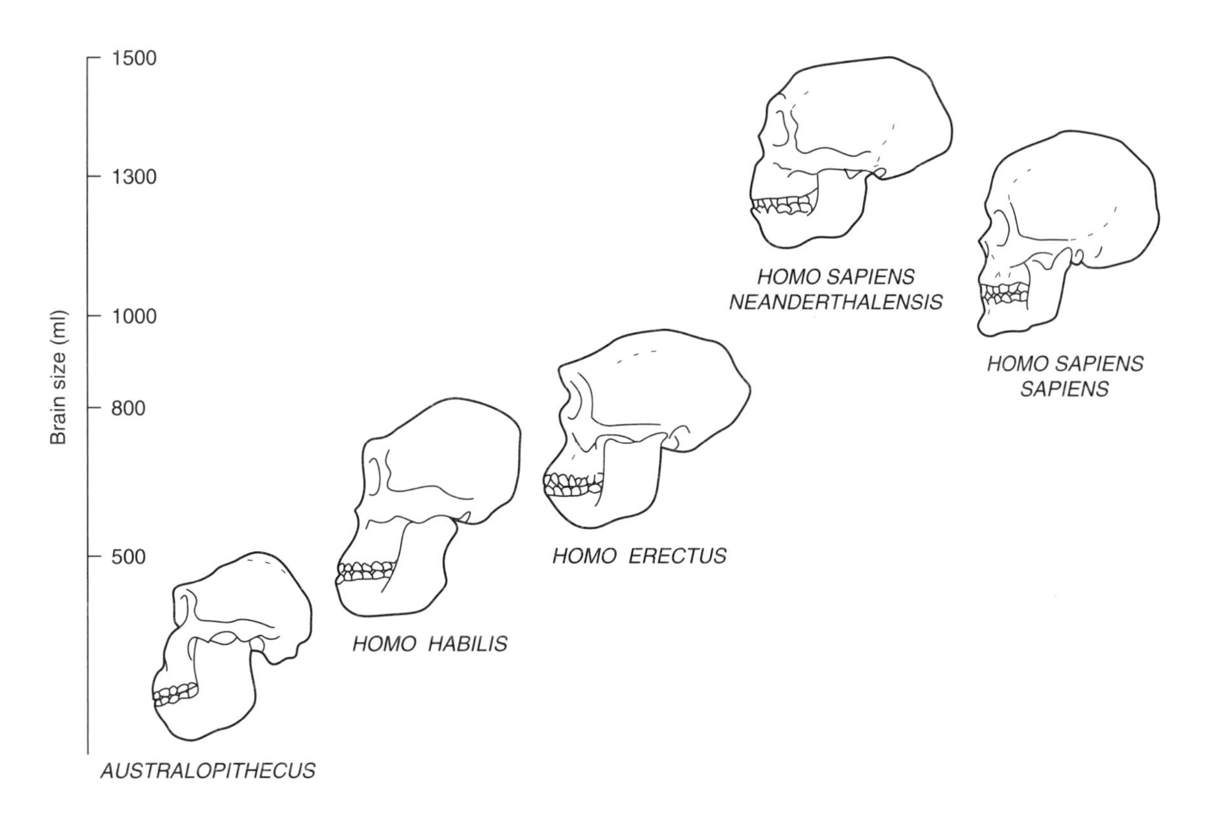

Figure 1.1 Changes in skull shape and the increase in brain size in the hominids. The finer jaws and reduced teeth are possible because of the use of tools to aid feeding. Much of the increase in brain size is in areas responsible for higher mental processes (thinking and memory) and speech.

ably consumed the remains of kills by other species. Perhaps the first tools, simple stone clubs, were used to break open these bones for their marrow.

But technology moves on. Hominid evolution goes hand in hand with the evolution of their tools: over hundreds of thousands of years developments from simple cutting edges through hand-axes to arrow heads mark the advent of different human species in the fossil record (Figure 1.2). Along with the switch to hunting and gathering, new technologies opened up new habitats: fire, clothing and the building of shelters provided protection and warmth, allowing early humans to expand northwards into less hospitable environments.

Our divergence from the other primates may have begun with our tool-making skills. One suggestion is that the coordination needed between hand and eye promoted brain development. So too would the forethought required to envisage an end-use for the tool. Equally important, perhaps, was the effective communication needed in hunting. Cooperating to hunt, or to ward off predators, would have been essential for these small, not very fast apes.

The social cohesion of the group and the induction of the young into its traditions and skills benefit from a precise language. Using words and putting them together in a meaningful way requires conventions accepted by all. These rules impose a discipline on those trying to make themselves understood. With a developed syntax and vocabulary, teachers can convey complex and abstract ideas. Tool-making and other skills can then be learnt by instruction, not simply by copying the behaviour of others.

Individuals can rehearse and structure their thoughts using these same rules. Our verbal messages are usually what we wish the listener to hear and we anticipate the reaction our words might provoke. However, like many other species, we also

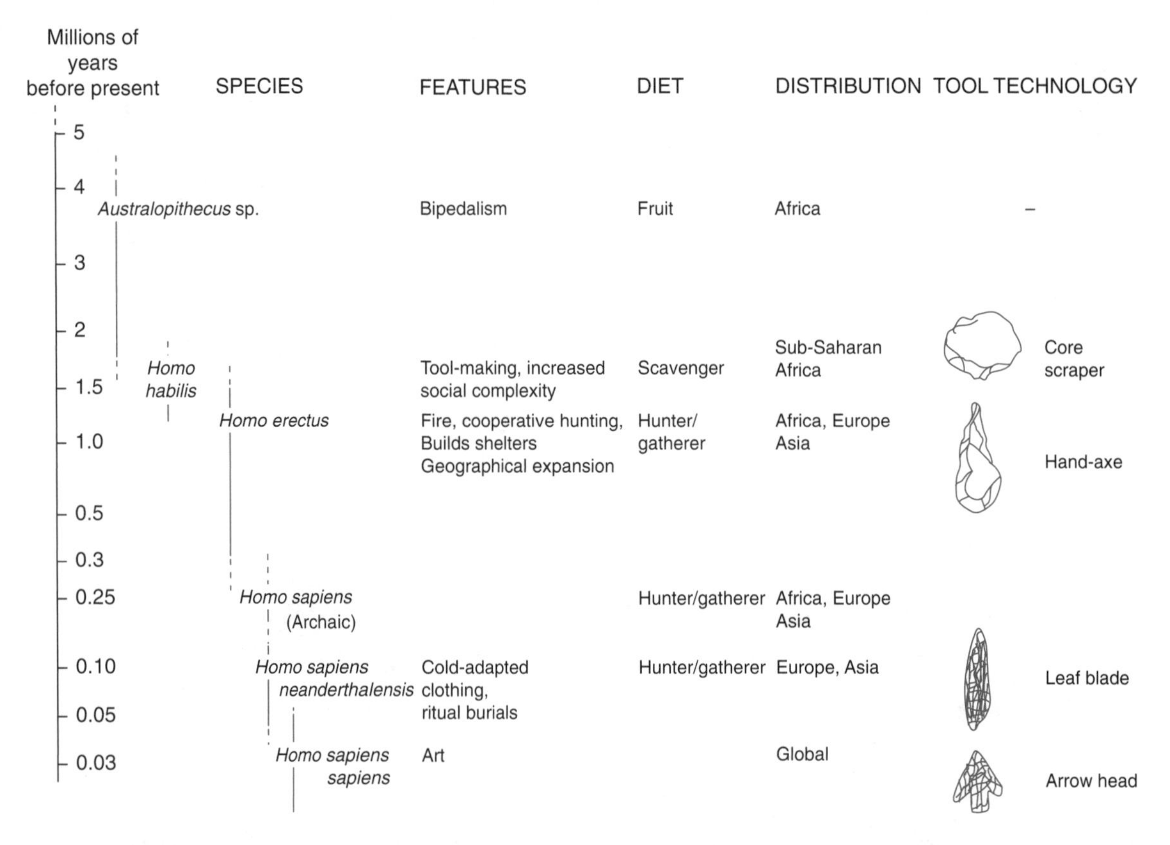

Figure 1.2 The origin of human beings, showing our immediate ancestors, which are distinct from the other great apes (chimpanzees and gorillas). The hominids began with *Australopithecus* and its capacity to walk upright, and whose fossils are found at a time when the forests in East Africa were shrinking. The next major development was a change of diet and way of life, from fruit-eating to being a scavenger/carnivore. Tools used to obtain food gradually improved and became more finely fashioned.

Whether *Homo habilis* or *Homo erectus* was on a direct line leading to modern humans is still debated. Equally, there is controversy about whether *Homo sapiens sapiens* arose from a single stock of archaic *Homo sapiens* in Africa, or developed simultaneously in several locations in Africa, Europe and Asia.

send messages to deceive. While the obligations to share food and to cooperate hold a social group together, there are circumstances (say, in competition for a mate) when the advantage is with those who think ahead, who predict outcomes and make trade-offs. Social intrigues and duplicities are known in groups of chimpanzees and some anthropologists have suggested that our mental evolution was spurred by having to second-guess social interactions.

Language demands memory and a capacity for imagination and abstract thought, and its increasing importance is recorded in the later fossils. In these skulls we find changes in areas of the brain associated with the machinery to process language. Also, the anatomy of the mouth and the position of the larynx, or voice-box, changed in these hominids. A larger space developed above the larynx, allowing a wider range of sounds to be produced. Unfortunately it also increases the chance of choking on food. It seems that articulating speech must have been advantageous indeed to risk this hazard.

You can see the significance of language and mental skills from looking at other aspects of our biology. Compared with other primates, we are born at an early stage of development, before the head grows too large to pass through the mother's pelvic girdle. Childhood itself is extended, an important period of induction into the experience and knowledge held collectively by our cultures.

Along with mother's milk we pass on the tricks of survival, feeding both the body and the mind.

The complexity of our language, the relative precision of its terms and its structure, puts the greatest distance between ourselves and the apes. Other animals are able to walk upright, some use tools, and most communicate with each other. Some, like chimpanzees, do all these things and live in highly integrated and duplicitous social groups. But it is our spoken language, and the culture it creates, that distinguishes the hominids. So central is language to human evolution that some linguists suggest it is less of an acquired skill and more of an inherited instinct (Box 9.4).

Their success allowed early humans to escape from their cradle in the East African Rift Valley, moving through the Middle East into Europe and Asia. Surviving in these different environments would have demanded adaptability and ingenuity, and the ice ages which followed two million years ago would have further tested human resourcefulness.

Yet the hominids continued to spread. Modern humans appeared around 100 000 years ago in East Africa, and again eventually moved out through the Middle East. In step with the glaciations, they migrated northwards to join their hominid cousins, the Neanderthals, in Europe and Asia. Neanderthals had been in the north before the evolution of modern humans (Figure 1.2) and had the anatomy of a cold-adapted people, with large chunky bodies. They also had larger brains.

For a period *Homo sapiens sapiens*, modern humans, lived side by side with the Neanderthals. Perhaps it was competition for resources between the two subspecies that led to the demise of the Neanderthals. Or perhaps there was a more direct conflict between them. Not for the first time, more than one type of hominid were alive together. Today we can only imagine what it would have been like to look into Neanderthal eyes, perhaps to speak with someone with a perception quite unlike any other human being.

Hominids appeared because of a series of environmental changes, to which their biology adapted. But this animal also adapted by using its mental abilities, an adaptation which is much quicker than any physiological or genetic change. In many respects, the changes in our physiology and anatomy are relatively minor, and we still share much in common with other great apes. It is the mind, and not the body, which makes the human species successful.

We have been describing how one species has developed from another, the sequence of changes that describe human evolutionary history. Compared with many other species, we have a relatively complete fossil record of our recent ancestry. We continue to search for remains but now we can also look inside our own cells at the history written in our genes. It is changes in this code, selected over the millennia, that have produced modern human beings. We now need to describe the mechanism by which changes at this molecular level have led to humanity.

1.2 EVOLUTION BY NATURAL SELECTION

The **theory of evolution by natural selection** proposes that species change over generations because of the selection of inherited characteristics. A trait which enables some individuals to reproduce more prolifically will eventually come to dominate a population. The success of one trait relative to another depends upon the selective pressures in the environment. We have suggested that the appearance of hominids was the result of a change in their habitat, when inherited characteristics, such as bipedalism, were better fitted to life in open grassland.

Over many generations, the accumulation of a series of small, gradual changes may lead to a population becoming distinct from its neighbours. When these differences prevent successful matings between populations a new species has arisen (section 2.5). We can explain this process most easily with an example.

Imagine a field of annual plants – that is, plants which die at the end of the year, but leave behind seeds to germinate the following spring. The seeds are viable for one year only, so each generation is derived solely from the parents of the previous year. When space or nutrients are scarce seedlings compete for available resources, and those best able to survive contribute the most seed to the next generation. Others fail to reproduce because they lose these competitive battles.

Now although they belong to the same species we can see differences between the plants. Some are larger, some are dull in colour and some produce more flowers. One reason for their variation lies in the soil. Perhaps the larger plants are found where nitrogen is abundant. Then all individuals would be larger if they had same supply of nitrogen.

Box 1.2

STORING THE INFORMATION TO MAKE A LIVING ORGANISM

Once upon a time, not so very long ago, the idea that we could write out the recipe for a living organism was the stuff of science fiction stories. Now scientists all over the world are currently reading the entire code in our chromosomes, as part of the human genome project. This set of instructions will not allow us to construct a living creature, but it will give us important information about human biology and disease.

A chromosome is a long strand of protein and **DNA** (deoxyribose nucleic acid) held in the **nucleus** of a cell. All higher organisms have several **chromosomes** held within a nuclear membrane, though bacteria and their relatives have a single chromosome not confined to a discrete nucleus. Most higher plants and animals are termed **diploid** because they have two sets of chromosomes ($2n$). Human beings have 23 pairs of chromosomes (n = 23), one of each pair derived from either parent.

The information is actually encoded by a series of **genes**, built up from subunits of the DNA called **nucleotides** (Figure 1.3). There are four chemical forms of nucleotides, and their sequence along the chromosome encodes the instructions. This is rather like a morse code message, except it has four characters rather than two. As with any language, much of the sense of the meaning is derived from the position of each unit of information.

The position of a gene on a chromosome is called its **locus**. A gene is 'read' in a process that leads to the production of a **polypeptide**, a chain of many amino acids joined together (Figure 1.4). Either on its own or in combination with other polypeptides, this can be folded to produce the three-dimensional structure of a finished protein. The activity of the protein is determined by both its chemical behaviour and its structure, according to the folding of the polypeptide chain.

Paired chromosomes carry instructions for the same functions and are termed **homologous**. Each gene from the father is matched by a gene from the mother. When the code is the same on both chromosomes, the individual is described as **homozygous** for that character; otherwise it is **heterozygous**. Alternative forms of the same gene are called **alleles**. Within a population there may be many different alleles for a character. The totality of genes in a population is termed its **gene pool**.

A key distinction is drawn between the genetic information carried by an individual, its **genotype**, and the traits it shows, its **phenotype**. The two will differ because some of the genetic information is not expressed in the phenotype. For example, a child with brown eyes may actually have the code for both brown and blue eyes, but the action of one gene masks any activity by the other. The brown gene is dominant over its allele. We cannot tell from looking at the phenotype (the brown eyes) whether the child is heterozygous or homozygous. Heterozygotes in the population thus carry some hidden genetic variation with the two alleles at this locus.

This is a physiological response to the nutrient and such differences are termed **phenotypic variation**.

On the other hand, some plants produce more flowers than their neighbours, irrespective of soil quality. We shall say that this is a trait they have inherited. This difference between the plants reflects information stored in their genes and is known as **genotypic variation**.

Thus there are two possible sources of variation between individuals: genetic differences inherited from the parents and phenotypic differences which are not. Phenotypic characteristics may be either reversible, such as wilting at times of water shortage, or irreversible. Irreversible differences are fixed during the development of the organism. Once plants have grown tall on the nitrogen-rich soil they do not shrink again when the nitrogen supply is depleted.

Our main concern here is with genetic differences. The instructions for building a new plant are stored in the **chromosomes** of the fertilized egg, in the code written into the chromosomal deoxyribose nucleic acid – **DNA**, a long molecule whose chemical structure carries the information of the genes (Figures 1.3 and 1.4).

Let us say that having multiple flowers gives a better chance of producing seed in very wet years. Perhaps the insects that pollinate the flowers only fly when there is no rain. We assume that these flowers are unable to fertilize themselves. A flower lasts a short time in wet weather, and few are pollinated before they deteriorate. A plant with a series of blooms is available for pollination for longer and is more likely to produce seed.

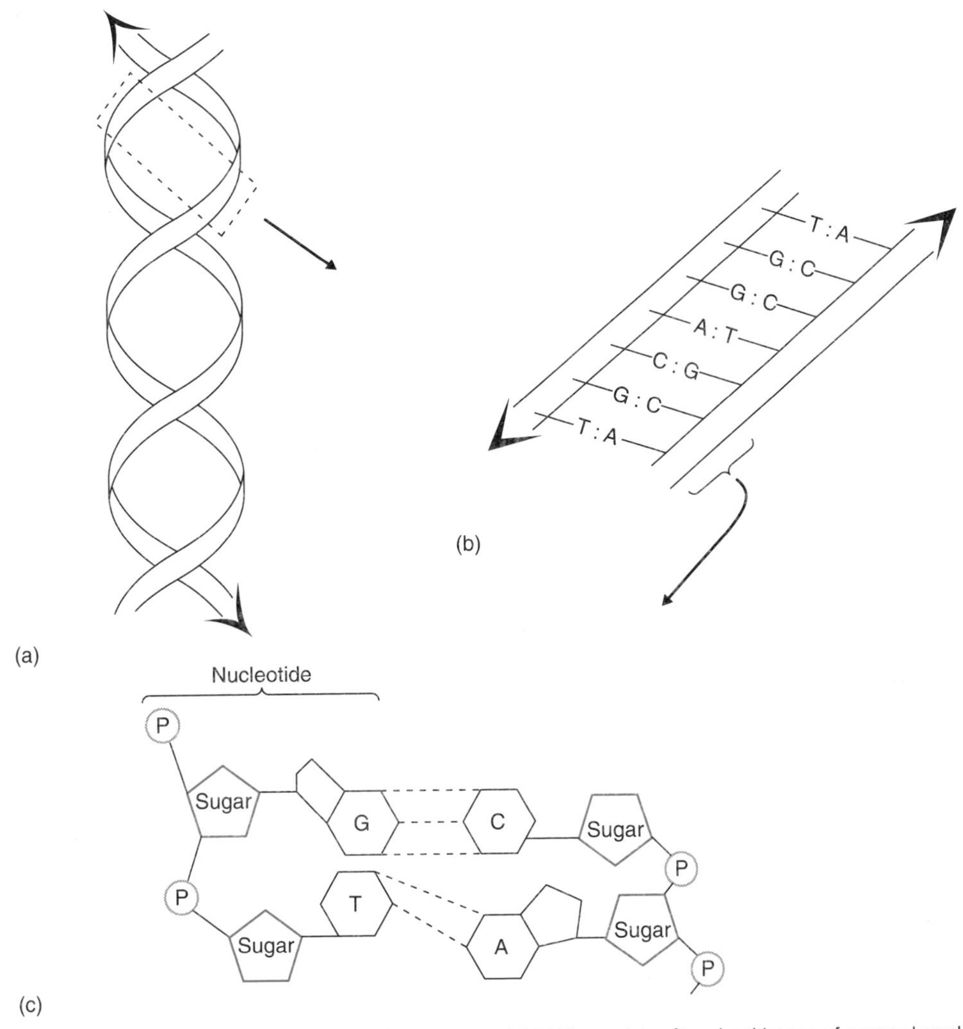

Figure 1.3 The storage of the genetic code in the chromosome. (a) DNA consists of two backbones of sugar-phosphate groups, coiling around each other in a double helix. (b) The information is actually stored in the sequence of bases (A, G, T, C) that bind the two strands together. The direction, sequence and position of these bases give the code its meaning, and is read in blocks of three bases. The bases come in two forms: purines (**adenine**, A, and **guanine**, G) and pyrimidines (**thymine**, T, and **cytosine**, C), distinguished by their chemical structure. Notice that only two combinations of bonds will form – A with T, and G with C. (c) This means that one strand of the DNA is a complement or 'mirror-image' of the other. The code is read by the two strands separating and the sequence of bases being read on the strand carrying the code being used.

The vast amount of information needed is packed into such a small space by winding the double helix around protein bundles, and then packing these 'beads' or nucleosomes into a sequence of loops that eventually comprise one chromosome.

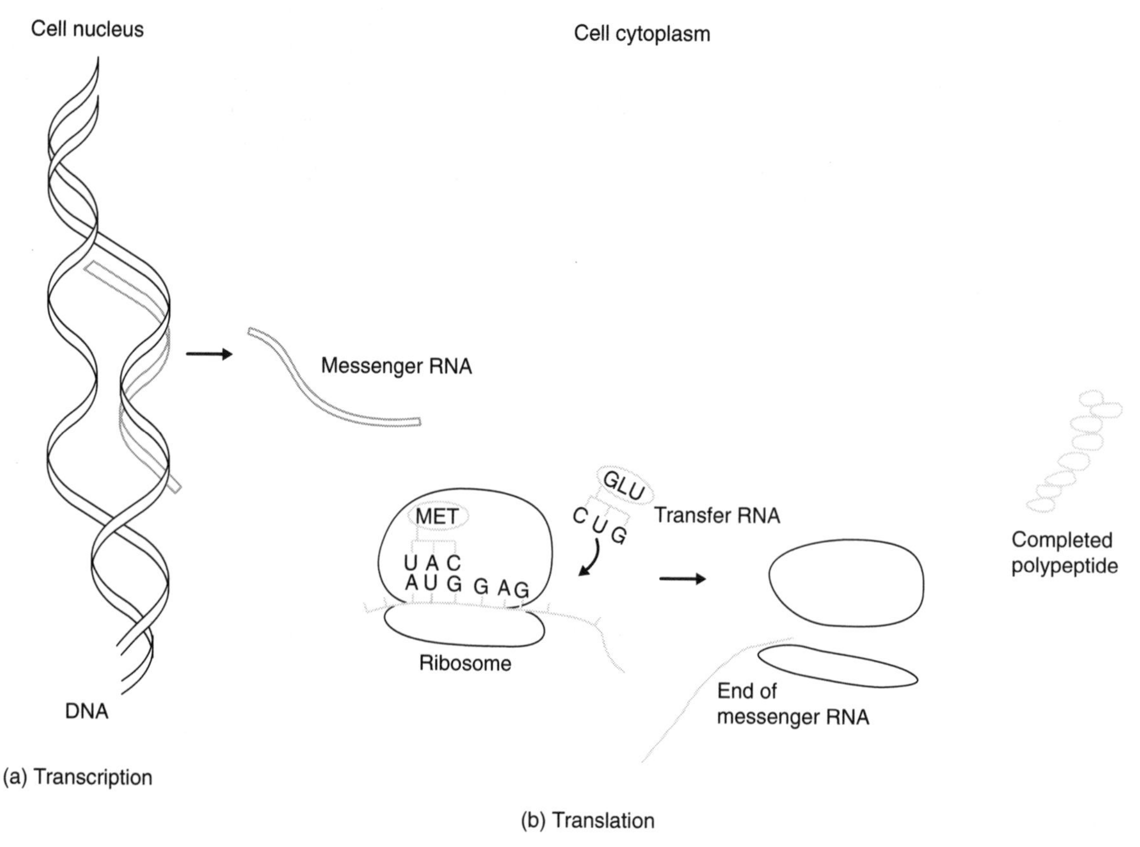

Figure 1.4 The language of the chromosome is translated into the language of proteins. The linear code of nucleotides on the DNA strands has to be translated into the linear sequence of amino acids to build the coded polypeptide. There are two stages in this process:

(a) **Transcription**, where the two strands of the DNA are parted so that free nucleotides can pair with their corresponding bases on the strand carrying the code. Note here that uracil (U) replaces thymine as the complement of adenine in all RNA. The read sequence is joined together with RNA to form a length of messenger RNA (mRNA), and the completed mRNA then enters the cytoplasm.

(b) **Translation** itself takes place in the ribosome. Here the mRNA is read sequentially, with three base sequences matched against their complementary bases attached to a transfer RNA (tRNA). Each tRNA molecule has an amino acid attached that corresponds to a particular triplet code. This makes the translation. As the appropriate tRNA molecule is selected by the complementary base-pairing, that amino acid is added to the line being assembled. Particular codes also signal the start and finish of the linear sequence needed to synthesize each protein.

In a wet year, the single-flowering plants are unsuccessful. Most die unpollinated and few produce any seed. Those with a sequence of flowers were pollinated in the early months, and now have a higher proportion of seed in the soil. Next year the field is a carpet of blue for the whole season, and most plants carry the gene for multiple blooms, as do most of their seed. This gene has increased its frequency within the population and our plant population has changed (Figure 1.5).

Despite being hypothetical, this example demonstrates the inevitability of evolutionary change in a changing world. If a selective pressure favours some individuals over others, and if the selected trait is inherited, then a change in the population must follow. Individuals best adapted to a particular habitat will produce more offspring than their competitors. Traits which improve reproductive success will proliferate within the population.

This is genotypic adaptation and its essential elements are inheritance, variation and selection. Consider the example again. It makes several important points about the nature of genotypic adaptation:

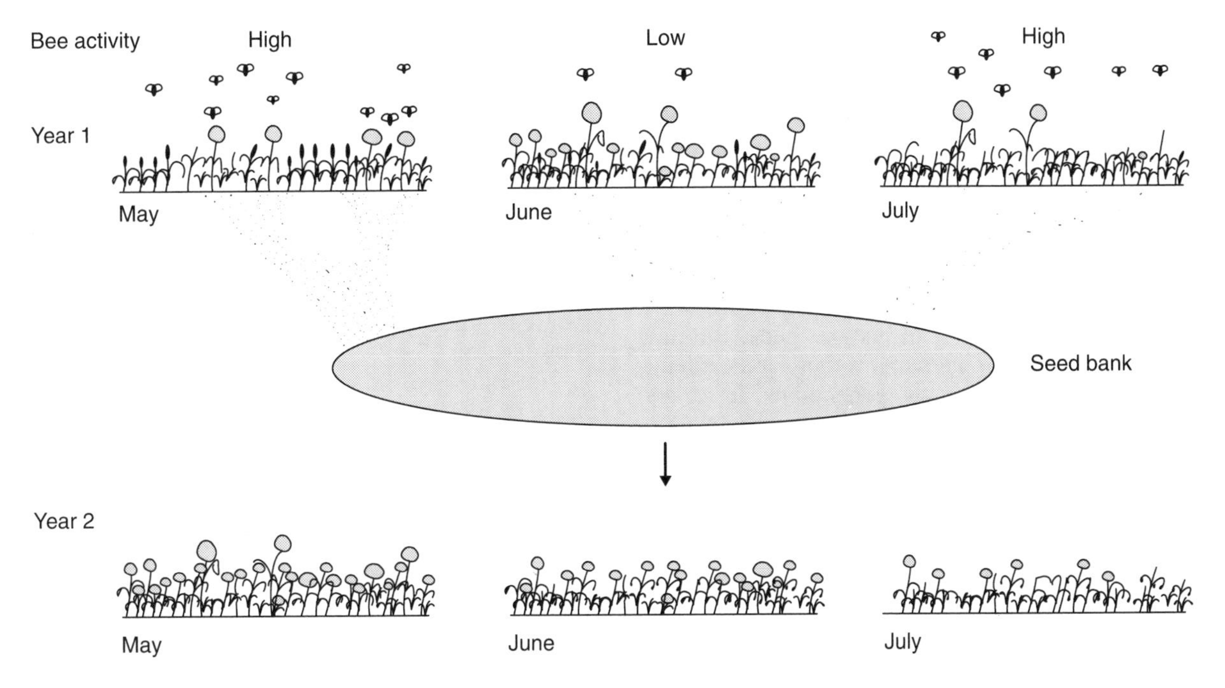

Figure 1.5 A simplified example of natural selection producing a change in a population. In this case, the selective pressure is exerted by bees, responsible for pollinating a flower and thereby ensuring its seed production; hypothetically, the bees here only fly during fine weather. In year one this happens only in May, so those plants which flower over a long period, producing several blooms during the season, are pollinated. Those plants that only flower once, perhaps in June or July, miss out when the weather is poor and the bees are not active. In this particular year the majority of the seeds in the seed bank contain the gene for multiple blooms. Next year this variety dominates, and the field is covered in flowers in May as well as June.

1. **The character selected must be inherited.** To become part of future generations a trait has to be written in the genetic code. Phenotypic adaptations are lost with each generation – and must develop anew each year.
2. **An individual carries the traits which were of selective advantage in its parent's environment.** The new generation is dominated by multiple flowers because such plants were more frequently pollinated in the wet conditions prevailing last year. Yet this year may be dry. It was the selection pressures operating upon the parents that determine the nature of the current population. An individual is adapted to its environment only as far as it matches the conditions under which its parents lived.
3. **One type dominates only while environmental conditions favour it, or there is no selective pressure against it.** If there was no selective pressure against multiple blooms in the drier years, their proportion in the population would not change. Only if they were at some disadvantage would their numbers decline with a series of dry years: for example, if their seed were viable for shorter periods or they germinated less readily under dry conditions. Then a series of dry years might shift the balance in favour of single blooms.
4. **Individuals need to differ for there to be selection.** Without variation, all plants would have the same chance of producing seed. With variation, those less able produce fewer seed. Success – the relative fitness of each type – is the proportion of offspring that each contributes to the next generation.
5. **The gene responsible for the adaptation was not induced by the environment.** Chance threw up the successful genotype; it was not produced in response to a wetter summer. Genetic change is blind to selective pressures operating in the environment. Most major changes in the genetic code are disadvantageous, or confer no advantage. We discuss this further later.

6. **The effects of natural selection are seen as changes in the proportions in the population.** A consistent shift to multiple blooms from one year to the next might lead us to suspect that this variety was being selected. This directional change in the population, over a number of generations, might lead to complete dominance or **fixation** of the favoured gene.

We have observed only a change in the preponderance of one variety of the plant, reflecting a change in the frequency of just one gene. But now imagine that the climate turns wetter for an extended period, so that, over the generations, fewer and fewer single-flowered plants survive. Eventually the gene for single blooms is lost. If the population remains isolated, not swapping genes with other populations, other genetic changes may follow. An accumulation of differences may lead to reproductive isolation (section 2.5).

The important difference between the plants is the availability of the flowers because this is the phenotypic character selected by the environment, in this case by the bees. Yet future generations will only reflect this selection as far as the character is written into the genes. However, indirectly, only the selection of genetic code can produce inherited change and thus generate new species.

This distinction between the phenotype and genotype is important. To understand its significance we need to recall the facts of life.

1.2.1 The blooms and the bees

The curious might reasonably ask why the plants should flower at all. Rather than hoping for a dry year and the attention of some rather fussy bees, why do they not simply produce some form of runner or tuber which could survive the winter?

With the picture we have painted so far, this might indeed be a good strategy. It would work fine if life in the field never varied. The plant need not change and an over-wintering stage might be a good strategy. The species could then avoid all the tribulations of sexual reproduction and a successful genotype passes unaltered into the next generation.

But if the environment were to change, perhaps with the climate getting wetter, such plants, despite all their runners and tubers, would find it increasingly difficult to survive. Relying solely on asexual reproduction, the capacity of a species to adapt genetically is very limited; once its phenotypic flexibility has been exceeded there is little prospect of accommodating further change. Sexual reproduction may be risky, but it does at least produce new combinations of genes and a possibility of increasing the range of tolerance.

Enter the bees. Inadvertently, they act as go-betweens, carrying the male genes in the pollen grains to the egg held in the flower. The flower is simply a device to attract the bee, a feeding station that advertises its presence. The cost to the plant is worth bearing given all the advantages of sexual reproduction.

Not only are the costs of sexual reproduction high; so too is the risk of failure. Each individual has to meet a partner, and two **gametes** (the sperm and the egg) must fuse to form a single cell, the **zygote**. For the female, who may protect and nourish the egg, that means sacrificing half of her genetic code and supporting the male genes. The zygote must then have a viable combination of genes. In multicellular organisms, it has to undergo a series of divisions and transformations to develop all the tissues of the adult. In the process it must produce special organs for sexual reproduction, such as flowers.

To survive, each offspring, each new genotype, needs to be well suited to the environment in which it finds itself. Producing large numbers of offspring, all slightly different from each other, improves the chance that some will survive, but this is a major cost to the parents. The luck of the draw will mean some will be less fit than their parents. Asexual reproduction avoids all these risks and costs and is a good bet when little changes.

The merits of sex are not obvious to biologists and there is a long-running debate about its selective advantages. One view is that sexual reproduction is favoured because uncommon genotypes may allow escape from parasites seeking to adapt to the internal environment of the host (Box 4.3). These benefits are indirect, however, arising from the genetic variation resulting in the next generation. This would not benefit individuals reproducing in this generation, only the genes whose code is passed on to the offspring.

Yet for most higher plants and animals sexual reproduction is the only viable strategy. A variable environment favours different traits at different times. The variation that comes from shuffling the genes in sexual reproduction means new genotypes which will, occasionally, translate into new phenotypes better fitted to the environment. Sex also provides the means by which deleterious variations might be lost, at meiosis and at fertilization (Figure 1.6), and via those individuals which fail to reproduce. Variation itself may confer some advantage, primarily for heterozygous individuals that have alternate alleles, perhaps providing them with a greater range of adaptive responses.

1.3 SOURCES OF VARIATION

The infinite variety possible from sexual reproduction provides for change in a changing world. These adaptations follow from alterations in the molecular sequences carrying the genetic code. In Chapter 2 we shall see how the gradual accumulation of such changes over many generations can produce new species. Now we look at how variation between individuals and across generations is created.

Traits may be inherited from either parent, but what is actually transferred by the sperm to the egg? By what mechanism is genetic information passed on? Charles Darwin could only speculate about the nature of inheritance and never knew that Gregor Mendel had begun to describe the mechanism just seven years after he had published *The Origin of Species*. Much later the **chromosomes** within the **nucleus** of the cell were identified as the site of its genetic information. The code is written as a series of discrete units, the genes, down the length of the chromosome (Figure 1.3).

The code is read sequentially, bearing the instructions either for reading the code (its 'punctuation'), or for building a protein (Figure 1.4). The protein is actually constructed at the **ribosome** from a series of smaller molecules, **amino acids**. This product is then used in a cellular structure or in cellular metabolism. Genotypic change occurs when part of the code reads differently. We detect this as a phenotypic change when the sequence of amino acids produces a protein with different structural and chemical properties.

In most higher plants and animals, chromosomes come in matching pairs, one from each parent (Box 1.2). A chromosome has to be copied every time the cell divides so that each daughter cell has a copy of the full genetic code (Figure 1.6). Each chromosome is replicated, producing an identical copy, and these separate, one to each cell. Thus, in this process of **mitosis**, the chromosome number is maintained in the daughter cells. This is not true of the gametes, where each cell has only a single set of chromosomes. Sperm and eggs have undergone **meiosis**, a process that firstly duplicates and then separates the paired chromosomes (Figure 1.6). When the gametes fuse at fertilization the full chromosome number is restored.

You look a lot like your parents, but you are not a perfect blend of the two. Genes carry discrete quanta of information and the proportion of traits you have inherited from your father and mother depend on the shuffling at meiosis and the chance unions at fertilization. A **crossing over** of sections of chromosomes may occur between pairs of chromosomes during meiosis (Figure 1.6), producing new gene combinations. The **segregation** of each pair of chromosomes, one to each gamete, and the union of sperm and egg give a vast number of possible combination of traits. An individual inherits only some of each parent's genetic code.

Most of the variation within a generation thus comes from the segregation and the shuffling of existing code during sexual reproduction. Entirely novel code requires that the sequence of bases along the chromosome be changed, for mistakes to be made in their replication. Genes are copied and read chemically in a highly reliable process, and cells have mechanisms to try to preserve the fidelity of the copying procedure. But mistakes do occur and these may be passed on with the genes in gametes that produce the next generation.

A **mutation** is either a change in the code at a single point, or deletions or insertions of lengths of code. A mistake that affects the code for a protein may change the sequence of its amino acids, or the way in which the molecule is folded. Sometimes this has little effect on its function, but it can cause more dramatic change. Most mutations are deleterious and some are lethal. Very rarely, a mutation improves performance or enables a protein to carry out a new function.

Chromosomal aberrations also result from the duplication and separation processes in meiosis. Large pieces of code can be swapped between non-homologous (non-matching) chromosomes (**translocation**), or chromosomes are duplicated or lost in meiosis. Again most chromosomal changes are likely to disrupt the development of the organism after fertilization.

Genetic and biochemical analyses show that a large amount of variation is normal in most populations. Much of the variation appears to have little selective advantage, so minor code changes have no

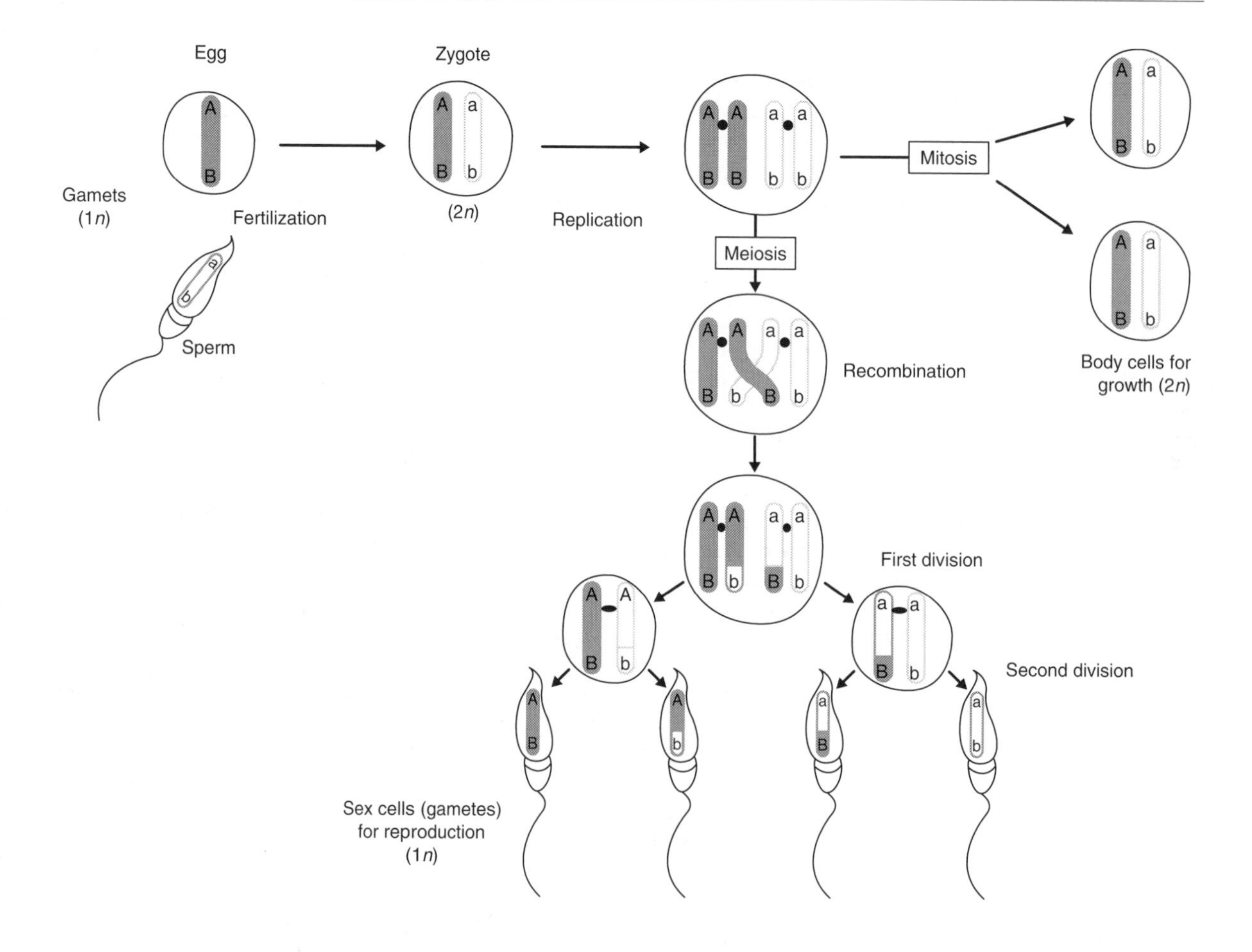

Figure 1.6 Chromosome replication and division. In most higher plants and animals, fertilization produces an individual with matching pairs of chromosomes, one from each parent. To simplify matters, here there is only one chromosome. Thereafter, chromosomes are replicated for either growth or the production of gametes.

Mitosis is the replication of the chromosomes to produce two identical daughter cells, each with a full complement of chromosomes ($2n$). The chromosome number (here $n = 1$) remains the same and the chromosomes themselves remain intact, so that the genetic code in each cell is the same.

Mitosis is responsible for cell division within the non-reproductive cells of an organism, and has no significance for evolution. However, notice that the $2n$ cell may contain recessive genes (represented by the lower case letters: a, b; the upper case letters A and B represent dominant genes) which are not expressed in the phenotype. This is a store of variation, retained in the population since it is never exposed to selection.

Meiosis produces haploid ($1n$) gametes carrying just one set of chromosomes. When these fuse at fertilization the diploid condition is restored ($2n$). Meiosis also involves replication, but now there are two division stages. The first division allows a chromosome and its replicate to remain together (termed sister chromatids). Homologous pairs of chromosomes separate into each cell at this division. In a second division parts the two chromatids separate into individual cells. This is the haploid state.

Notice that when homologous chromosomes form pairs, they become attached at various points. In some cases, pieces of the chromosome may cross over, and become transposed between non-sister chromatids. This is **recombination**, an important way in which genetic information can become shuffled, producing new combinations and variation. Meiosis does not change gene frequencies in the gene pool, but it does produce new combinations of genes.

Variation will also occur between individuals because gametes do not contain equivalent combinations of chromosomes. This is termed **independent assortment**. Gametes differ from each other because of both independent assortment and recombination.

apparent impact on reproductive success. This is termed **neutral variation**. Some biologists believe, however, that a high level of variation may, in itself, contribute to the vigour of a species.

Variation also resides in the code 'hidden' or not expressed in the phenotype. At a heterozygote locus, where one allele dominates another, the recessive gene is not subject to selection (Box 1.3). It is only ever exposed to selection in a homozygous individual, when it is expressed in the phenotype. Recessive genes not lost from the population can be an important source of additional variation at recombination.

In some situations 'ancient' code can re-enter the population. With our hypothetical flowers, we carefully arranged for all seeds from one generation to be derived from plants alive the previous year. If, instead, some seeds were several years old, their code would come from a slightly different gene pool, reflecting the environmental conditions prevailing when their parents were alive. Germination of very old seed (and some seeds can remain dormant for centuries) can reintroduce variation into the gene-pool.

Variation can also be lost. Over the generations, a directional change in the code may occur as the most fit alleles are selected for a particular trait. Rapid change in the population can lead to a gene becoming 'fixed' at its locus. Then there is little or

Box 1.3

UNITS OF SELECTION

What does natural selection operate upon? The DNA of the chromosome is not directly subjected to the selective pressures of the environment. Selection operates only on the traits expressed in the phenotype; an individual represents the expression of thousands of genes and the phenotype is their combined effect. Natural selection and chance decides which of these phenotypes reproduce, whose gametes produce offspring. However, it is only the code, the information, written in the genes that survives into the next generation. Genes replicate themselves; individuals do not.

Some biologists thus emphasize the role of genetic code in the process of evolutionary change. Richard Dawkins distinguishes between the genes as 'replicators' and the 'vehicle' in which they travel, the individual. Individuals are communal vehicles for a collection of replicators. A gene improves its chances of survival by endowing the vehicle, the phenotype, with a trait that better fits it to the environment. But since only genes survive into the next generation, natural selection can only select between replicators.

A gene is, in effect, in competition with its alleles. Actually, Dawkins does not want to identify the 'gene' as the sole unit of selection but rather any length of DNA, any part of a chromosome, which survives intact into the next generation. Replicators can thus be whole sequences of genes. The length of the replicator depends on the frequency at which they are fragmented when the gametes are produced after meiosis (Figure 1.6). While individuals and populations change over the generations, the replicator remains the same.

The alternative and more traditional view is put by Ernst Mayr, who emphasizes that it is individual phenotypes which are adapted to their environment. It is the phenotype, the collective expression of the genotype, that succeeds or fails to reproduce. Mayr contends that the effect of a gene can only be understood in the context of the whole organism, in its environment. Under some conditions a gene may be lethal; under others it is advantageous.

This is largely undisputed by the other side. However, they contend that the phenotype and its collection of genes is part of the environment in which a replicator finds itself. It succeeds or fails in this context. Combinations of genes which work – mutually compatible replicators – are more likely to reproduce and remain together. The gene as replicator will only survive as long as selection does not act against it.

no variation in this characteristic. This may happen in small and localized populations where chance and circumstance mean a small gene pool and breeding is between individuals, relatives perhaps, who share much of the same code (section 2.5). Then they are said to have undergone **genetic drift**. Isolated from other populations, with no genetic exchange between them, these changes may become sufficient to create a new species.

1.4 RATES OF EVOLUTION

The idea of species changing over time was not particularly revolutionary amongst the scientists of Darwin's generation. They knew that the fossil record included unknown and strange animals that had died out millions of years previously. Remains of many familiar species appeared in much younger sediments. Giant mastodonts and mammoths recalled living elephants, similarities suggesting that modern species might be derived from ancient relatives. The experience of breeding domesticated plants and animals demonstrated the range of variation that can exist within a species and how readily traits could be selected. What was needed was a natural mechanism for selecting inherited traits to produce new species.

The same answer suggested itself to two British naturalists, both of whom had experienced the immense diversity of tropical life. Within a few years of each other, Charles Darwin and Alfred Russel Wallace independently originated the idea that the selective force was imposed by the demands of the environment. Given competition for limited resources, and the capacity of all organisms to produce large numbers of offspring, only the fittest, the best adapted, would survive to reproduce. Traits selected over many generations, the accumulation of small-scale changes, would generate sufficient differences to distinguish a new species.

Today we can map differences in the code of individuals and populations and describe in great detail how they change with time. In the middle of the last century the commonly held belief was that species were ordained by a supernatural power, following a fixed design. Darwin and Wallace showed how generations would change under selective pressure. Not only were species not from a fixed template; the disturbing implication was that humans had evolved from an ancestor shared with the great apes. The human design, once deemed to be perfect, even divine, was now seen as a compromise born out of chance.

There was much that the theory did not explain. Species did not seem to change around us so the process appeared too slow and the Earth too young to produce the great diversity of plants and animals. We now know much more about the time-scale of events. Today we can trace life back perhaps 3.5 billion years and have found evidence of several bursts of speciation in the fossils embedded in the rocks of different eras.

Several times slates have been wiped clean and new species have evolved to exploit the opportunities as competitors have disappeared (section 9.4). Darwin knew that his theory discounted a single day of creation but would have been surprised how frequently and how rapidly species have arisen in the geological past. Massive upheavals in the ecology of the planet have led to both the extinction and the evolution of species. In between times, the fossil record shows long periods with few discernible evolutionary developments. Even today it is difficult to know how rapidly species were changing at different times, given the imperfect record in the rocks. Certainly there were quiet times but also times of rapid change, perhaps following some evolutionary innovation, such as a novel body structure.

The study of **speciation** is a branch of evolutionary biology in its own right. A species forms when populations become isolated and genes cease to pass between them. This reproductive isolation can develop for reasons other than just the physical separation of the populations, a subject we take up in Chapter 2. Rapid speciation will follow rapid environmental change, but the mechanism remains the same – small, gradual changes, building over many generations to produce the differences that will separate populations into new species.

At the cellular level, there is great uniformity in biological systems: all living organisms share similar mechanisms of metabolism and reproduction. These essential features are very conservative in their detail, so that even the flowers and the bees share much of their cellular machinery. The differences we see between species result from their interaction with the environment, from their adaptations to their way of life.

It is the information in the chemical code, often minor differences, that distinguish one species from another. Life became possible when molecules able to copy themselves were large enough to carry information that aided their own replication. Chromosomes are immensely complex macromolecules with elaborate mechanisms for reproducing themselves and for reading the signal they carry. Replication and transcription are chemical processes, requiring energy to drive the bonding of the molecules. If there is a secret to life, it is how such a system could have originated.

1.5 THE ORIGINS OF LIFE

Natural selection does not explain how life originated on the planet, but it does suggest where we might look for some clues. It implies that organisms had evolved from ancestral and less complex forms. As we uncovered its chemical basis, speculation began about the conditions under which chemical processes could give rise to life.

A living cell fixes and liberates energy to grow and reproduce, replicating its chemical machinery. The complex molecules of the cell – the nucleic acids, proteins, carbohydrates and lipids – are built up from smaller and simpler chemical units. A source of energy, in the form of electrons, is needed to make the chemical bonds that hold these units together.

But in no sense are these chemicals alive. Living systems replicate themselves and they change or evolve. That means occasional failures in the copying process – chemical changes that produce variation. Together with a system of energy management (metabolism), replication transforms inanimate matter into an evolving, living system.

When life first appeared on the planet 3.5 billion years ago conditions were very different from today. Many primitive life forms, and indeed some cellular processes, only function in the absence of oxygen and so could only have originated in a low-oxygen atmosphere. This is not surprising for oxygen is an extremely corrosive, highly reactive gas, and most carbon-rich (organic) compounds in living systems would be destroyed, oxidized by abundant free oxygen.

Biochemists have experimented with a variety of atmospheric conditions under which organic molecules may have formed. Several combinations of gases can be made to generate an impressive range of organic compounds. All 20 amino acids found in modern proteins have been synthesized, and one or two others. Sugars, lipids and other important organic molecules have also been produced. Different energy sources have been used to mimic the electrical discharges of primeval thunderstorms or the intense ultraviolet radiation of this time. Overall, it seems the formation of these simple compounds, the **monomers** needed to build the **polymers** of proteins and nucleic acids, can be achieved relatively easily.

How could the monomers be made to join up, to polymerize into chains? Bonds between them are formed with the loss of a hydrogen ion (H^+) and a hydroxyl ion (OH^-) as a molecule of water. Assuming that the monomers were formed in an aquatic environment, we need a mechanism which lowers the concentration of water for this to happen. Several theories have been suggested. One possibility is that monomers became concentrated by evaporation, perhaps around volcanic vents, or small pools in a tidal zone, where drying occurred temporarily.

Alternatively, the monomers may have become concentrated at the edge of ice sheets, or on minerals that could bind them. Clays have a massive capacity for binding charged molecules. We know that metals bound to the clay surface may aid the polymerization of amino acids, again in the absence of water. There are clays that only bind certain forms of amino acids and sugars: amino acids that have left-handed configuration and sugars that are right-handed. Today all living systems on Earth are based only on left-handed amino acid and right-handed sugar molecules. One fascinating possibility is that the biochemistry of life on Earth results from monomers bound to these clays. Perhaps these were the monomers most able to replicate themselves.

The energy for such polymerization could have come from heat from the Earth's core, though some reactions can take place under freezing conditions. Some energy sources could actually destroy the polymers: ultraviolet radiation is likely to break the chemical bonds, so the the polymers may only have accumulated in relatively sheltered sites, out of the glare of the sun.

Polymerization alone does not give us a collection of molecules able to copy themselves. Nucleic acids can do this, but they need enzymes to replicate – that is, proteins which aid the chemical reaction. There is a dilemma here: modern cells manufacture their enzymes using code written in their nucleic acid. Which came first, the replication of the nucleic acid or the replication enzyme it was coding? It seems these two components must have developed together, with the code for the enzyme promoting the replicating nucleic acid and vice versa.

RNA (ribonucleic acid, Figure 1.4) almost certainly preceded the appearance of DNA, and might answer this conundrum. Because of its simpler form, RNA has long been proposed as the first self-replicating molecule, a potential precursor of DNA. Recent work has shown that RNA has the catalytic properties of an enzyme, so that it promotes some chemical reactions inside the cell. In addition, fragments of RNA spliced from larger molecules can act catalytically to produce new RNA molecules from a pool of nucleotides. While RNA may be able to self-replicate, more problematical is creating RNA in the first instance. Not only is it difficult to synthesize in the laboratory; the environmental conditions on a primeval Earth would not have been conducive to RNA production.

Perhaps it needed a simpler replicating molecule as a precursor of RNA. One possible site for its formation is inside the microscopic droplets which form naturally in water that is rich in long-chained organic molecules. Proteins, carbohydrates and lipids will collect together as concentrates and may even form a membrane-like boundary layer. These droplets scavenge organic molecules from the surrounding water and bud off daughter droplets when they pass a critical mass. Inside the droplet further polymerization reactions are likely. Entirely the result of chemical and physical processes, these droplets are nevertheless an important early stage in the self-organization needed for a living system. What these aggregations lack, of course, is the inheritance that a self-replicating molecule confers upon its copies.

We know replication reactions can take place inside droplets. Replication requires only that a molecule should bind components from its surroundings and configure them into a copy of itself. The chemistry of the parent molecule simply ensures that the binding sequence leads to a copy, or to a complementary molecule (a 'negative') that can then make a copy of the parent. When completed, the newly formed polymer breaks away.

Copies will then bind components in their own synthesis reactions. From a collection of replicating molecules, some may be more effective at scavenging monomers than others. These would become the most prevalent. In turn, it is the most prevalent molecule that makes the most copies. These replicators thus come to dominate. If there are occasional errors in the copying process, slight variations that are passed on to the copies, there is the possibility of evolutionary change. Changes which improve polymerization will tip the balance in favour of the new template. In any competition for resources, for monomers, selection will favour the most efficient replicator. This is chemical evolution, where the prevalence of one molecule depends upon its capacity to replicate itself. We can see genes, or replicating DNA, as an extension of the same process (Box 1.3) and, in turn, the evolution of living organisms.

Droplets able to scavenge monomers quickly, perhaps because they have rapid polymerization, will grow at the expense of competitors. Rates of synthesis are increased by a source of energy: a synthesis reaction that produces energy-rich molecules, or that aids the synthesis of the polymers, will mean the droplet is more likely to survive. Indeed, it may grow large enough to bud-off daughter droplets, so that the 'species' proliferates.

But these droplets are not living. They have no orderly mechanism for storing and replicating hereditary information, essential if all daughter droplets are to be identical, with the same biochemical abilities.

This picture is highly speculative, and there are still large gaps in our knowledge. It seems likely that RNA played a central role in this early evolution, if only because of the range of duties it carries out in modern cells. Not only does RNA translate the code of the more complex DNA; it also binds the amino acids to be assembled into proteins in the ribosome, which itself is largely composed of RNA (Figure 1.4). A system of such complexity must have evolved over a long period of time.

Systematic protein synthesis must have begun without the aid of ribosomes. RNA-mediated polymerization will occur outside ribosomes and we know RNA will bind amino acids weakly, long enough for them to be linked together. Any RNA molecule that coded for a polypeptide which, by chance, aided its own replication would be favoured. Again, the prevalence of this polypeptide and this RNA molecule is increased by the presence of each other. Perhaps inside a droplet such favourable associations would lead to further interdependence with other molecules. We might then at last have a self-replicating system catalysed by an enzyme.

We now have to show how particular amino acids become associated with certain nucleotide codes on RNA and how these were matched against DNA sequences.

We do not know how life began but we have reduced the problem to one largely of chemistry and the early environment. Of course one short cut would be for a replicating, metabolizing life to arrive from space. Indeed, some elaborate organic molecules have been isolated from meteorites. However, this does not solve the problem; it only shifts it to another planet (of which we know even less).

From the fossil remains of primitive bacteria-like cells, life appeared about a billion years after the Earth was formed, and these cells were able to exploit a variety of energy sources. To this day different bacteria can be found in extreme environments, often devoid of oxygen, splitting chemical bonds to derive energy. Some, primarily the cyanobacteria (blue-green algae), developed the means of capturing energy from sunlight and a capacity to use water as a source of electrons. In splitting the water molecule in their photosynthesis (Box 5.1) they added oxygen to the atmosphere. Gradually the chemical environment was being changed.

With large amounts oxygen in the atmosphere the conditions under which non-living synthesis of large organic polymers could happen came to an end. From around 2.5 billion years ago, new polymers would be produced only by living systems, within the protection of the cell. Indeed, a new cellular organization followed as oxygen was used in metabolic processes. The eukaryotic cell with its well-defined nucleus was to be the fundamental building block of all higher plants and animals.

1.6 ORIGINS OF THE MAJOR GROUPS

The system for naming and classifying species created by Linnaeus in the eighteenth century helps us to pick out the ancestry of multicellular organisms (Box 2.1). In its original form the system used mainly morphological characters to classify around 10 000 plants and animals. In its hierarchy, Darwin could see clues to the ancestry of many species.

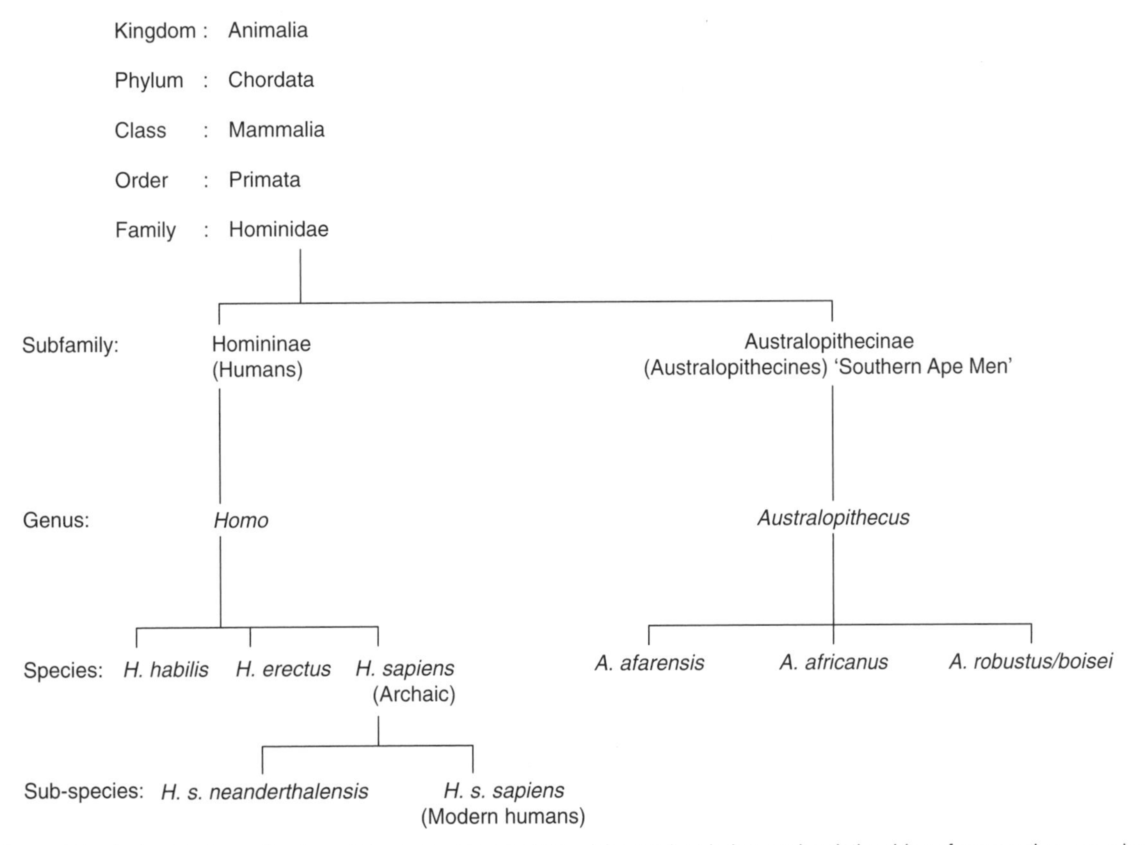

Figure 1.7 A simplified classification of Hominids. Note that this picks out the phylogenetic relationships of our species – species that are closer together on the phylogenetic tree share more of their ancestry than those far apart.

We look at the modern classification system in the next chapter, but all classifications collect species together according to shared characteristics. This produces a hierarchy of groups, with increasing similarities as we approach to level of the species (Figure 1.7). This helps us to recreate the evolutionary history or **phylogeny** of a species. Originally based largely on external morphology, today we also draw upon the evidence written in the cells, in sequences of genes and the structures of key proteins.

For example, a muscle protein, myoglobin, differs by one amino acid only between humans, gorilla and chimpanzee. Baboons show five amino acid differences to those of humans. The evidence from DNA sequences taken from subcellular organelles (mitochondria) suggest that *Homo sapiens sapiens* originated from a single population in Africa. Alternatively, some palaeoanthropologists, using different criteria, argue that modern humans have evolved from the archaic genus *Homo* several times in different parts of the world.

Studies of cellular and biochemical organization have also expanded our ideas about how the major groups of living organisms have arisen. Some molecular biologists have attempted to reformulate the whole classification of life based only on the genetic sequence in one type of ribosomal RNA. This divides all organisms into just three major categories: the bacteria, the eukaryotes and a group called the **archaebacteria** that share an ancient ancestry with the eukaryotes and which are largely confined to very extreme environments. Many biologists have questioned whether we can base an entire classification on just a few phenotypic characters.

A more conventional classification using a range of criteria defines five Kingdoms (Figure 1.8), with the main division being between the **prokaryotes** and the **eukaryotes**. Prokaryotes, including the bacteria and blue-green algae, have no nucleus and a simple looped chromosome. The cells of eukaryotes, which include all higher plants and animals, wind their DNA around proteins, form them into chromosomes and hold them within a nuclear membrane. Other organelles within the eukaryote cell are equally well packaged, including the mitochondria, whose enzymes drive much metabolism, and the chloroplasts of photosynthetic plants.

According to one view, this compartmentalization of the cell followed from a series of invaginations of the cell membrane, isolating the organelles in their own membranes. An alternative view has the eukaryote cell formed from the association and assimilation of a collection of prokaryote cells. For example, chloroplasts show much similarity to blue-green algae (cyanobacteria) cells, and indeed, have their own DNA. One suggestion is that they have been assimilated into a host cell.

Cooperation for the mutual benefit of two species is termed **symbiosis**. The hypothesis of **symbiogenesis** suggests that an association becomes so interdependent that each partner can only replicate and survive by remaining in such a partnership. Lynn Margulis argues that this is how the eukaryotes originated (Figure 1.8). In her **endosymbiont theory**, Margulis suggests that chloroplasts and mitochondria were derived from prokaryotic cells that formed symbiotic associations with a host cell.

A variety of evidence supports this theory, based on the similarity of these organelles to prokaryotic cells: they are the same size, they share a sensitivity to oxygen, have their own stores of DNA, and have RNA sequences similar to prokaryote polymers. However, the choloroplasts and mitochondria are synthesized using the host's nuclear DNA, which means their code had to find its way into the chromosome. Although this is a complication, there are known mechanisms by which this could happen.

Margulis points out that this makes the eukaryotic cell a community of structures with different genetic histories, which have together evolved to a high level of integration. In this sense, we might regard them as miniature ecosystems. The endosymbiont theory recognizes that evolution occurs through not only competition but also cooperation, when the replication of each partner is enhanced by the association. Margulis argues that symbiosis has been responsible for a level of cellular organization. In this case, genes from more than one source have contributed to the evolution of plants, animals and humans.

SUMMARY

The origin of humanity from an ancestral stock shared with the great apes followed a series of changes in the environment of Africa around 5 million years ago. The advent of bipedalism in these apes, prompted by the advancing grasslands, freed the hands for carrying and for tool-making. The

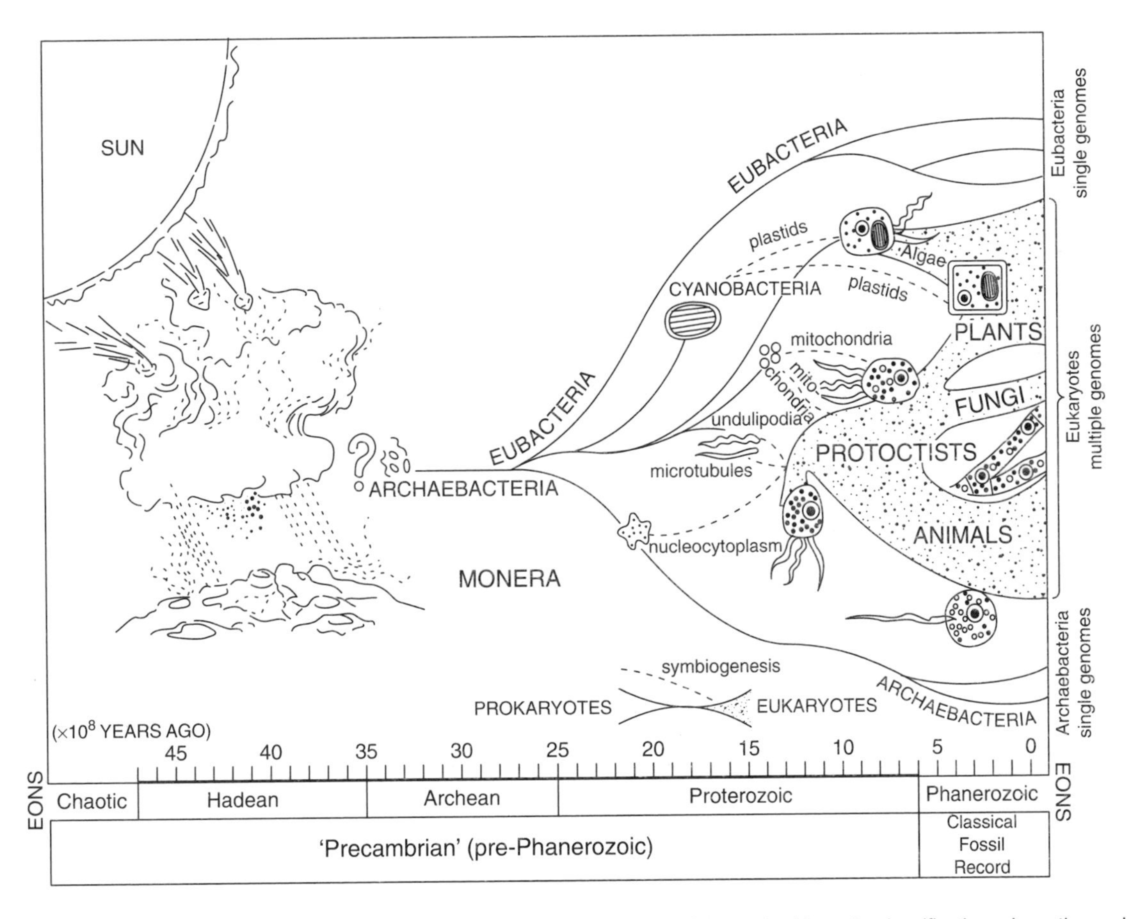

Figure 1.8 The origins of the five Kingdoms (shown in bold capitals). This follows the Margulis classification where the major division is between the prokaryotes (Monera), without a well-defined nucleus and with only one source of genetic code, and the nucleate eukaryotes which have more than one. Associations have been formed through evolutionary history so that symbiogenesis (shown by the hatched lines) produced structures we observe today inside the cells of higher plants and animals but which evolved independently in the first instance. At various stages the additions of microtubules, mitochondria and plastids have meant the advent of animals, fungi and plants.

techniques of making tools and of working within groups would benefit from a precise language, all of which fostered mental development. In time, the adaptability of mental skills made for a more adaptable species.

Such developments had their origins in shifts in the genetic code stored in the chromosomes. Natural selection determines which individuals are able to pass their genes to the next generation. A gradual accumulation of genetic changes can lead to a population becoming reproductively isolated from its relatives, and evolving into a new species.

Evolution operates on the code stored in the genes. Replication of the molecules carrying this code requires a means of storing and releasing energy. Replication and metabolism must have developed closely together. Many of the early steps in the origin of life, starting with a chemical evolution, are relatively easy to reproduce in the laboratory. The major features of the cell may have followed when structures and chemical processes united in symbiotic relationships. These represent large steps forward in the evolution of life on Earth, reflected in the five major Kingdoms commonly recognized today.

FURTHER READING

Dawkins, R. (1986) *The Blind Watchmaker*, Penguin.

Jones, S. (1993) *The Language of the Genes*, Flamingo.

Leakey, R. and Lewin, R. (1992) *Origins Reconsidered*, Abacus.

EXERCISES

1. Why do we not expect polymerization reactions (building the large macromolecules necessary for life) to be very prevalent outside of living cells on Earth today?
2. Where might you look to find clues to the metabolism of the early life on the planet and why?
3. The idea that modern humans are all derived from a single population, somewhere in Africa, is based on the base sequences in the DNA carried by our mitochondria, outside the nucleus. Why are these sequences more indicative of our ancestry than those carried in our chromosomes? Why is this referred to as the Mitochondrial Eve hypothesis and not Mitochondrial Adam?
4. Why might possession of a genotype that is common within a population make an individual more susceptible to parasitism?
5. Complete the following paragraph by inserting the appropriate words, using the list below (some words may be used more than once, some not at all).

Any replicating system produces copies of itself. For the system to change, the copies cannot be perfect; there must be ____________ between them. Those ____________ may have no significance: unless they confer some advantage or disadvantage, the ability to make copies, any variation would be ____________. If, however, the ____________ favours some varieties over others, so that some are more likely to ____________, then their ____________ in the population will rise. The population will have changed and ____________ will have occurred. Thus ____________, ____________ and ____________ are the three elements needed for evolution to occur.

adaptation, number, differences, proportion, environment, replication, evolution, reproduce, inherited, selection, neutral, variation

Tutorial or seminar topic

6 Examine the ways in which the language of a people carries its culture.
- Are ways of speaking ways of thinking?
- Has the discipline of language promoted our ascendency as a species? And if so, how?

- Does our language change or evolve with our understanding of the world? If so, does this have an adaptive significance?

The binomial system of classification – species and their various definitions – classification by biochemical and genetic techniques – species and ecological niche – intraspecific competition and the process of speciation – humans as agents of speciation.

2 Species

That which we call a rose
by any other name would smell as sweet

Romeo and Juliet, William Shakespeare

Classification comes naturally to us. Faced with something unfamiliar we look for points of comparison with objects or events which are already part of our experience. Recognizing similarities allows us to place it into a category and assume it shares the properties of that group. Establishing categories and describing their characteristic properties is central to the way in which we learn.

This is a good adaptive strategy. It means that we can make predictions and perhaps anticipate situations that we have never encountered before. Indeed, pattern recognition is common to all animals with enough neural machinery to store the information. Identifying key patterns is so central to the behaviour of many higher animals that some plants have developed means of exploiting it. The male bee looking for a mate is readily fooled by the orchid with the correct signals associated with the category 'female'. The flower may have no wings, no legs, no genitalia, but it does have what a male looks for in a female bee, or at least enough for him to change his behaviour.

One difference between us and the animals is that we use our language to name the categories. Names are abbreviations that imply the properties of a category, a term that serves to summarize a complex description. The general term 'dog' implies all that we take to be characteristic of this category from being hairy with a leg at each corner to the fine detail that distinguishes dog from cat. Even then we are sometimes surprised to learn that our category 'domestic dogs' contains everything from a chihuahua to a great dane.

The danger, as Romeo reminds us, is to mistake the name for the category. We have to draw a distinction between the agreed term for an object – its name – and its essential properties. Sometimes the distinctions we draw are not a true reflection of the real world. At other times we are not sufficiently discerning to pick out the fine detail of nature. This is certainly true of biology, where we have repeatedly had to refine our ideas about the variety and diversity of life.

All sciences begin by creating categories and classifications to help us understand the relations between their important elements. This is invariably a hierarchy, ordered into levels of increasing similarity. In biology this is termed **systematics** and it groups species according to their shared properties. In the process, biologists can pick out the evolutionary lines linking species, more formally called their **phylogenetic relationships**.

In this chapter we introduce systematics and also the conventions used in describing, naming and classifying organisms – **taxonomy**. The basic unit in biology is the species, which we defined earlier as individuals sharing a common gene pool and able to produce viable fertile offspring (Box 1.1). Although this is a good functional definition, we shall see that the boundaries between some species are often poorly defined, sometimes because we are witnessing the process of speciation in action, sometimes because we cannot measure the distinguishing traits and sometimes because no real distinctions exist. We go on to look at how new species arise and the finer and finer distinctions we have to draw to represent accurately the variety of life.

2.1 WHAT'S IN A NAME?

The closest we get to a universal language is in the sciences. Even if nations do not use the same names or even the same alphabet we invariably share the same symbols. Chemists speak to each other using one or two letter symbols as short-hand for the elements of the periodic table. Biologists have the convention of referring to species with two Latin names, the **binomial system** (Box 2.1). This was developed in the eighteenth century by Linnaeus, who actually streamlined earlier systems that used several names.

The most important property of the binomial system is that it establishes a unique combination of two names for each species. When Europeans and

Box 2.1

KEYS TO THE KINGDOMS

Modern classification is a hierarchical system based on Aristotle's concept of the Kingdom as the highest category into which living organisms can be grouped. This was further developed into the binomial system by a Swedish botanist, Carl Linne, who latinized his own name to Carolus Linnaeus (Figure 2.1). His system consisted of a hierarchy of seven nested categories:

- Kingdom (originally either plant or animal)
- Phylum (suffix -phyta for plants)
- Class (suffix -phyceae for plants)
- Order (suffix -ales for plants)
- Family (suffix -aceae for plants and -idae for animals)
- Genus
- Species

We have already seen how this system is used to classified hominids in the previous chapter (Figure 1.7).

In Linnaeus's time it was not always easy to identify the precise relationships between species in the intermediate groups (Class, Order and Family) and so genus and species tended to be the most reliable and precise way of defining an organism by name. The hierarchy still works well but we have found it necessary to extend it considerably, dividing and subdividing these main categories. Nevertheless, it retains its important property – that species grouped together at the lowest levels have more in common than those only linked together further up. Plants or animals belonging to the same genus or the same family share more characters than others in the same order or class. In effect, the hierarchy helps us to pick out the evolutionary ancestry of a species.

By the nineteenth century problems were starting to emerge with the Two Kingdoms system. The development of the microscope revealed an unexpected world populated by microorganisms. These could not be easily classified as either plants or animals and so in 1866 Ernst Haeckel proposed a third Kingdom, the **Protista**, to include the protozoa, primitive algae and fungi. These groups were lumped together with the prokaryotic bacteria and the Kingdom was little more than a biological dustbin for groups that did not fit anywhere else. Eventually a further Kingdom, the **Monera**, was created to contain the bacteria and blue-green algae (more properly known as cyanobacteria). Finally in 1969, Robert H. Whittaker separated the Fungi from the Protista into a Kingdom of their own, so producing the Five Kingdoms system which is generally accepted today (Figure 2.2). On top of these, Lynn Margulis has introduced two Superkingdoms, the **Prokaryota** and **Eukaryota** to acknowledge the fundamental difference in cellular structure between these groups (Figure 1.8).

Figure 2.1 Carolus Linnaeus (1707–1778).

There are problems, however, as some life forms do not seem to fit into any group. Viruses are so varied and so different from anything else that Lynn Margulis half-joked they were more closely related to their hosts than to each other. To this day, viruses remain outside mainstream classification, little more than replicating molecules that hijack the replication machinery of living cells.

Taxonomists have an altogether different problem when it comes to lichens. Although these may look like individual organisms, they are in fact a close association between algal and fungal cells. Lichens are the only true naturally occurring **chimeras** – organisms which have two genotypes (section 4.2). There are relatively few algal species involved in these associations (sometimes several different algae occur in a single lichen) but each 'species' of lichen has a different fungal species and so they are classified by their fungal partners.

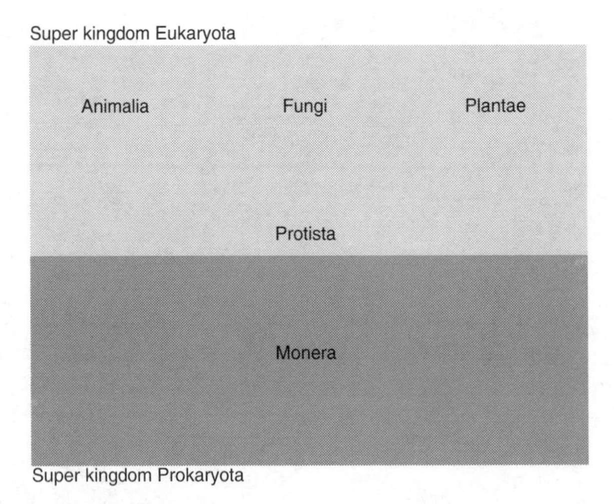

Figure 2.2 The Five Kingdoms divided between the two Superkingdoms. All organisms without a discrete nucleus – bacteria, blue-green algae (cyanobacteria) and others – comprise the Monera and are the sole members of the Prokaryota. Protozoans, algae and slime moulds together form the Protista. The rest – higher plants, animals and fungi – each have their own Kingdom inside the Superkingdom Eukaryota.

Americans use the term 'blackbird' they are invariably referring to two different species with different Latin names. *Turdus merula* is the European blackbird and belongs to an entirely different genus to the North American blackbird (*Agelaius phoeniceus*). The same is true for plants: the British bluebell is *Hyacinthoides non-scripta*, of the lily family, whilst the plant called a bluebell in California is *Phacelia tanacetifolia* (Figure 2.3, Plate 3) is from an entirely unrelated family, the Hydrophyllaceae.

Notice here the detail of using a Latin (or scientific) name:

- Each consists of two names (hence the binomial system).
- The first name is the **generic name**, the name of the genus. This always begins with a capital letter.
- The second name is the **specific name**, the name of the species. It always begins with a lower case letter.
- A scientific name is always printed in italics or, if handwritten, underlined.
- It is the combination of the generic and specific which is unique. So two species may belong to the same genus, and therefore share the same generic name, but will share different specific names – e.g. *Mustela erminea* (stoat) and *Mustela nivalis* (weasel). Two species belonging to different genera may share the same specific name – e.g. *Primula vulgaris* (primrose) and *Calluna vulgaris* (heather).
- After first being cited in full, the name may subsequently be abbreviated if there is no possibility of confusion – e.g. *P. vulgaris* and *C. vulgaris*.
- Often the name is descriptive, even for those with just a little knowledge of Latin or Greek. For example, *vulgaris* means 'common'.

Sometimes the citation also gives the name of the **authority** immediately afterwards; for example, *Helix aspersa* Müller (the garden snail). The authority is the first person to describe that species in its currently accepted classification and who gave it the name. Scientific papers usually cite the name of the authority when the Latin name is used for the first time. There are strict rules about naming and giving authorities for species, with well-known authors having their names abbreviated. In *Homo sapiens* L., the L refers to Linnaeus, the authority behind our own species name.

These rigorous rules make the binomial system universal. Species names are agreed by an international committee which operates the International

Classification				
European Blackbird	American Blackbird	**Category**	European Bluebell	Californian Bluebell
Animalia (animals)	Animalia (animals)	**Kingdom**	Plantae (plants)	Plantae (plants)
Chordata (chordates)	Chordata (chordates)	**Phylum**	Trachaeophyta (vascular plants)	Trachaeophyta (vascular plants)
Aves (birds)	Aves (birds)	**Class**	Angiospermae (flowering plants)	Angiospermae (flowering plants)
Passeriformes (songbirds)	Passeriformes (songbirds)	**Order**	Liliales (lilies)	Polemoniales (Jacob's ladders)
Muscicipidae (warblers)	Icteridae (icterids)	**Family**	Liliaceae (lily family)	Hydrophyllaceae (phacelia family)
Turdus	*Agelaius*	**Genus**	*Hyacinthoides*	*Phacelia*
merula	*phoeniceus*	**Species**	*non-scripta*	*tanacetifolia*

Figure 2.3 The full hierarchical classification of European and North American blackbirds and bluebells.

Codes of Biological Nomenclature. Animals are dealt with by the International Code of Zoological Nomenclature (ICZN), which recognizes organisms down to subspecies level, whilst the International Code of Botanical Nomenclature (ICBN) goes further, recognizing not only subspecies but also varieties, subvarieties, forms and subforms (Box 2.1).

A taxonomist's work is never done. Classification of most groups is under constant review, not only as new species are found and named, but also as we learn more about the evolutionary history of known groups.

2.2 THE SPECIES

Ideally, our classification system would accurately reflect the important differences between organisms. Then each species, as the basic element in the phylogeny, would represent a unique collection of adaptations, the current end-points of two different evolutionary histories. However, the boundaries may be blurred where two species have only recently diverged or if the character which distinguishes them is not obvious.

Distinctions between orders, families or genera are usually easily made. For example, it is easy to differentiate between a grasshopper and a butterfly. Whilst we recognize that they are both insects, the differences in their anatomy (the enlarged hind-legs of the grasshopper and the broad, dusty wings of the butterfly) readily place them in different orders. As we descend down the hierarchy, deciding where a family or genus begins or ends requires closer and closer observation (Figure 2.3). At the level of the species, the difficulty is in deciding whether the differences we observe between individuals represent significant adaptations. Are the variations in the coloration of the butterfly's wings part of the natural variation within a species, or a characteristic that helps to define a distinct species with some adaptive significance? Often, variations in colour mark only a different race, a variety confined to a particular area, but still capable of interbreeding with the rest of the species.

Shared morphology remains the most common means of classification and the basis of most identification keys for higher plants and animals (Figure 2.4). In a key, a series of descriptions are offered at each level in the hierarchy. An unknown specimen is compared against each description and placed in the category which has the closest fit. This continues, running down a sequence of alternatives with directions to the next stage. By choosing the most

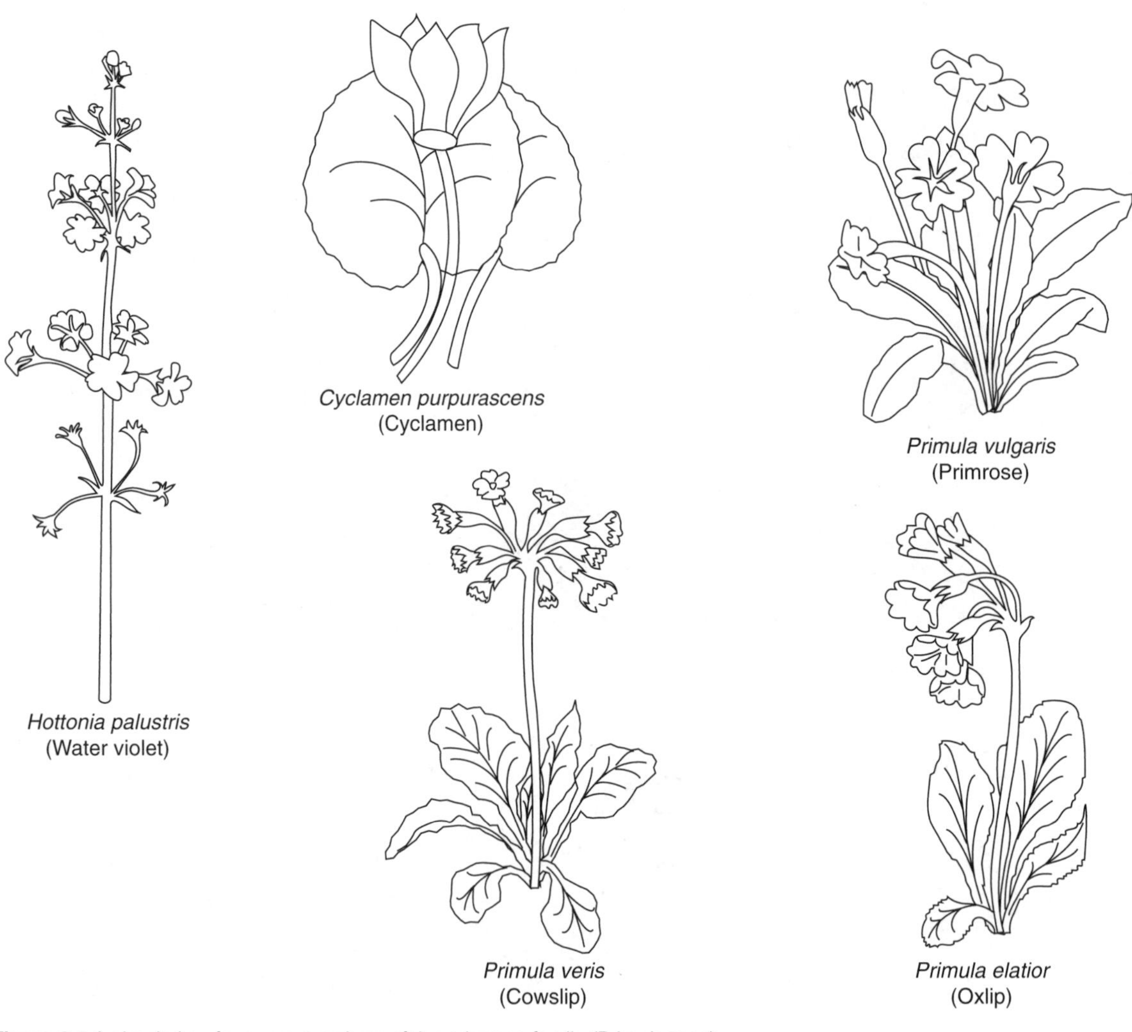

Figure 2.4 A simple key for some members of the primrose family (Primulaceae).
1. Leaves in basal rosette ..... go to 2
   Leaves in rosette and along stem ..... water violet (*Hottonia palustris*)
2. Petals flat/turned forwards (*Primula* species) ..... go to 3
   Petals turned backwards ..... cyclamen (*Cyclamen purpurascens*)
3. Flowers in a cluster ..... go to 4
   Flowers solitary ..... primrose (*Primula vulgaris*)
4. Flowers fragrant ..... cowslip (*Primula veris*)
   Flowers not fragrant ..... oxlip (*Primula elatior*)

appropriate description at each step, the key leads to a identification that gives a complete description that should match the specimen.

Classifying on looks alone can be misleading. Morphological variation caused Edward Poulton considerable difficulty when he tried to work out the classification of the African mocker swallowtail butterfly (*Papilio dardanus*). The males were obvious enough, but there seemed to be no females of the species. Females of the genus were known, but there were at least three different types, so variable that they resembled neither the males nor each other. Each group of females was therefore classified in a species of its own. In fact, the females come in three forms, each of which mimics one of three species of unpalatable butterfly (*Amauris* species) (Box 4.2). The males, on the other hand, are uniform because they need to be recognized by the female for her to allow them to mate. Poulton found that different female forms shared the same anatomy despite being polymorphic. Notably, those females which are intermediate between the three forms, and which are not close mimics to the unpalatable species, are most likely to be preyed upon.

Another complication in classifying from morphology alone is that natural selection can produce very similar structures from very different starting points. This is known as **convergent evolution**. One of the most spectacular examples is the thylacine or marsupial wolf. Probably now extinct, this was, to all outward appearances, a member of a dog family, but as a marsupial it was more closely related to possums and kangaroos (Plate 4).

Such structural resemblance is termed **homoplasy** and is particularly common amongst plants, possibly because they are relatively immobile and are forced to adapt to local conditions. Again, similar selective pressures will tend to produce similar results. Mediterranean-type plant communities are found in five distant regions of the world, but although they have very different plant groups, many species have evolved thick, tough leaves and a short shrubby stature, adaptations against a prolonged summer drought and periodic fires (section 7.1).

A more general example of homoplasy is tendrils developed by plants to reduce the cost of raising themselves up toward the light. Tendrils enable the owner to cling to other plants and use the support of their substantial stems and trunks. Various plant families have evolved tendrils, but as modifications of different structures (Figure 2.5) including leaves, stems, petioles (leaf stalks) and stipules (scales).

2.2.1 The biological species

The complications of morphological variation and convergent evolution have led biologists to treat morphological species with a good measure of caution. Today, most biologists use a more functional definition of a species, termed the **biological species concept**. A biological species comprises

Figure 2.5 *Clematis* uses its petioles (leaf stalks) as tendrils.

those individuals that can interbreed successfully to produce fertile and viable offspring (Box 1.1) forming a gene pool and sharing a common set of adaptations. So while there may be differences in morphology, like the butterflies, and while the genetic code differs between individuals, or between males and females, they are a species if they can produce fertile offspring together.

An advantage of this definition is its recognition that species are variable and changeable. By referring to an interbreeding group, it does not preclude the possibility of change in the future, perhaps with some becoming excluded from the breeding population. Over time, species evolve and new breeding populations form.

However, the biological species concept can be less readily applied to those organisms that reproduce asexually. The dandelion (*Taraxacum officinalis* agg.) has fewer than 10 sexual species, with the rest (around 2000) reproducing asexually. They are **apomictic**, meaning that they produce seed without pollination, without the fusion of a male gamete (pollen) with the ovule (Figure 2.6a). Apomictic species (also known as **micro-species**) are genetically isolated. Since the flow of genes from one individual to another has stopped, different micro-species are often found growing very close together, but remaining distinct from each other. John Richards discovered that as many as 100 micro-species of dandelion could be found within a single hectare. Brambles (*Rubus fruticosus* agg.) have also abandoned sex to reproduce apomictically, with Britain having over 400 and North America 381 recognized micro-species (Figure 2.6b). Micro-species are notoriously difficult to classify and to avoid the complication we simply use the abbreviation 'agg'. (for aggregate) after the specific name.

Many plants do not conform to the principle of the biological species and they exchange genes across species boundaries. Sometimes, often under rather special circumstances (such as captivity), the barriers that normally keep some animal species distinct may also be insufficient to prevent interbreeding. Any offspring from such unions is called a **hybrid**. Interspecific hybrids are rare in most animal groups and these are rarely fertile. The sterile mule (the hybrid from a horse and a donkey) or the ligers and tigons (crosses between tigers and lions sometimes produced in zoos) are genetic dead-ends.

Nevertheless, there are examples of fertile hybrid animals, some of which have been taken for different species. Indeed, we have only learnt about their true ancestry after looking at their genetic signature (Box 2.2). One such surprise was the discovery that the red wolf (*Canis rufus*) was actually a fully fertile hybrid between two distinct species, the coyote and the grey wolf (Plate 5).

The red wolf underwent dramatic decline after 1900 due to hunting, loss of habitat and increasing hybridization with the coyote, which had been extending its range. Robert Wayne and Susan Jenks compared the mitochondrial DNA (mtDNA) of the few red wolves collected together in a captive breeding programme with that of coyotes and grey wolves. They also collected mtDNA from six pelts of red wolves that dated before hybridization was considered to have begun on a large scale. The mtDNA and its cytochrome *b* sequences revealed that the red wolf is a hybrid of coyotes and grey wolves, and that their range represented a zone of hybridization between the two separate species. Indeed, Wayne and Jenks suggest that some of the wild animals they sampled in the wild would have been classified morphologically as grey wolves but actually approximate more closely to the red wolf genotype.

In several plant groups fertile hybrids are quite common. Many species of orchid belonging to the genus *Dactylorhiza* will readily cross with each other. Some hybrids appear to be a perfect blend of the two parent species, while others resemble one or other parent. This makes them particularly difficult to identify (Plate 6).

The results of crosses between fertile hybrids also make it difficult to trace phylogenies, their genetic ancestries. For these groups, the idea of a species as a discrete entity is misleading and we have to accept a more fluid classification, of a highly variable group of individuals passing genetic code readily between themselves. Nowadays, molecular biology and genetics allow us to explore these differences in increasing detail (Box 2.2).

2.2.2 Defining species by their chemistry

We could also take Romeo's advice and identify a rose by its scent. Flowers are recognized for their ability to produce complex chemical signals to attract insect pollinators. Some scents even mimic an animal's sex pheromone, the chemical signal

Figure 2.6 Apomictic species: (a) dandelion (*Taraxacum officinalis* agg.); (b) bramble (*Rubus fruticosus* agg.).

Box 2.2

READING THE MOLECULAR CODE

Looking at the genetic code, the blueprint for a species, may seem the obvious place to look for the real differences between one species and another. We have, however, only been able to read gene sequences for a few years. Before that, molecular biologists had to look at the products transcribed from the code (Figure 1.4). This meant comparing the composition of proteins, especially those responsible for key functions, to reconstruct the phylogenetic relationships of a species.

The greater the genetic distance between two individuals, the less likely they are to share proteins with exactly the same structure. Even in the absence of a selective pressure, genotypic differences and chance variations will change the amino acid composition or sequence and we can quantify this relatively easily. **Immunological methods** use the remarkable capacity of mammalian immune systems to detect 'non-self' proteins. Antibodies produced against the foreign protein are highly specific and the degree of binding between an antigen (the protein under test) and its antibody can give us a gross measure of the structural similarities of proteins from different species.

Gel electrophoresis separates proteins according to their electrical charge and molecular weight. The greater the charge across a protein molecule and the smaller its size, the faster it will migrate toward one pole of an electrical field. Two protein samples migrating the same distance through a gel are likely to have the same configuration and amino acid composition. This technique is widely used because it can pick up subtle differences in protein structure and is particularly good at identifying hybrids that otherwise look the same.

Sequencing the amino acids in a protein, using a series of enzymic scissors and assays, has paved the way for more detailed analysis of proteins. One protein, **cytochrome**, has been of particular interest because it is found in all living organisms and is therefore very ancient. Cytochrome plays an important role in the energy-generating electron transport chains of mitochondria and chloroplasts (Box 5.1).

Analysis of the 104 amino acids within cytochrome has allowed biologists to group species according to these sequences. All share a sequence of 33 amino acids in the same position along the protein chain, but the remainder fall into groups – a hierarchy of sequences that mirrors the phylogeny of living systems. All the vertebrates, for example, fall into a group of their own, as do the invertebrates, plants and fungi. The cytochrome of human beings and chimpanzees is identical.

The most direct way of deciding the genetic differences between individuals and species is to read the genetic code itself. Again, technological advances in the last 20 years have meant that we do this with increasing reliability and speed.

Part of the double strand of DNA is unzipped when the genetic code is being read (Figure 1.4). Strands can also be separated by heat treatment and will re-associate as they cool. In **DNA hybridization**, DNA from two sources is heat treated, mixed together and allowed to cool. Some strands from each source will associate to form hybrid strands. The hybridized DNA is then separated again by heat treatment, but the closer their match, the higher is the temperature required to dissociate the strands. Using the temperature scale over which complete dissociation takes place between hybridized and non-hybridized strands, we can calculate the percentage similarity for an entire genome.

More precise information can be gained from **gene sequencing**, especially when we wish to look in detail at a particular part of the genome. DNA fragmentation techniques use specific enzymes to break the strand of DNA at known places. We can then examine the gene sequences within each fragment and relate this to their known coding instructions. Once again, the more fragments shared between two individuals, the longer is their common evolutionary history.

These techniques can be used wherever DNA occurs (including the remains of some long-dead organisms). We can also distinguish the DNA of the nucleus from that of the mitochondria or the chloroplast. **Mitochondrial DNA** (mtDNA) is only inherited from the mother with the cytoplasm supplied in the egg cell (section 1.6), making it particularly useful for working out relatedness within parthenogenic species where females produce young without having to mate.

It also helps us to sort out the parents of the hybrids. Using this technique, Steven Carr was able to unravel the complex mating system of white-tailed deer (*Odocoileus virginianus*) and mule deer (*Odocoileus hemionus*) in Texas. He found that mtDNA of the hybrids was the same as *O. virginianus*, indicating that crosses between the two species invariably had a mule deer male and a white-tail female as their parents.

Classifying species and working out their origins by genes alone can be problematic. Changes in DNA may not accurately reflect evolutionary change. DNA molecules are more changeable in some groups than in others. This makes it difficult to construct accurately timed phylogenies on the basis of molecular evolution alone. For example, plants within the sunflower family, the Asteraceae, show little variation compared with other flowering plants. This may mean that sunflowers are a recent evolutionary development or simply that the DNA within their chloroplasts is slow to change. There are also risks in classifying an entire species based on the genetic material of a small number of individuals, especially in a very variable species.

used to attract a potential mate. Until recently, a group of Panamanian orchids defied all attempts at classification. Defeated, taxonomists simply bundled the whole lot into single species, *Cycnoches egertonianum*, and referred to them as the 'Egertonianum complex'. Unable to classify the plants by traditional means, Katherine Gregg looked at the sex lives of the bees that pollinate them. After they have attempted to mate with the orchid, male bees collect the delicate lemon scent and reuse it as an ready-made pheromone to mark out their mating sites. A number of different species of bees use this strategy and field observations showed that they were extremely fussy about their choice of plant. Fascinated by this behaviour, Gregg ran a chemical analysis of the scent and found that it was composed of up to 18 separate compounds. She also found that differences in the composition of the scent/pheromone reflected links with different pollinators. By reference to the species of bee and type of scent she was able to make sense of the Egertonianum complex, dividing it into four 'chemotypes', which are now recognized as four separate species.

Although the bees might have got there first, our use of chemistry to classify organisms also has a long history. In fact, the standard test for lichen identification was developed by Nylander back in 1866 and enables lichens to be grouped into any of six chemical races. The molecular classification of plants is also well developed because of our detailed knowledge of the secondary plant metabolites (section 4.4) which have medicinal or economic value. Genera like *Eucalyptus* hybridize freely and it is not always easy to work out which two species were the parents (especially when one of the parents may itself have been a hybrid). A particular group of these compounds – the terpenoids – have proved invaluable in sorting out such plant paternity cases. The same technique has been used to work out the origins of hybrids in other aromatic species such as junipers (*Juniperus* species) and pines (*Pinus* species).

Although morphological characteristics remain the basis of most keys to multicellular plants and animals, chemical and biochemical techniques enable us to make finer distinctions. Our goal is to arrive at a classification and phylogeny that reflects real adaptive differences. The complication is the great variability from one individual to another.

2.3 VARIATION WITHIN A POPULATION

Variation between individuals is usually fairly obvious. We use these differences to distinguish one person from another and to recall their name, at least most of the time. Under other circumstances, we might wish to know what significance these differences have for an individual's survival and reproductive success.

If we plot some measure of fitness (say, growth rate or abundance) against some key environmental factor (such as temperature or water availability), we invariably find that a species has an optimum range (Figure 2.7). Usually this produces a characteristic curve, termed a **normal distribution**. Most individuals are found close to the mean value and numbers rapidly tail off with increasing distance away from the mean.

This tells us something about the species adaptation to that parameter. Away from the optimum, conditions would be less favourable and individuals will incur physiological costs to maintain themselves. For a warm-blooded animal (or endotherm) this means using energy to keep warm or to keep cool. If the costs are high, or have to be sustained for a long time, an individual will have fewer resources to devote to growth or reproduction (section 4.1). Those forced to live at the margins of their physiological range may thus fail to reproduce or grow.

However, some individuals may be adapted to living outside the optimum range for the larger population. This is the nature of variation within a species: not every individual responds in the same way to a selective pressure and there will be some who favour conditions some way from the population optimum.

Of course, not all of the variation we observe will have an adaptive significance and some may differ simply because of the chance combination of genes they were dealt at fertilization. This is neutral variation (section 1.3). Much of the variation in proteins (Box 2.2) appears not to be adaptive. Even so, in new or changing environments, some traits may, again by chance, equip some individuals to survive and reproduce with a higher frequency than their neighbours. As we saw in the last chapter, these differences allow for natural selection and evolution.

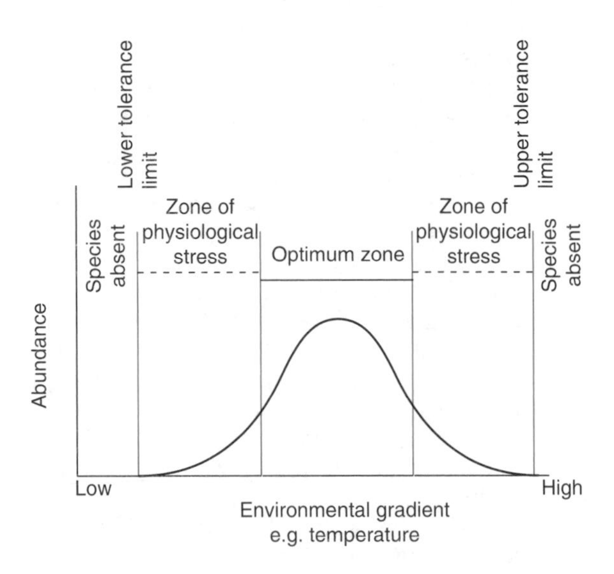

Figure 2.7 A plot of abundance of a species against some environmental parameter, in this case temperature. This shows the range to which a species is best adapted and the same sort of pattern would be found for any important parameter, such as water availability, nutrient level, salinity and so on. Outside the optimum range, numbers decline rapidly because individuals have to expend energy or other resources to adapt physiologically to the poor conditions. At extreme temperatures, conditions are too harsh and the species is not found. We could make the same sort of plot for any other measure of a species performance, such as reproductive success or growth rate.

2.4 ECOLOGICAL NICHE

If a species is defined by its collection of adaptations, which selective pressures have been most important in shaping it? And what selective pressures have led to it becoming separated from its close relatives?

Major morphological adaptations, such as those needed for flight, have a long evolutionary history and will separate animals high up in the hierarchy, at the Class level (Class Aves, the birds) or Order level (Order Chiroptera, the bats). Smaller adaptive changes – variations in colour, for example – may not require major physiological or structural changes. They distinguish closely related species which have diverged relatively recently. Even more recent are the distinctions between varieties of plants or subspecies of animals that perhaps are only separated geographically rather than genetically.

What about closely related species that share the same geographical range, but apparently remain distinct? Often, it is the details of their life history, or the habitat they use, which keeps them separate. For example, two micro-moths, *Lampronia rubiella* and *Lampronia praeletella*, share a similar range, and as adults have similar markings. Their caterpillars, however, are leafminers, and one attacks strawberry leaves (*L. praeletella*), the other raspberry fruit stalks (*L. rubiella*). They have evolved to use different resources and, indeed, live their larval stage in different habitats. This separation at one stage in their life cycle means they are not in direct competition with each other. The larval food plant is one of a large range of factors to which each species has to adapt, any one of which can serve to distinguish them from other species in the same genus.

For any environmental factor, we could define the optimum range to which each species is adapted (Figure 2.8), both abiotic and biotic. The performance and distribution of a species depend on how well it is suited to its environment and the combined effects of all the selective pressures. This totality of factors to which it has adapted is called its **ecological niche**.

Niche is a complex idea that tries to describe the fit between a species and its environment. An ecological niche is not a place is the normal sense of the word and it is also much more than a species' habitat. It is represents the interaction between a species and its habitat, describing both where a species lives and how it lives there.

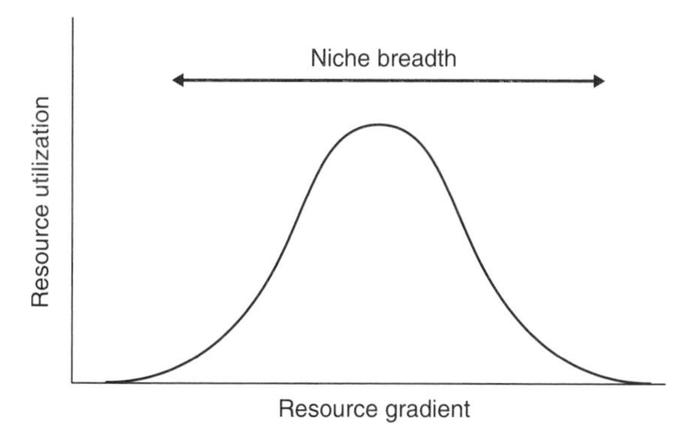

Figure 2.8 Niche, defined by one environmental gradient – in this case, a resource being used. For every environmental parameter we can define an optimum (Figure 2.7), but the full range exploited by a species is termed its niche breadth. For a plant, such a gradient might be light levels of different intensities, whereas the niche breadth for an insectivore would be the size-range of insects on which it feeds.

Niche is sometimes described as a species' role in its community. That role depends on the interaction with other members of the community and so can include everything from being, say, a strawberry leafminer to a prey item for a bird or bat, or a host to a parasite.

Although niche is difficult to define, it does have a very real ecological meaning. We can see this in convergent evolution, where species filling equivalent niches in different areas develop the same adaptive solutions. The thylacine (Plate 4) and the dog is one example. Anteaters in the three southern continents have elongated snouts and long probing tongues – equivalent adaptations to the same selective pressure. Like the thylacine, one of these (*Myrmecobius*) is a marsupial, another (the echidna) is an egg-laying mammal, while the pangolins of Africa or the tamandua of South America are placental mammals. These fundamental differences in reproductive physiology place their separation high up in the hierarchy (at the level of the subclass), indicating a large phylogenetic distance between them. Yet each has evolved equivalent adaptations to collect ants.

The match is not perfect and differences between most convergent species are obvious on even the most casual inspection. This emphasizes that a niche is defined by the organism and its adaptations, interacting with its environment. Rather than simply being a role that might be played by a number of species, a full description of a niche includes the detail of the species which occupies it. So, part of the niche occupied by *Lampronia rubiella* is defined by the raspberry stalk it uses when a larva.

Clearly there are practical difficulties in describing every significant factor in a species' niche. Usually studies have concentrated on those which are easily measured and which seem to be important in separating species. Two principle niche measurements are made. **Niche breadth** is the range of a factor over a which a species is found. This could be abiotic factors like temperature, or (more often) the range of a resource which a species uses (Figure 2.8).

An insect that feeds on the leaves of a variety of plants is said to have a larger niche breadth than one which feeds only on a single species. Species with a large niche breadth are **generalist species**, able to exploit a wide range of resources. **Specialists** are adapted to a narrow range of a resource, but attempt to out-compete other less well-adapted species by being more efficient in its use. Being a generalist is a good strategy in an unpredictable habitat, when switching to a different part of the **resource spectrum** may be necessary at times of shortage. Being a specialist is only viable when the resource supply is very reliable, in constant or predictable habitats (section 3.5). We would therefore expect to find frequently or recently disturbed habitats dominated by species with broad niches.

Niche overlap is the part of a resource spectrum which two species occupy (Figure 2.9a). Sometimes overlap indicates that two species are competing with each other for the resource. At other times it can mean the opposite. To understand how, we have to look at the two forms competition can take.

2.4.1 Niche and competition

Competition between individuals of the same species is termed **intraspecific competition** (section 3.2) and is one of the prime forces driving natural selection. Those most able to secure resources – food, space or shelter – will be more able to reproduce and ensure that their genetic code enters the gene pool of the next generation. Ordinarily, we would expect those individuals close to their optimum range to have the greatest reproductive success (Figure 2.7).

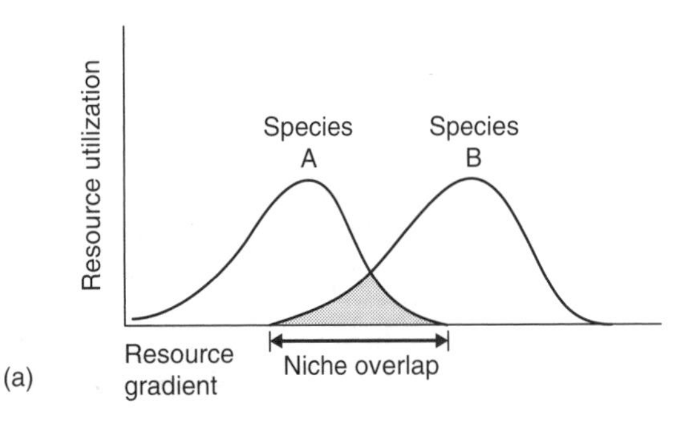

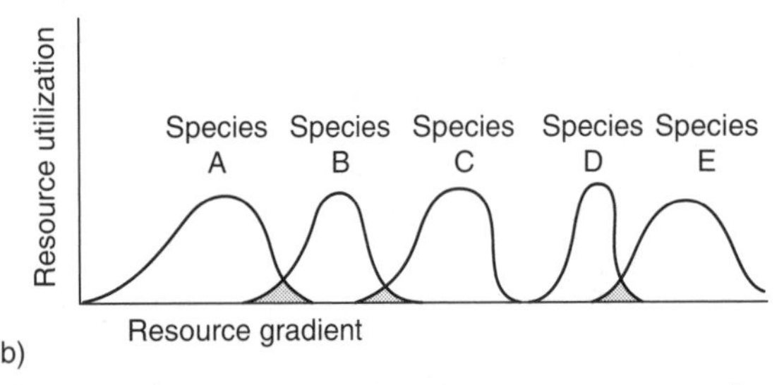

Figure 2.9 (a) Two species that exploit the same part of a resource gradient (or 'spectrum') are said to show niche overlap. In some cases, this may indicate competition between them for that resource. (b) A collection of species exploiting the same resource can only coexist if they each exploit different parts of the gradient.

However, genetic change might adapt some individuals to a different part of the environmental gradient. Perhaps a change at one locus means a key enzyme is more effective at lower temperatures. This new genotype may then be able to thrive at a new optimum and its numbers increase (Figure 2.10). Mutation has produced a variety that eventually might lead to a new species, but at the moment, it can still breed with the rest of the population. Over time, the entire population may become dominated by the new genotype and its optimum range shifts. An alternative is that the population fragments and a new species forms.

If we examine the use of a resource we often find that different species concentrate on different parts of the range. They are spread out along the resource spectrum (Figure 2.9b). Two species sharing a large zone of overlap may indicate a region where the two species are competing for the same prey or the same nutrients in the soil.

It can, however, indicate the opposite. A resource which is very abundant may not limit the growth of either species and there may be enough for both. If the overlap persists over many generations, we may conclude that there is little competition, at least along that resource gradient.

By the same token, a lack of overlap between two species only indicates no current competition for this resource. It tells us little about their past battles or their ongoing fight for other resources. Perhaps we observe the outcome of competitive battles fought long ago, after each has become adapted to a different range.

Finally, we may witness a competitive battle in progress. Where a resource limits the growth or reproduction of both species, utilization by one species depletes the supply for the other. **Interspecific competition** (between species) means that either species would perform better in the absence of its competitor. Then niche overlap does indeed denote competition.

Under these circumstances, ecologists distinguish two types of niche. The range which a species could occupy in the absence of interference from

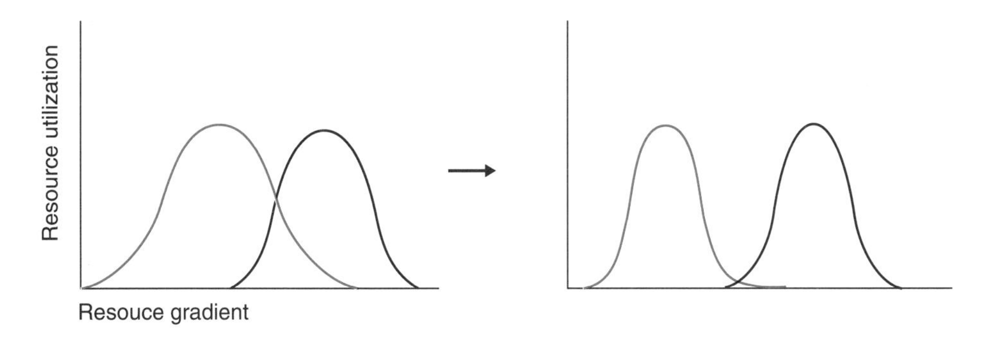

Figure 2.10 Where overlap represents a high competitive pressure on each species, one or both of them will, through natural selection, gradually change to reduce the overlap, shifting their niche. This is termed character displacement and may be represented by some change in physiology or morphology.

other species is its **fundamental niche**. The range to which it is confined by competitors or predators is its **realized niche**. Under severe competition, a species may only use a very narrow part of a resource spectrum and have a small realized niche. Then natural selection will be very intense, favouring those individuals able to make best use of what is available. These will be the most successful reproducers and will soon dominate the population and gene pool.

In this way, a species becomes highly specialized, often showing distinct morphological or other changes to adapt them to use that resource most effectively. This is known as **character displacement** (Figure 2.10) and is seen when two closely related species or races begin to diverge.

Dolph Schluter has shown how rapidly competition can lead to significant character displacement. Schluter and colleague Don McPhail work on the three-spined stickleback (*Gasterosteus aculeatus* complex), a variable group of species whose taxonomy has yet to be fully worked out. These fish have been isolated in a series of small coastal lakes in British Columbia since the retreat of the ice sheet 10 000–13 000 years ago. In some lakes, two different species are found: a larger and rounder stickleback that feeds on the invertebrates on the lake bottom ('benthic' species) and a smaller, slender species that feeds on the phytoplankton closer to the surface ('limnetic' species). In lakes where there is only one species an intermediate form is found, able to exploit both food sources. These species will readily interbreed with each other.

Schluter set up an experiment to examine the pressure for character displacement when two species share similar morphologies and the same diet. He bred three intermediate forms which were either close to the limnetic form, close to the benthic form or true intermediates. These were then grown in experimental ponds in the presence or absence of wholly limnetic species. By measuring the performance of each species (in this case by measuring their growth rate over a season) Schluter was able to work out the response of different intermediate forms to the presence of a limnetic species (Figure 2.11).

As expected, the intermediates closest to the limnetic species grew less well in the presence of limnetics. In contrast, the same intermediates in control experiments showed no depression in their growth rate. The greater the similarity between an intermediate and the limnetic form, the greater the niche overlap and the reduction in growth rate. Eventually this might translate into the loss of intermediates as only those using the different niche would survive. Over a number of generations, under interspecific competition, the intermediates would be expected to show character displacement, presumably approximating to the rounder larger form of benthic sticklebacks.

It seems that the niche separation we see in some lakes today is the result of past displacements. Perhaps, with time, the intermediate forms in single species lakes may show the same character displacement. Schluter reports signs that the process is already under way, with intermediates either taking most of their food from the phytoplankton or concentrating on the invertebrates. Again, it is the bigger fish that are predominantly bottom feeders. It may be that the same niche separation has occurred independently in five lakes over the last 13 000 years. Where single species intermediates are found, it seems to be happening again.

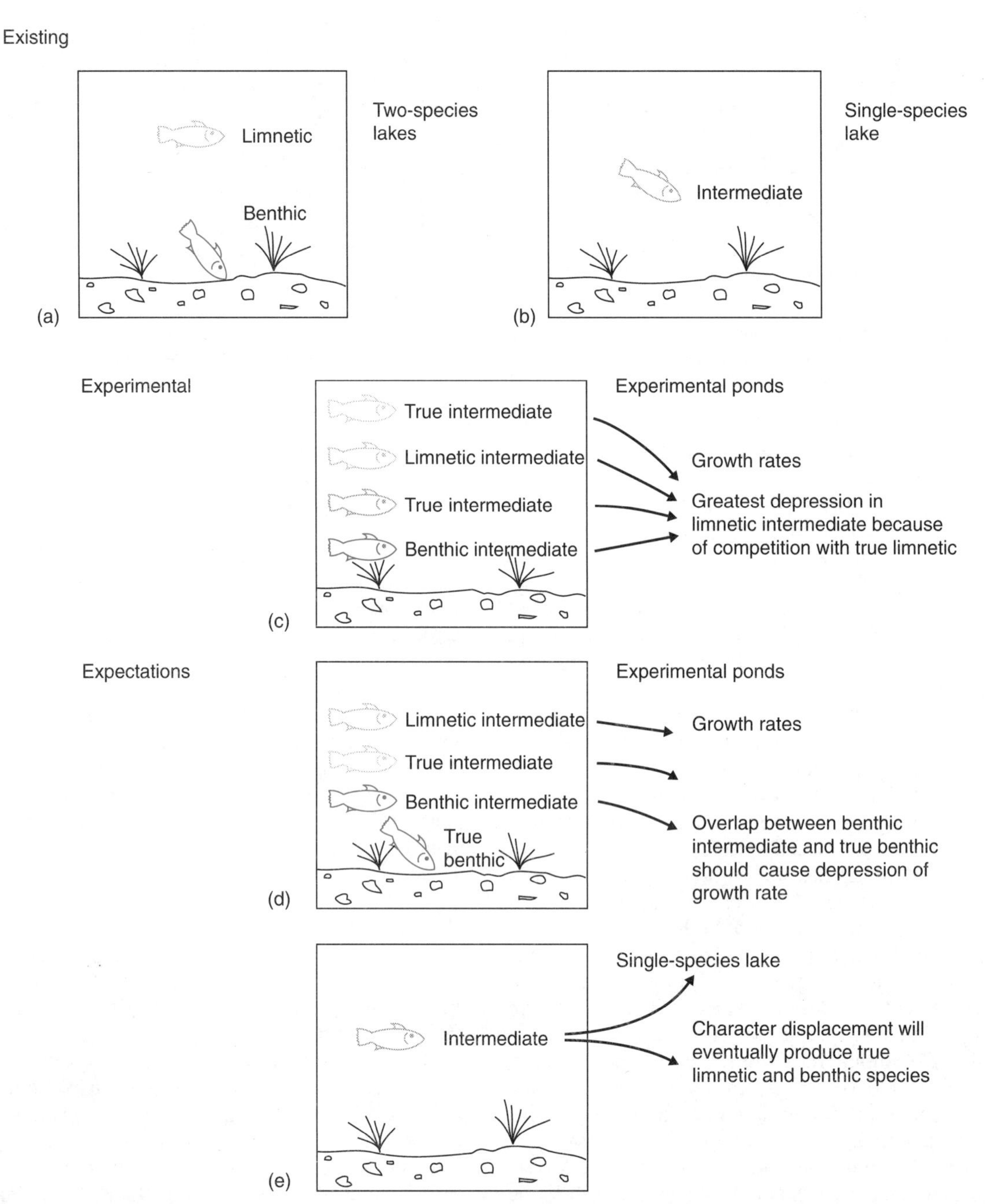

Figure 2.11 Schluter's experiments with three-spined sticklebacks (*Gasterosteus aculeatus* complex) demonstrating character displacement. Lakes in British Columbia have either two species of stickleback with different morphologies or one which is intermediate between the two. (a) One form (the 'benthic' species) feeds on bottom-dwelling invertebrates and has a large mouth and rounded body; the other feeds on phytoplankton in open water and has a smaller mouth and slender body (the 'limnetic' species). (b) Intermediates feed in both categories. (c) Schluter has shown that intermediates with forms closest to the limnetic species have the greatest growth depression. (d) Similarly, we would expect intermediates in closest competition with benthic species to grow less well. (e) Over many seasons, these intermediates would be expected to show further character displacement, feeding more and more in the opposite category of the resident specialist. Lakes now populated by intermediates are likely to differentiate into the two forms given sufficient time.

Plate 1 Biosphere 2, a microcosm of the Earth's biosphere built in the Sonoran Deserts of the south-western United States.

Plate 2 The Earth's biosphere taken during the *Apollo 17* lunar mission. This view includes the Mediterranean Sea, Africa and the Antarctic polar ice cap. According to the Gaia hypothesis, the temperature and composition of the atmosphere, waters and solid surface forming the surface of the Earth shown in this photograph are controlled by Life.

Plate 3 Blackbirds and bluebells. Photographs (a) and (b) show the European blackbird (*Turdus merula*) and bluebell (*Hyacinthoides non-scripta*); (c) and (d) show their North American namesakes, *Agelaius phoeniceus* and *Phacelia tanatcetifolia*.

Plate 4 The marsupial thylacine is remarkably similar to placental dogs, or perhaps a blend of dog, cat and hyaena. As one of the few major carnivores in Australia throughout the Pleistocene, it probably filled the role of all three, and shows how evolution can converge on similar adaptive solutions.

Plate 5 The red wolf of North America. Classified as *Canis rufus*, the red wolf is actually a hybrid of the coyote (*Canis latrans*) and the grey wolf (*Canis lupus*).

Plate 6 (a) The southern marsh orchid (*Dactylorhiza praetermissa*) and (b) the common spotted orchid (*Dactylorhiza fuchsii*). These two species were hybridized to produce an interspecific hybrid orchid (c).

(a)

(b)

Plate 7 Distyly in *Primula vulgaris*. The flowers occur in two forms: (a) the long-styled pin-eye form and (b) the short-styled thrum-eyed form.

Plate 8 The Bialowieza National Park, Poland.

Plate 9 Regenerating woodland, dominated by large numbers of small saplings.

Plate 10 Elephant browsing acacia, Masai Mara, Kenya.

Plate 11 Elephant damage to acacia trees, Masai Mara.

The alternative to character displacement may be oblivion. If the competition for a resource between two species is very intense, one species may lose out completely. Given sufficient time, any difference in their capacity to use a resource, or to interfere with the other's exploitation of it, will mean one species flourishes at the other's expense.

This is the **competitive exclusion principle** and simply states that two species may not occupy the same niche at the same time in the same place. One niche can only be occupied by one species (section 4.3).

However, not every niche may be filled. We see this occasionally when an alien species manages to establish itself in a new habitat. Sometimes, a invader can squeeze between two other species, by being better adapted to resource space that neither fully occupies. For any newcomer to persist, it needs to be sufficiently different from its new neighbours to carve a new niche for itself. There are also lots of examples of the alternative outcome, where an introduced species outcompetes the native niche-holder, so the latter is lost (section 4.4).

Competitive exclusion and the partitioning of resources between species means there are probably limits to the number of species that a community can hold. The number of species that could be packed along a resource spectrum depends on how different they need to be to avoid excessive competition. This has proved very difficult to measure, partly because their separation on one environmental gradient is often tempered by their interactions on other gradients (section 9.3).

Niche theory is useful because it links the evolution of species to their use of resources and the number of species that a community can contain. It also links the range of resources used by a species to the constancy of their habitat, and explains why we might expect more species and more specialist species in constant or predictable habitats, a theme that we explore in some detail in the last chapter.

2.5 SPECIATION

Dolph Schluter's sticklebacks are in the process of making new species, but are not yet perfectly separated because they can interbreed and produce intermediate hybrid forms. The reproductive barriers that would define them as new species have not yet formed. If two populations remain isolated for long enough, genetic differences emerge which eventually prevent them from mating at all. Their gene pools are then separate and any changes that become established in one pool are not transmitted to the other.

Consider the allele we described earlier, coding for an enzyme that had a different temperature optimum. Under character displacement, individuals with different alleles flourish in different places where the temperature regime is most suitable for each type. This means they are physically separated – perhaps in space, say by living on different sides of a hill, or perhaps they are active at different times of day. Either way, the effect may be the same: the two populations rarely meet or interbreed with each other and only exchange genes with their near neighbours.

Differences between the two populations are then never diluted by gene flow. Genetic code is not shared between the populations and they begin to diverge. Over a long enough period of isolation, their genetic differences may preclude mating, even if they are subsequently reunited.

Speciation which follows populations being separated in space or time is termed **allopatric speciation**. In this case, gene flow is reduced by their physical separation. With **sympatric speciation**, populations may live together, but gene flow is intially restricted because of some genetic change.

2.5.1 Allopatric speciation

Spatial separation can occur when a population fragments, perhaps as some individuals become adapted to more marginal environments. Only a small number of individuals may show an adaptive change, but if this proves successful their numbers quickly build. If they can survive where others cannot, they benefit from a lack of competition and may have part of a resource spectrum to themselves. Their separation also means that they mate primarily with each other, rather than with the main population. As a result, their adaptive traits soon become fixed within their subpopulation.

Sometimes, a small section of a population is cut off from the main population by a major disturbance, such as a flood or a forest fire. Again, gene flow with the parent population is restricted. In the same way, individuals invading a new habitat (perhaps an island) may only breed with those that have colonized the new area.

Notice that in all these cases the new population is breeding from a small collection of individuals, either because they are colonists (or survivors) or because they are the first to show the new adaptive trait. These few individuals will have only some fraction of the genetic variation of the parent population. This has two important consequences for future genetic change.

First, traits that might occur at a low frequency in the main population may be more common here. The few individuals establishing the subpopulation can be unrepresentative of the genotype of the larger population, if only by chance. From then on, their future generations may maintain and exaggerate these differences. This is known as the **founder effect**, and it means that small gene pools can rapidly diverge from the parent population. Given the different genetic stock each has to draw on, the two populations may quickly differentiate.

Second, in a small population rare alleles are easily lost by chance. With large numbers rare alleles will be found in a few individuals and may therefore persist down the generations. However, the same proportion in a small population implies very few individuals, so the allele is readily lost. In this way, the frequencies of some alleles can change very rapidly in a small population, a process called **genetic drift**. Isolated, a smaller gene pool becomes more and more distinct with time. Again, with no exchange with the parent population, the two gene pools quickly diverge, perhaps to reach a point when they can no longer interbreed.

2.5.2 Reproductive barriers

Of course, the populations which have become physically separated can be reunited again, and gene flow can be resumed. For physical separation to produce a new species, it must last long enough for the populations to diverge such that mating between them is no longer possible, creating a reproductive barrier between them.

Reproductive barriers take one of two forms according to whether they operate before or after fertilization: **pre-zygotic** and **post-zygotic barriers** (Table 2.1). Pre-zygotic barriers are either mechanical, physiological or behavioural. Many insects have intricately sculptured genitalia, so that the male fits the female rather like a key inside a lock. This seems to be one way in which the female insect ensures that she is fertilized by the correct male, especially in congeneric species (species of the same genus).

A similar situation exists in the primrose (*Primula vulgaris*). These plants have flowers of two forms: pin-eye flowers have a large prominent style while in thrum-eye flowers it is the male stamens that are prominent (Plate 7). Both forms reduce the chance of self-fertilization but it may also be that one group serves as functional females (with the prominent styles) while others take on the male role (with the prominent stamens). However, we find that entire populations are dominated by one form or the other. An alternative explanation is that pin-eye forms are favoured in areas where there is an intense form of herbivory. Slugs will feed on stamens so that thrum-eye forms are readily emasculated where slug numbers are high. There, perhaps, pin-eyes will have a selective advantage.

Even when sperm are introduced to the egg, fertilization may still not happen. Often there are physiological barriers between the gametes of different species, and at the cellular level the two cells have to recognize each other. Failure of an egg cell to recognize a sperm (or vice versa) prevents fertilization between species, and sometimes within species.

Behavioural barriers are more easily observed. Often the female is highly selective, choosing a mate according to the signs and signals he uses to demonstrate his identity and worth. Within a species, females will often select between males according to some indication of the quality of their genotype. Males boast of their prowess by their plumage, or by their song or dance routine, and are selected by the female (though in some species the roles are reversed). This is termed **sexual selection** and explains why male birds of paradise or peacocks sport such elaborate plumage. Unless the signals are recognized by the female, the male may never get to pass on his genes. Part of the massive variety of fruit flies (*Drosophila*) in Hawai'i is due to sexual selection. Males of some species are required to dance for the right to mate with a female. By their dance so they are known, and getting the steps right signals to the female that the male belongs to the correct species.

Post-zygotic barriers may operate at various stages in the development of the zygote. Sometimes a fertilized egg will not develop. A mismatch between the number of chromosomes in the sperm and the egg means that the development of the offspring is not likely to proceed very far. Even amongst closely related species, matching (homologous) chromosomes may differ according to their gene sequence and such incompatibilities usually cause meiosis to fail.

Sterile hybrids unable to produce effective and viable gametes are often the result of interspecific

Table 2.1 Reproductive barriers

Stage	Barrier	Description
Pre-fertilization (pre-zygotic barriers)	Ecological isolation	Populations are separated by distance or barriers (such as mountains or water bodies).
	Temporal isolation	Populations may be reproductively active at different times; they may flower at different times or have different breeding seasons.
	Behavioural isolation	Without the correct signals to initiate reproductive activity, males and females of different populations may never interbreed.
	Mechanical isolation	Reproductive organs need to complement each other for the exchange of gametes. Anatomical differences can thus prevent fertilization (Plate 7).
	Gametic isolation	Unless the sperm and the egg recognize each other, fertilization may be prevented by a failure of them to fuse.
Post-fertilization (post-zygotic barriers)	Hybrid inviability	Embryonic development may be impaired so a hybrid never reaches the adult stage.
	Hybrid sterility	Offspring are produced but they are infertile, producing either dysfunctional gametes or no gametes at all.
	Hybrid breakdown	Although the offspring are fertile and may reproduce, their young fail to develop properly, cannot reproduce or are poorly adapted to new habitat.

crosses. Even where a hybrid develops to full maturity, there are a number of ways in which it may be prevented from breeding, collectively called **hybrid breakdown**. Many hybrids are poorly adapted to the habitat. One example is the intermediate forms of *Papilio dardanus*. Because these are a poor match with any of the distasteful butterflies mimicked by the rest of the species, they are readily taken by predators. In the same way, hybrids are likely to lose competitive battles with the parent populations, more closely adapted to specific niches. This is what Schluter was able to demonstrate with the hybrids he entered into competition with the limnetic sticklebacks.

Reproductive barriers are the feature that makes animal species the most well-defined taxonomic unit. Whereas the rest of the hierarchy is a classificatory convenience for us, many species are distinct and operate as functional units. Reproductive barriers prevent closely related species from blending into each other by a mingling of their genes. The faster these reproductive barriers are erected, the quicker the identity of a species becomes established. The longer the barriers have been in place, the less likely it is that hybrids will form.

2.5.3 Sympatric speciation

In this case, species are formed by becoming reproductively isolated through genetic change, even though they are living side by side. Gene flow is halted not by a physical barrier but by individuals becoming divided by a genetic change.

This is thought to happen most readily in highly fragmented habitats, where significantly different conditions are found within a short distance. For example, adjacent valleys can differ in the amount of sunlight they receive or in their geology and soil type. Similarly, the adaptations needed to survive in one pond may differ from a second pond perhaps only a few kilometres away. In these cases, local conditions will select traits in their populations. This may mean that an individual moving to the next pond or the next valley has a lower chance of survival in its new habitat, to which it is less well adapted. Similarly, crosses between individuals from the two habitats may be less fit for either location. Natural selection then favours genetic isolation, so that offspring bred within each local population are better adapted to that habitat and, once again, the intermediate hybrids are not fit for either pond.

This means that we should expect greater speciation in very heterogeneous habitats. This may explain the higher species richness of some mediterranean-type communities (section 7.1) and also the tropical forests (section 9.3). The same sort of separation can occur temporally, where populations become specialized for flowering or feeding at different times of the day or during the year. As we saw earlier, a population adapted to a different temperature optimum may have a restricted period of activity, reducing its gene flow with the rest of the population.

The other main mechanism by which sympatric speciation can occur is through **polyploidy**, most especially in plants. This happens when normal diploid ($2n$) individuals produce gametes with multiple copies of their chromosomes. A gamete that fails to undergo meiosis will stay diploid so that when it unites with a standard haploid gamete a triploid ($3n$) zygote is formed. Tetraploids ($4n$) form when two diploid gametes combine.

Whilst polyploidy rarely results in viable or fertile offspring in animals, it is much more common in plants. This may be because plants tend not to have sex chromosomes – save for a few notable exceptions such as cannabis (*Cannabis sativa*), hops (*Humulus lupulus*) and white campion (*Silene alba*) (Figure 2.12). In fact, with 95% of all plant species being hermaphrodite, this is perhaps the best strategy for an organism which is largely immobile.

Polyploids are often bigger than their diploid counterparts and survive despite, and even because of, their unusual genetic make-up. Polyploidy has been an important cause of speciation in crop plants, especially cereals (Box 2.4).

The importance of sympatric speciation is still a matter of considerable debate, especially regarding animal species. Many evolutionary ecologists question whether we should regard small changes of habitats as effective barriers, albeit small-scale, making sympatry just another form of allopatric speciation. Some of the best evidence that speciation can indeed follow genetic change alone comes again from freshwater fish, this time the remarkable cichlids of Africa.

A sequence of lakes occurs within the African Rift Valley; some of them are freshwater and some are highly saline. The lakes change dramatically in water level from one year to the next, and over thousands of years have expanded and contracted considerably. Some lakes have been cut off from each other for many thousands of years, during which time their fish communities have speciated into a large number of forms. Even within a single lake, differences in habitat type and the habitats created

Figure 2.12 White campion (*Silene alba*), one of the few plant species to have sex chromosomes.

by changing water levels have promoted speciation. Overall, 580 species of cichlid fish are found in African lakes.

Outside the Rift Valley there are some lakes that are never connected to other watercourses. Two volcanic crater lakes in Cameroon have been studied by Ulrich Schliewen and his co-workers. These have species unique to each lake (11 and 9 endemic cichlids species, respectively), despite being very small and with little differentiation of habitats. From analysis of mtDNA cytochrome *b* sequences it seems each group was probably derived from a single colonization event in each lake.

Since the lakes are very uniform, have no effective inflow from surrounding rivers and have not been connected by rising water levels, the speciation was almost certainly sympatric. Indeed, by looking at the phylogenetic tree for the species of each lake, Schliewen, Tautz and Pääbo suggest that their divergence was prompted by niche differentiation which again, like the sticklebacks, was based on feeding behaviour.

Some ecologists also recognize **parapatric speciation**. In this particular form of sympatric speciation, gene flow is restricted to individuals occupying a small area because the species is sedentary. Genetic change which follows this inbreeding often leads to highly adapted local populations, in distinct habitats. The immobility of plants especially restricts gene flow and parapatric speciation is perhaps most important for them, producing local races or **ecotypes**. A good example is the metal-tolerant ecotypes of various grasses (Box 2.3).

In this case, natural selection has allowed some individuals to colonize a marginal habitat, one which is poisonous to most other members of the parent population. Here they can grow in the absence of competition. We often find that highly adapted ecotypes are less competitive forms, so that they invariably lose out in intraspecific battles with the main population. It seems that the costs they have to meet to withstand the high levels of stress (in this case the metals in the soil) put them at a disadvantage when growing in normal soils.

Box 2.3

THE EVOLUTION OF METAL TOLERANCE IN PLANTS

Large-scale extraction of metal ores has created a series of distinctive habitats in which most plant species cannot flourish. Yet some plants have colonized these tips and have provided evidence on how rapidly natural selection can operate. In Britain, several sites date back to Roman times but many were first exploited 200 years ago, and within this time they have been colonized by varieties of grasses able to tolerate toxic metals. While there are naturally occurring soils with large amounts of toxic metals, many of these grasses appear to have been evolved from local populations where no tolerance is evident. Adaptations to lead, copper, zinc and others have arisen in different species and some varieties are tolerant to combinations of metals.

Tolerance seems to evolve fairly readily in certain grass species. Sowing normal grass seed on a spoil will usually produce one or two seedlings that are able survive the poisonous substrate. This is an indication that the genetic information for tolerance occurs as part of the background variation in the population. Interestingly, while a range of plants are known to have such alleles, albeit at a low frequency, others never show the capacity to develop metal tolerance.

Such locally adapted varieties are termed ecotypes. Metal-tolerant ecotypes are also found among some animal groups, especially soil-dwelling invertebrates. The grasses have proved particularly useful in restoring spoil heaps, where little else will grow (section 6.5).

It seems that, within the grasses, the genetic change needed may not be that large. We know that some plants produce special proteins that can bind these metals and prevent them passing from the roots to more sensitive parts of the plant. It may be that the most important adaptive change is simply an increase in the amount of this protein that is produced.

Others actually accumulate the metal and store it in parts of the cell where it will do the least damage. This includes vacuoles within the cell or storage in the cell wall. Several plants are so effective at accumulating metal in this way that they are being developed as means of concentrating precious metals from wastes (section 6.5).

Some clues to the nature of the tolerance mechanism in grasses were gleaned from studying non-tolerant ecotypes of Yorkshire fog, *Holcus lanatus* (Figure 2.13). These will grow on arsenic-contaminated soil provided they are supplied with additional phosphate. Mark MacNair and Quiton Cumbes found that *Holcus* was unable to distinguish between the poison and the nutrient, simply because the size and shape of their two ions are so similar. By swamping the soil with excess

Figure 2.13 Yorkshire fog grass (*Holcus lanatus*). Varieties of this species have evolved a tolerance to toxic metals.

phosphate, the uptake of arsenate could be reduced to a level low enough for the plant to survive.

Tolerant ecotypes of *Holcus* have modified their nutrient uptake systems so that they distinguish between the nutrient and the poison. Not only does this tell us how this tolerance mechanism functions; it also shows how non-tolerant plants may become poisoned.

Very often we find that tolerant ecotypes pay a price for their capacity to live in marginal habitats. In normal conditions – say, growing in a unpolluted soil with normal plants – the ecotypes grow less vigorously than their non-tolerant neighbours. It seems the costs they incur in being tolerant (perhaps the costs of producing metal-binding proteins, for example) mean they are less well adapted to this habitat. For this reason, their genotype is only found at a low frequency in unpolluted habitats.

2.5.4 Man-made species

Over the past 10 000 years or so humans have been the major selective pressure in the lives of thousands of plant and animal species. By means of domestication and cultivation, we have selected those most suited to our purpose (Box 2.4).

Box 2.4

THE ORIGINS OF WHEAT

Wheat was among the first plants to be cultivated and domesticated and continues to be one of the staple foods of many peoples today. The evolution and spread of wheats and other cereals follow those of a number of civilizations (Box 7.1, Table 5.2).

Archaeological evidence from the Near East indicates that wheat was being grown in the Jordan Valley, Jericho and Damascus around 10 000 years ago but the first use of wheat may have been a thousand years before this. Perhaps at that time the grain was simply collected from the wild; eventually, some seed was saved and sown to ensure the size of the harvest the following year. In so doing, human beings began its domestication, on the one hand ensuring the success of these wild wheats over their competitors and on the other selecting those with the most useful characteristics. This first agriculture probably began in the 'Fertile Crescent' – an area which reaches from the Balkans to Mesopotamia.

The early wild wheats fell into two categories: the diploid einkorn (*Triticum boeoticum*) and tetraploid emmer wheat (*T. araraticum* and *T. dicoccoides*). Einkorn is thought to have been the first species in cultivation but after generations of selection by the farmers a cultivated form of emmer (*T. monococcum*) arose and quickly spread to central and western Europe. Not only was emmer a tetraploid and therefore larger and more vigorous; it was also free-threshing, allowing its seed to be readily separated from the ears. By 9000 BP, cultivated emmer (*T. dicoccum*) had supplanted einkorn to become the single most dominant wheat in cultivation. In less than 2000 years it spread as far as Europe, Ethiopia and India and, along with barley (*Hordeum distichon*), was a staple cereal of the Neolithic period, (Figure 2.14).

Around this time a newcomer appeared, a species which would eventually come to dominate world agriculture for the next 9000 years. This was bread wheat (*T. aestivum*), which is hexaploid; that is, it has six sets of chromosomes (6*n*). *T. aestivum* is thought to

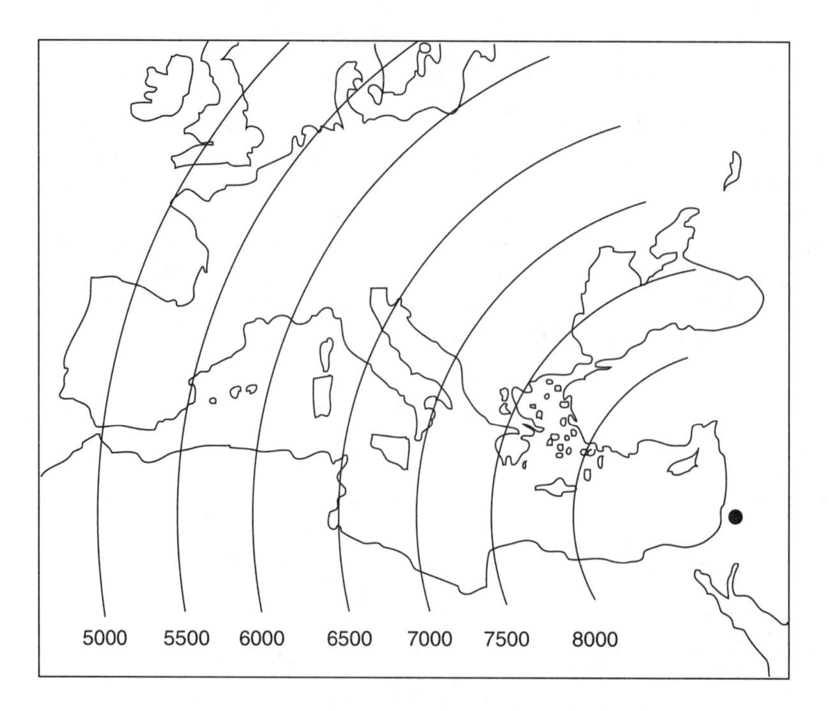

Figure 2.14 The rate of spread of modern wheat (*Triticum aestivum*) and barley (*Hordeum distichon*) from their origins in the 'Fertile Crescent'. Lines show the date of arrival in each area in years before present.

have resulted from intergeneric hybridization between cultivated emmer (*T. monococcum*, which is 4*n*) and an unknown species of goat grass, possibly *Aegilops squarrosus* (2*n*).

Although bread wheat has become the most widely grown of all wheats (Figure 2.14), with hundreds of different cultivars adapted to different soil types and climates, other wheats are still in cultivation. Chief among these is durum wheat (*T. durum*), otherwise known as macaroni wheat. Durum, a descendant of the tetraploid emmers, is extra-hard with a high protein (gluten) content. It is also rich in the pigment beta-carotene, which helps to give pasta its firm texture and golden colour. This has ensured its continued use in traditional agriculture and cookery. This, along with the other cereals, has had an important role in the civilizing of humanity.

There are perhaps around 70 000 cultivated plant hybrids derived from 1100 wild species and some have an ancestry that might include as many as 35 species. Not surprisingly, all this makes them particularly difficult to classify. In the case of hybrids, we simply list the most recent parents using a multiplication sign (×) to indicate the cross.

In the past, cultivated plants were known as varieties, but this term is now used to describe naturally occurring variation within species. Plants bred by humans and dependent on cultivation for their continued existence are therefore called **cultivars**. According to the International Code of Nomenclature of Cultivated Plants a plant must be listed according to its genus and species which is then followed by its cultivar name (usually signalled by the abbreviation, cv.). For example Red Ace, a cultivar of shrubby cinquefoil is usually written as *Potentilla fruticosa* cv. Red Ace.

Hierarchical classification tends to break down at the level of hybrids and cultivars, primarily because we have often blended very different genotypes to create new cultivars. Even so, there is considerable value in understanding the evolutionary history of domesticated plants. Many modern crops will only thrive with an abundance of water and fertilizers and with the help of pesticides to reduce competition from other species. Deprived of their 'technological fix', many cultivars are of limited use and fail in competitive battles (section 5.4).

Their wild-type relatives often hold genes which make them tougher and more resilient, and we draw upon this to produce more resistant cultivars. Potatoes are plagued by aphids and the cost of insecticides to protect them is considerable. The wild hairy potato of Central America has been used to produce new aphid-resistant cultivars which impale aphid attackers on short, sharp hairs.

Unfortunately, pests evolve too and adapt to our new varieties. Pest-resistant cultivars therefore have only a limited useful life. For example, strains of wheat resistant to various fungal ('rust') attacks last little more than five years before the fungi itself has adapted to the change in its niche. Within a decade the build-up of pests and diseases associated with a cultivar may make it uneconomic to grow. Crop breeding programmes can thus represent a continuing evolutionary battle between ourselves and the pests (section 4.5).

While we select our defences from the genetic variation in wild species, we have to be aware that these too are under threat. Their loss of habitat under agriculture means that we risk losing much of the genetic variability needed to sustain breeding programmes.

The stakes are even higher now, as genetic engineering has taken some of the guesswork out of crop and stock improvement. By incorporating selected genes from other organisms into the genome of economically important species, we can enhance their productivity or confer protection against pests and disease. Gene technology offers the opportunity to move genetic code across species boundaries and even between phyla. This is evolution by design rather than by chance.

SUMMARY

The binomial system for classifying and naming living organisms provides a convention by which scientists can refer to different species, based on generic and specific names. It also provides an indication of the phylogenetic relations by grouping species into a series of hierarchical categories.

Until recently, most classification has been based on morphology, but with our increasing capacity to read the genetic code and its products, we can measure genetic distances between different groups and often revise the phylogeny for a species. Variations in body form mean that morphology can be a poor guide to species differences. The biological species concept, based on a group of interbreeding individuals, also has to be used with some caution, especially in those plants and animals which hybridize freely.

A species represents a set of adaptations to life in a particular ecological niche. The niche is the totality of factors, biotic and abiotic, to which a species has adapted, and therefore describes its relations with other organisms. For this reason, niche is sometimes described as the role of a species within the community. Niche also describes the resource utilization by a species: specialists have a narrow niche whilst generalist species have a broad niche, and use a larger part of the resource spectrum. Two species may not occupy the same niche and competitive exclusion will mean that one species will be lost. Intense competition for a resource can also lead to character displacement, where a species adapts to use a different niche. Populations which fragment and diverge by character displacement may eventually lead to the formation of new species.

Speciation occurs where gene flow ceases between two populations. This can occur because the subpopulations become physically separated (allopatric speciation) or are isolated from each other by a genetic change (sympatric speciation). Speciation not only follows from natural selection but also occurs as a consequence of selective pressures applied by humans in their breeding of plants and animals.

FURTHER READING

Jeffrey, C. (1977) *Biological Nomenclature*, Edward Arnold, London.

Price, P.W. (1996) *Biological Evolution*, Saunders, Fort Worth.

Weiner, J. (1994) *The Beaks of Finches*, Knopf, New York.

EXERCISES

1. Listed below are the names and categories associated with the classification of wild clematis. Rank the categories according to their hierarchy and then match the names to their respective categories.

Order
Ranunculaceae (buttercup family)
Clematis
Species
Plantae (plants)
Phylum
Kingdom
Angiospermae (flowering plants)
vitalba
Ranalės (buttercups)
Family
Trachaeophyta (vascular plants)
Class
Genus

2. Using Figure 2.4 as an example, devise a key to classify the following insects according to the characteristics listed in the table below. Seek to use the minimum number of steps.

Table Ex.2.1

Characteristic	*Insect Type*					
	Wasp	*Beetle*	*Butterfly*	*Fly*	*Grasshopper*	*Ant*
Wings present	Y	Y	Y	Y	Y	N
Number of wings	4	4	4	2	4	0
Hard wing case	N	Y	N	N	N	N
Large dusty wings	N	N	Y	N	N	N
Large jumping legs	N	N	N	N	Y	N

3. Complete the following paragraph by inserting the appropriate words, using the list below (some words may be used more than once, some not at all).

An ecological niche is the sum of a species adaptations to both the _________ and _________ environment. It is defined by the use of _________ such as food, water, light etc. and these can be plotted along a _________ _________. The overall range of resources a population exploits is known as its _________ _________. This is _________ for generalist and _________ for specialists. Where niches of two species overlap, the organisms concerned _________ for the resources concerned. Complete niche overlap leads to competitive _________ of one or other of the species.

abiotic, axis, biotic, breadth, character, compete, displacement, exclusion, fundamental, hypervolume, narrow, niche, population, realized, resource, resources, species, wide.

4. From the list below, select the **most** correct answer to complete the following sentence:

Species which specialize ...

(a) have a narrow niche breadth.
(b) are bad competitors.
(c) have a narrow niche breadth and are good competitors.
(d) have a broad niche breadth.
(e) are good competitors.
(f) have a broad niche breadth and are bad competitors.

Tutorial or seminar topic

5. What are the problems with the species concept? When does the category of species fail to classify a group of organisms?

Models of population growth – harvesting populations and sustainable yield – two simple fishery models – life history strategies – the dangers of small population sizes – the threat of extinction in small and fragmented populations.

3 Populations

Part of the attraction of ecology is its appeal to our sense of order and balance. Many people see in ecology a set of principles which explain a constancy and regularity in living systems: the cycle of life and death, of eating and being eaten, of continuity between generations. The natural world seems to be ordered by simple rules that maintain some sort of 'balance of nature'. We capture these ideas in familiar phrases such as ashes to ashes, big fish eat little fish, and the circle of life.

Perhaps there is a natural order out there, but few modern ecologists would try to write out the rules of the game. It is difficult to show that natural populations and communities are actually stable or unchanging for long periods. A variety of ecosystems appear to be similarly organized and regulated, yet differences start to emerge when we look closely at each one in detail.

Close up, chance and variation blur the sharp outlines of these principles. To use another common phrase, the devil is in the detail. Our ideas might make intuitive sense, but their simple logic fails to reflect the complexity of the living world. Ecology, the science, often muddies the waters, showing that nature does not always come as clean as we might think it.

Nevertheless, most of us have some conception of an order in our environment, a common-sense view of the economy of nature. We see that nature works by cycling nutrients, that food and the energy it provides fuel living systems and govern the growth of individuals and populations. Such ideas are part of the knowledge we all need to exploit and survive in our various environments.

A simple example is relevant to this chapter. A fisherman sitting by a pond knows that there are a limited number of fish in the water. With no stream bringing new individuals, the fish caught can be replaced only by the reproductive efforts of those remaining in the pond. The frugal fisherman understands that the population must be allowed time to recover.

Consider the elements in this simple system. The pond has a population of a limited size and the fish can reproduce at only a limited rate. The population is defined solely by the boundaries of the pond, with no stream allowing fish to enter or leave. To maintain a constant population the fisherman can remove fish only at the rate at which they are replaced by reproduction.

The intuitive sense of this example underpins the first part of this chapter. Here we use simple models to examine population growth, the limits on population size, and also to estimate how many individuals might be caught without damaging the population's reproductive potential. We shall find the simplicity of these models is clouded by the detail of real fisheries, when data on species' life histories are needed to manage their exploitation properly.

Later, the same models are used to examine why some populations and species face extinction. The loss of a species seems inevitable when its reproduction does not match its death rate. Again, we find that the detail of the ecology of each species tempers the use of general rules in conservation.

3.1 MODELLING

Models are used extensively in ecology and in this chapter, so it is worth saying something briefly about their nature and their purpose.

All sciences use models to do one of two things. First, they can simplify a complicated system by removing non-essential information or 'noise'. Think of a road map. This is a model in which most of the background detail of the real world has been removed to highlight the important information – the pattern and direction of roads. We model ecological systems in the same way by isolating the key factors in a study. The model helps to clarify our understanding of a system by defining its important elements and their relationships. These are **analytical models**.

A second type of model simulates the behaviour of a system. Again, much of the spurious information of the real world is pared away, but now sufficient detail remains to produce accurate predictions. A flight simulator is a physical model (or a computer program) for training pilots that provides enough information to give a realistic impression of flying an aircraft. A **simulation model** attempts to make realistic predictions about a system, rather than analyse its workings. Ecological simulations make predictions about populations, communities and ecosystems and, like flight simulators, they allow us to try out different manoeuvres without any risk of destroying the real thing.

Models abbreviate reality, enabling us to analyse the workings of a system, or to make predictions about its behaviour. Here we begin by making simple analytical models of population growth, using the minimum of information. Later, we add details that improve our predictions and help us to manage both exploited populations and those in danger of extinction.

3.2 SIMPLE MODELS OF POPULATION GROWTH

The number of fish hatched in the pond at the end of the breeding season will depend on the number of adults breeding there at the start. In describing the relationship between these two numbers we create a simple population model. This is pared back to just two elements and has only one function: to make predictions about population growth in future years.

This population is defined by the boundaries of the pond (Box 3.1), and we assume there is no migration in or out. We also assume no deaths during the breeding season, which is perhaps unrealistic but by no means impossible. Our model is for a fish with a single breeding season each year, true of many species in temperate waters.

We make one other simplifying assumption: that all fish live for just one year, with adults dying after spawning. If there are 100 adults each responsible,

Box 3.1

INDIVIDUALS, POPULATIONS AND METAPOPULATIONS

Ecologists counting individuals and treating them as a population need to define their terms very carefully. Criteria that most people would use to define the individual become problematic when we look closely at different species.

Consider organisms that increase their numbers by budding or some other form of asexual reproduction. Both botanists and zoologists study species that grow by forming modules that might separate and grow independently. This is true of many plants and several groups of colonial animals. For these we have to distinguish between individuals that are genetically distinct (having arisen from different zygotes), which are termed **genets**, and asexual clones that are genetically identical, which are termed **ramets**. In some species (for example, potatoes) the count of individuals will include both genets (seedlings) and ramets (tubers).

We then have to decide whether we include all stages in the life cycle in the count: for example, do we include both tadpoles and adult toads, or seedlings and overwintering tubers? Our decision will depend on the organism and the purpose of the study.

Even then, defining a **population** is not always straightforward. A good functional definition is a group of individuals of the same species occupying a particular area at a particular time. Notice that we define notional boundaries which delimit individuals who can actually breed with each other – for example, the elephants of the Serengeti in 1992. Elephants in the Serengeti would not normally mate with those many miles to the south. Nor, very obviously, could they breed with the elephants in the Serengeti in 1902. Even so, these spatial and temporal boundaries are only truly distinct in the mind of the ecologist, as a necessary convenience to define the terms of a study.

Most species have a collection of populations alive at the same time, separated spatially. This population of populations is called a **metapopulation**. While some populations may swap individuals fairly regularly, others may have little or no contact with each other. The degree of interchange within the metapopulation is often crucial to the survival of some endangered species.

You will also come across the term **population density**. This is used where we have no need or have made no attempt to count every individual in a population (for most plants and animals this is impossible). Instead, we count their numbers in a specified area or, in some circumstances, a known volume.

on average, for two offspring, the population in the next generation would be 200 fish. All reach adulthood next year, so we calculate the **reproductive rate** (R_0) as:

$$R_0 = N_g/N_0$$

The number of adults at the start (N_0) was 100. The number of offspring reaching reproductive age in the next generation (N_g) is 200. Therefore R_0 = 2.0.

Note that R_0 is the **average** reproductive rate per individual in the population; each adult does not produce exactly two offspring.

Here the generations do not overlap and the adults do not survive long after reproducing. More realistically, fish populations include individuals that survive for several years and take part in several breeding seasons. Keeping a count is then more difficult, requiring us to note the age of reproduction and of death for each individual in the wild. Instead, we can measure the reproductive rate by taking a census at certain times. Counting the number alive before and after a period of time allows us to calculate the **net reproductive rate** (R_N) over that interval:

$$R_N = N_t/N_0$$

The number of individuals alive (N_t) at the end of the time interval (t) is divided by those counted at the start (N_0). If the number alive at the end (survivors and offspring) is greater than at the start, the population has increased ($R_N > 1$). The population declines when the number of deaths is larger than any additions ($R_N < 1$).

R_N is called the net reproductive rate because it includes both deaths and births over the sampling interval. Again it is the rate of change per individual averaged over the entire population.

Can you see when R_0 and R_N will equal one another? They are the same when the census interval is equal to the generation time. Both then measure the average rate of change per individual over one generation.

We can follow the change in the population size on a graph. If we set R_N to a fixed value, and then plot the size of the population with time, we get a characteristic curve (Figure 3.1). In this example, we start with 100 individuals and set R_N to 1.2. After one breeding season there will be 120 fish in the pond. Note that some may have died in the meantime, but the net increment is 20 individuals.

Next time around, there are 24 additional fish, and in the third generation 29. You may like to derive a second set of data to check the shape of curve produced with a different value of R_N.

This example illustrates an important point about the pattern of population growth. Despite R_N staying constant, the increase in population size is not the same from one generation to the next. Instead, the size of the increment itself increases and population growth is accelerating. More individuals are added in each successive generation because N, the number of adults, rises with each generation.

This gives the curve its characteristic shape (termed **exponential growth**). This shape is true of any population with an R_N greater than one. Even with R_N values only slightly above 1 the curve is the same, though its rise is shallow. With time it will still grow to a very large population. Similarly, an accelerating decrease results when R_N falls below one, and the population races toward extinction.

3.2.1 Populations with overlapping generations

This simple picture becomes more complicated when we extend it to populations where births or deaths may occur continously. Many species of fish grow and reproduce for several years. We have all heard stories about the ageing giant that got away. But death can occur at any time and, however unlikely, there is some chance that this brute will one day be caught by the fisherman.

For some species reproduction is also a more or less continual process, not confined to a particular season. Then both the birth rate (or natality rate) and the death rate (or mortality rate) vary continuously. The difference between them is expressed as r, the **rate of change per individual** for a particular point in time:

$$r = b - m$$

When the natality rate (b) is greater than the mortality rate (m) the population is increasing at that time. The model of population growth becomes:

$$dN/dt = rN$$

If we fix r at a particular value, the population grows exactly as described before. The change in the population size (dN) after the time interval (dt) equals the average rate of change per individual (r) multiplied by the number of individuals (N). It differs from our previous example because it derives the change over a very small interval of time (dt).

As before, this model produces exponential growth as long as the birth rate is greater than the mortality rate. However, the rate of population

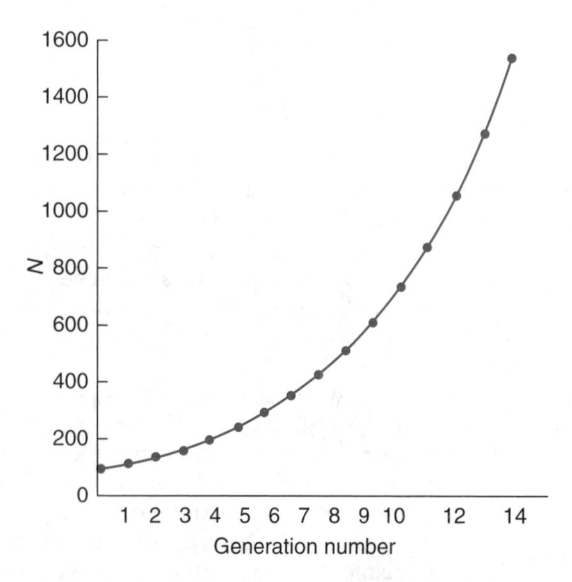

$N_0 = 100$ $R_N = 1.2$

Generation number	N	Increase in N
1	120	20
2	144	24
3	173	29
4	208	35
5	250	42
6	300	50
7	360	60
8	432	72
9	518	86
10	622	104
11	746	124
12	895	149
13	1074	179
14	1289	215
15	1547	258

Figure 3.1 Model of population growth in an organism with discrete generations, starting with a population (N_0) of 100 and a constant reproductive rate (R_N) of 1.2. The results used to plot the graph have been rounded up or down to give whole numbers.

growth (r) can change as conditions change, and b and m fluctuate.

3.2.2 Factors affecting the rate of population growth

The maximum value of r (r_{max}) describes how rapidly a population would grow under ideal conditions when b is maximized and m is minimized. This differs between species according to the details of their life history. Species that take a long time to reach reproductive age, or that produce few offspring on each occasion, have slower rates of population growth.

Elephants and other large-bodied animals typically have low r_{max} values because they produce few offspring over a long time. The gestation period in the elephant is relatively long (22 months), followed by a juvenile stage lasting 12–14 years, before the individual becomes sexually mature. They may live for an average of 30–40 years and produce just one offspring at a time. An African elephant has r_{max} value of 0.06 per year. At the other extreme, mice live short lives, quickly becoming sexually mature, and have multiple births after a brief (21-day) gestation period. Many small mammals, including mice, have a high r_{max}, ranging from 0.3 to 8.0 per year.

Why is r_{max} rarely achieved in the wild? It is usually because conditions for growth and reproduction are far from perfect. Natality is reduced with poor nutrition or disease, or when partners simply fail to meet. Rates of mortality depend on a wide variety of factors, and any rise in m depletes the numbers reproducing.

3.2.3 Populations in limited environments

In one famous calculation, Charles Darwin estimated that a single pair of elephants would have 19 million descendants in just 750 years, if all reproduced at their maximum rate. Yet we are not overrun by elephants and common sense tells us that some factor in their habitat must limit their population growth.

Rapid population growth can be sustained only for as long as conditions allow. Resources may become scarce, space may become limited or waste may accumulate in the habitat. These checks on growth, collectively known as **environmental resistance**, place an upper limit or **carrying capacity** (termed ***K***) on the population size.

What will happen in a population close to its carrying capacity? Any shortage of resources will mean competition between individuals, with some getting less than they need. A lack of food, for example, might prevent or delay some individuals becoming sexually mature, while others might die. Either reduces the number of offspring. Towards the carrying capacity, more individuals compete for fewer resources and the intensity of competition grows.

In some cases competition between individuals of the same species (**intraspecific competition**) is important in limiting population size. When the degree of environmental resistance depends upon the number of individuals growth is said to be density-dependent. Numbers increase rapidly as long as resources are abundant but as the density of individuals rises so growth slows down.

The next step is to include these checks from environmental resistance in our model, using the carrying capacity. For example, if our pond can accommodate 500 fish (K) and 100 are already present (N), there is 'space' for just 400 more. Put another way, of the total capacity, one-fifth is occupied and four-fifths remain:

$$(K - N)/K = (500 - 100)/500 = 4/5 = 0.8$$

This measures the degree of environmental resistance. The less spare capacity, the more resistance there is. We can build this into our model very easily:

$$\mathrm{d}N/\mathrm{d}t = rN(K - N)/K$$

Figure 3.2a, derived from this model, shows how increasing environmental resistance slows growth by its impact on r. A shortage of food, for example, may limit egg production (b falls) and/or increase mortality (m). Growth is gradually checked as the population settles smoothly at its carrying capacity.

We can see the effect more easily by plotting net incremental growth ($\mathrm{d}N/\mathrm{d}t$) against population size (Figure 3.2b). At first each increment rises as N increases. The greatest increase (23) is at half of the carrying capacity when N and r are both still large. Thereafter r declines because competition lowers b and increases m. Incremental size now falls, until, at the carrying capacity, $b = m$ and no net additions occur.

Consider what happens if the carrying capacity changes. Any expansion in space or resources

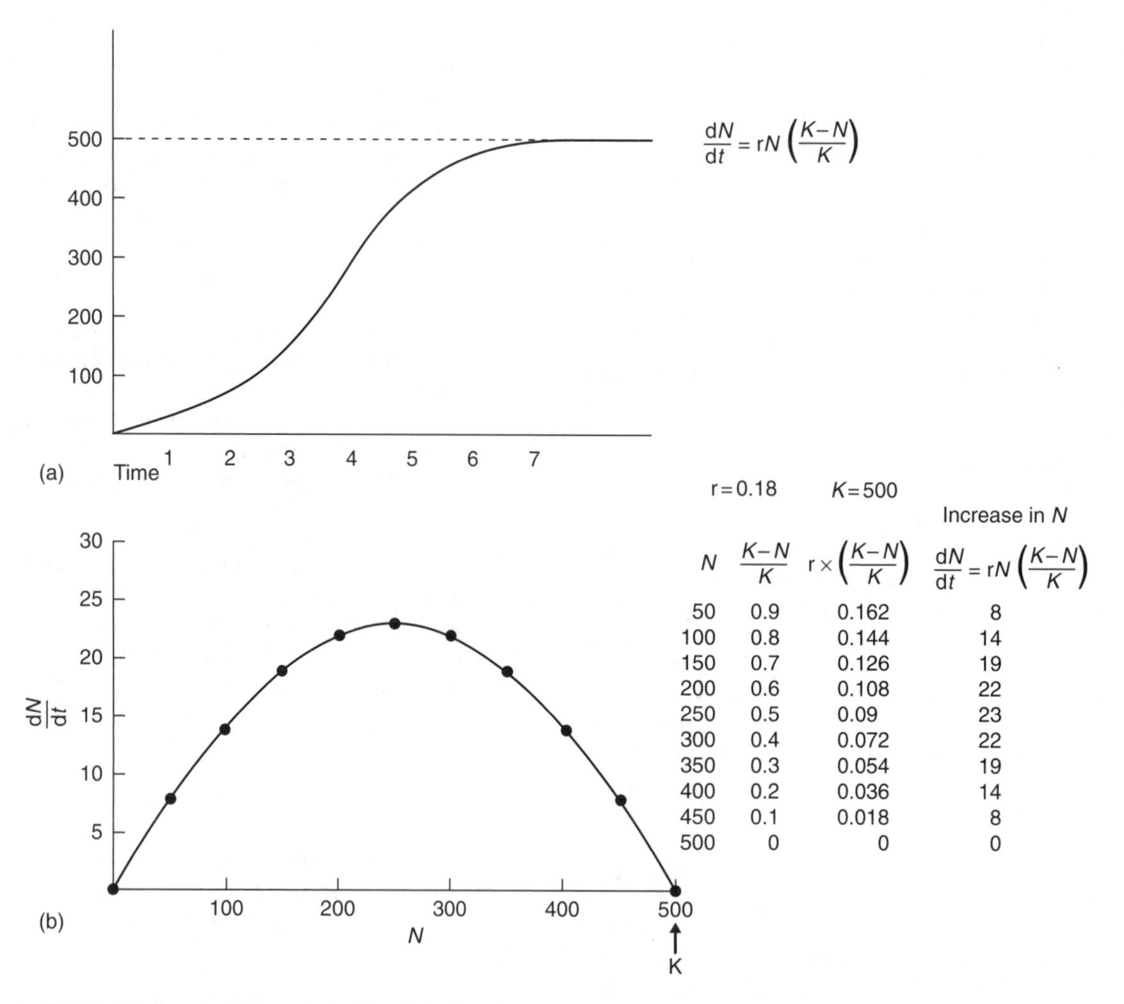

N	$\frac{K-N}{K}$	$r \times \left(\frac{K-N}{K}\right)$	Increase in N: $\frac{dN}{dt} = rN\left(\frac{K-N}{K}\right)$
50	0.9	0.162	8
100	0.8	0.144	14
150	0.7	0.126	19
200	0.6	0.108	22
250	0.5	0.09	23
300	0.4	0.072	22
350	0.3	0.054	19
400	0.2	0.036	14
450	0.1	0.018	8
500	0	0	0

Figure 3.2 Model of a population growing in a limited environment.

(a) In this case we model overlapping generations, with the carrying capacity (K) set to 500. The rate of increase per individual in the population (r) is 0.18. Population growth is reduced by the increasing effect of environmental resistance as the carrying capacity is reached.

(b) We can show the effect of environmental resistance by calculating the size of increase at particular population sizes. For each N, the environmental resistance is calculated as the proportion of capacity still available, using the equation $(K - N)/K$. For 50, this is $(500 - 50)/500 = 0.9$. Multiplying this by 0.18 (r) gives the actual rate of increase ($0.18 \times 0.9 = 0.162$). Finally, multiplying the result by N gives the net increase at that population size ($0.162 \times 50 = 8$).

Notice that the largest increment is when $N = 250$, half of the carrying capacity. This declines as the habitat fills, so that no increase is possible at K.

allows b to rise or m to fall, and the population grows to a new ceiling. Alternatively, a lowering of K leads to a reduction in the population as mortalities outstrip births. In this simple model, the population is regulated to stay close to its carrying capacity.

This s-shaped pattern of population growth (Figure 3.2a) is described as **logistic**. The model shows the population rising smoothly and resting perfectly at its carrying capacity. In the real world, time-lags and a variety of other factors cause populations to overshoot their K and oscillate around it.

Not all plants and animals have populations limited by intraspecific competition. For some, competition with different species (**interspecific competition**) is more important (section 4.3). Predation or parasitism, where one species is consumed by another, may limit the population growth of both consumer and consumed, sometimes in a density-dependent manner. Alternatively, the abundance of key resources means competition is rarely significant for some animals, such as insects.

3.2.4 What determines the carrying capacity?

Many insects show little constancy in their population size. Their habitats change too rapidly for their numbers to settle at an equilibrium. Frosts, floods, fires and other, larger-scale catastrophes cause major changes in their habitats. These shifts in the carrying capacity bear no relationship to the size of the resident population. They are density-independent changes. Not surprisingly, species living under these conditions undergo major fluctuations.

Even without dramatic changes in their environment, some plant and animal numbers fluctuate wildly from one year to the next. Various pests occasionally undergo explosive population growth often without an obvious upheaval in their habitat. Several species of locust have periodic population outbreaks, when massive swarms sweep across thousands of miles of Africa.

The simple model we have described is most relevant to species closely dependent on the long-term availability of resources in their environment – particularly larger plants and animals that grow slowly and live longer. Their numbers do not fluctuate wildly and the idea of a carrying capacity has some meaning.

What determines the maximum number of individuals that a habitat can support? For relatively stable populations we can make a reasonable prediction based on average body size. A large adult mass needs sufficient resources to grow and sustain itself. Meeting this demand requires space: a sufficient volume of soil from which to extract water or an area over which to forage for food. The amount of space needed to support each individual plant or animal thus determines an ecosystem's carrying capacity for the species.

Consequently, the average density of many large species at their carrying capacity shows some correspondence with adult body size. As we see later on, this has important consequences for the large plants and animals we seek to protect from extinction.

3.3 HARVESTING A POPULATION

Back at the pond, we will assume that growth of our fish population is density-dependent and limited only by intraspecific competition. This pond is a very stable environment and the fish population is close to its carrying capacity. We want to maintain its potential to provide catches from one year to the next. We also want to take the maximum possible number of fish each year, the **maximum sustainable yield** (**MSY**). If the population is to maintain itself, however, sufficient reproductive individuals have to remain to replace those caught.

What is the maximum number we can harvest? This will be the same as the maximum number that can be replaced each year. We already know this number from the graph of population increase in a limited environment (Figure 3.2b) – the greatest increment was at half the carrying capacity. At this N, the population produces the largest number of offspring, before r has been reduced significantly by intraspecific competition.

3.3.1 Fishery models

Calculating a maximum sustainable yield is feasible if we know the carrying capacity of the environment. That may be easy in a pond, but is highly problematical in the open sea. Besides K, we need to know the size of the catchable stock – those fish large enough to be caught in our nets and which are, in effect, the population (N). Many disputes about fishing quotas revolve around interpretations of such data and the methods used to estimate the MSY.

Of course, the fish are never actually counted. Commercial fisheries are interested in the yield in weight, and fishery models are based on biomass rather than numbers in the catch. Nevertheless, the principles and basic ideas of the models remain the same – we simply replace individuals with units of weight.

Assuming there is no net migration into the fish population, biomass is restored by the addition of new individuals and also by the growth of new tissues by existing fish in the stock:

$$F + M = G + R$$

Here the biomass caught by the fishery (F) and the losses due to natural mortality (M) are balanced by tissue growth (G) and by recruitment (R) of new individuals. Fish are only recruited to the stock when they are large enough to be caught.

In most fisheries we have no control over natural mortality. We can only regulate our fishing effort so that fishing mortality (F) does not exceed $G + R - M$, and the catchable stock (N) does not decline.

The maximum sustainable yield is achieved when the largest harvest is taken without the catchable stock falling from one season to the next. This is the basis of the **surplus yield model**, so called because yield equals the level of growth and recruitment surplus to that needed to maintain N. In effect, the population remains close to a constant size by replacing losses through growth and recruitment. This is the model that the fisherman at the pond has been using, if only intuitively.

This yield is maximized when the population is half of K, when $G + R$ are at their greatest rate of increase (Figure 3.2b). The model estimates whether the yield is optimal by looking at the relationship between yield and fishing effort: any additional effort producing no increase in catch means the optimum has been passed. The catchable stock is then being depleted and each additional unit of effort brings a diminishing return (Figure 3.3).

For a long time, the size of the catchable stock was estimated from catches landed at the quayside. A decline in the catch per trip (or more accurately the catch per unit fishing effort) over a number of seasons implied that the stock was declining. Fishery managers knew that a pattern of falling yields indicated overfishing. Catches then included a proportion of the population whose reproduction and growth were supposed to replace losses to F and M.

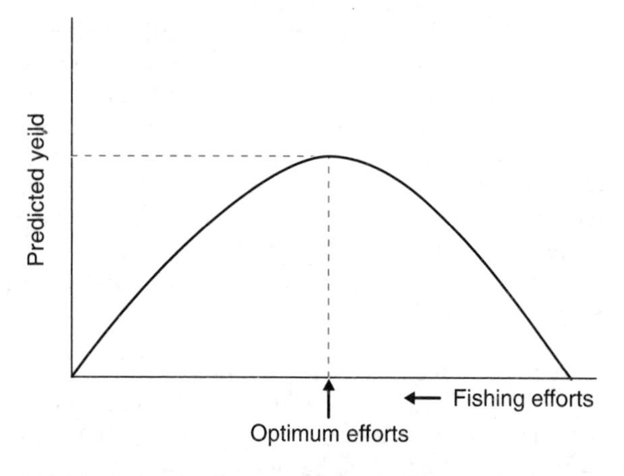

Figure 3.3 The estimate of the maximum sustainable yield from a fishery, based on the fishing effort. This is derived from Figure 3.2b. Fishing yields above the optimum will decline with increasing effort because we are now catching the stock needed to maintain itself. Below the optimum the scope for higher yields are indicated by increases in yields with more effort.

In fact, such patterns are notoriously unreliable. Many commercial fisheries are based on highly variable populations, whose size fluctuates with changes in weather, the movements of ocean currents and a range of other factors. Estimating the stock is very difficult under these circumstances. Equally, G and M can be highly variable, as can, most importantly, recruitment (R).

Measuring recruitment rates is crucial in intensively fished populations because these often determine the size of catchable stock, more so than the carrying capacity. Such populations are far below their K value, and any depressive effect of competition on r will be small. The critical problem is to sustain recruitment.

We can see that our model is only effective up to a point. Its analytical simplicity does not include important details governing tissue growth and recruitment rates. Because we are primarily concerned with the weight being caught, fish should be harvested only after they have had time to grow, not as soon as they have entered the catchable stock. We need to estimate of rates of weight increase and of recruitment.

A second group of models, **dynamic pool models**, do just this. They use growth rates to calculate the biomass harvested from each age group. Many fish grow at a relatively constant rate throughout their life, so size is a direct reflection of age. Average weights can be estimated from the length of time an individual remains in the stock. We can then work out the optimal survival time for an individual to deliver a maximum yield.

Just as a viable population must remain for growth in numbers, so individuals should be given time to grow new tissues. At high fishing intensities, few fish escape capture, and few survive long in the catchable stock. Even if numbers are replenished, biomass is not because fish have just a short time to grow (Figure 3.4). At the optimum fishing effort, residence times allow individuals to reproduce and to increase their weight. The model estimates the appropriate fishing intensity to give the maximum sustainable yield, based on recruitment and weight.

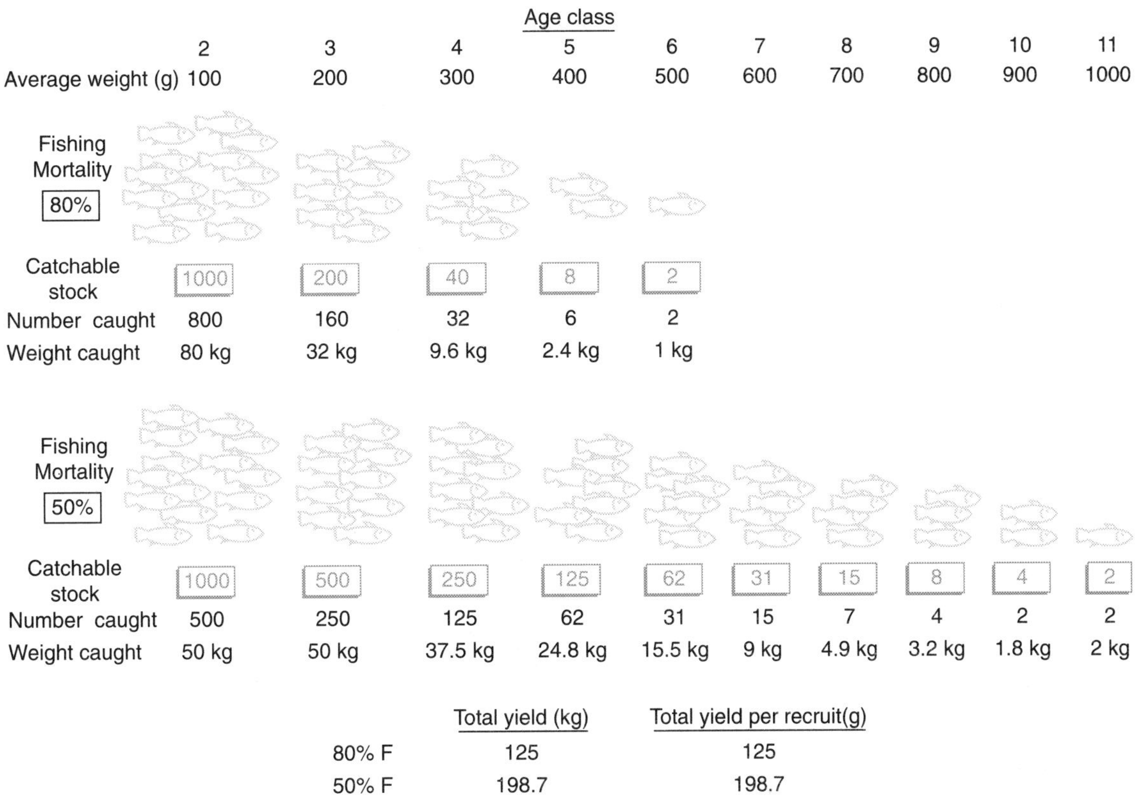

	Age class									
	2	3	4	5	6	7	8	9	10	11
Average weight (g)	100	200	300	400	500	600	700	800	900	1000

Fishing Mortality 80%

Catchable stock	1000	200	40	8	2					
Number caught	800	160	32	6	2					
Weight caught	80 kg	32 kg	9.6 kg	2.4 kg	1 kg					

Fishing Mortality 50%

Catchable stock	1000	500	250	125	62	31	15	8	4	2
Number caught	500	250	125	62	31	15	7	4	2	2
Weight caught	50 kg	50 kg	37.5 kg	24.8 kg	15.5 kg	9 kg	4.9 kg	3.2 kg	1.8 kg	2 kg

	Total yield (kg)	Total yield per recruit(g)
80% F	125	125
50% F	198.7	198.7

Figure 3.4 The effect of fishing intensity on yields to a fishery. Here we have two examples, each starting with a cohort of 1000 fish, but with differing fishing intensities. In the first, 80% are removed during each fishing season. The cohort is fished out in 6 years. At the lower intensity the cohort continues for 11 years. (*F*, biomass of fish caught.)

The weight of a fish depends upon its age and, at 80% *F*, most individuals are taken when the fish are youngest. In age class 2 for example, 800 fish are caught weighing on average 100 g, giving a yield of 80 kg. At 50% *F* the yield is only 50 kg, but more fish are left to grow bigger. The net effect is a larger yield overall at 50% *F*.

A dynamic pool model provides plots of yield per recruit against fishing intensity and allows us to find where the maximum yield is achieved – where the balance between growth and catches is optimum.

3.4 GROWTH RATES, AGE AND RECRUITMENT

Dynamic pool models are a derivation of the surplus yield model, but they divide the population into age groups, and then work out the biomass each contributes to the yield. Such models require a lot of detail to make realistic predictions. We need data on the number of individuals being recruited, the rate of mortality in each age group, and the average weight of each age class. Ecologists summarize such data in **life tables**.

A life table estimates the chances of an individual of a particular age dying, or surviving for a length of time. You may be familiar with such statistics from comparisons of life expectancies for various age groups or for people in different countries. Ecologists have borrowed many of these techniques from insurance actuaries, who use life tables to calculate the risk of our dying while covered by their policy.

Life tables come in two basic forms: **cohort** and **static**. In the first, a group of individuals born at about the same time (a cohort) is followed and the numbers surviving in each age class are recorded. From this we calculate the survival probability and life expectancy for an individual of a certain age. A static life table derives the same data but from a census of the whole population, looking at the proportion in each age class, on one occasion.

Life tables tell us much about the dynamics of a population. With the data for several years, we can decide whether the mortality of a particular age group is significant for determining total population size. For example, larval mortality limits recruitment for many fish and this determines the size of the catchable stock.

Life tables also make plain the **life history strategies** adopted by an organism. All species try to maximize their population growth but differ in how they allocate resources between the stages of their life cycle. Some produce large numbers of offspring, of which few survive to mate. At the other extreme are species producing just one or two offspring each time, though these are more likely to reach adulthood.

The structure of a population – the proportion of individuals in each age class – reflects these strategies. It also tells us much about the potential of a population for growth. Rapidly growing populations are typically 'bottom heavy', dominated by the new arrivals in the youngest age classes (Box 3.2, Figure 3.5). Alternatively, a population close to its carrying capacity, when its birth rate is matched by its death rate, has a more even age distribution, with a larger proportion in the older cohorts. Few are being added, since competition favours the old and the large (Figure 3.6).

Box 3.2

AGE STRUCTURES

A population with overlapping generations will have a distinct age structure – simply the proportion of individuals in each age class.

The distribution of individuals over the age classes has a very direct effect on population growth. We recognize two basic age structures: one associated with a rapidly growing population, and the other in a population close to its carrying capacity.

When there is large scope for population increase and the environment remains relatively constant, the population will increase at a constant rate per individual (*r*). It then assumes a **stable age distribution** with birth rates and death rates constant in each age group (Figure 3.5). Such a population will be growing exponentially, with fixed proportions in each age group. However, the highest proportion of individuals is in the youngest age groups, and the distribution is characteristically 'bottom-heavy'. This pattern remains stable as long as there is little net change due to migration and the environment does not change significantly.

When close to its carrying capacity, a population has little capacity for further growth, and the birth rate is equal to the death rate. Again, the environment must be largely unchanging, but now the proportion of individuals in each age class is more evenly distributed – the birth rate is lower, and the younger age classes do not dominate the population. This is a **stationary age distribution** (Figure 3.5b).

Different human populations approximate to both these types. Stationary age distributions are a feature of those countries where the population has remained relatively constant, or even declined slightly. In those nations undergoing very rapid population growth, a stable age distribution is found.

In animal populations, these distributions are not always indicative of the population growth rate. One obvious case is the fish we harvest. These typically have an age structure dominated by younger individuals because our fishing techniques aim to catch only the older, larger fish. These are not populations increasing rapidly.

Figure 3.5 Age distributions. (a) A stable age distribution has a fixed shape – the proportion of individuals in each age group remains unchanged. This is because additions and losses to each age group are constant. Overall, the whole population has a fixed rate of births and deaths, and is growing geometrically. For this reason, it is dominated by the younger age groups. (b) A stationary age distribution is also fixed, but now the birth rate and the death rate balance each other, and the population is close to its carrying capacity, *K*. This is largely a theoretical distribution (partly because the shape is bound to change as the environment changes), but now, with resources scarce, we see a more even distribution between the age classes.

You can get a sense of this difference by walking through temperate woodlands of different ages. An undisturbed woodland, several centuries old, is typically dominated by a small number of large, old trees with a few young saplings waiting for a gap in the canopy (Plate 8). In contrast, an abandoned field that has begun to revert to woodland has large numbers of saplings, derived from the seed of a few nearby trees (Plate 9). This is a population growing rapidly, when resources, primarily light, are plentiful.

Life table analyses can be crucial for optimizing yields in both fisheries and woodland management. Knowledge of how quickly an individual grows and of its life expectancy allows us to estimate the best time to harvest the largest biomass. The dynamic pool model uses these kinds of data with sub-models to calculate rates of growth and chances of survival.

The same principles are used in forestry management, though here we benefit from a much more complete data set. Managers can measure competition directly, as the effect of tree density on individual growth. They also have more control over the population, including its recruitment (by their planting strategy), rates of growth (through the application of fertilizers) and competition (through controlling the density and the species composition of the woodland).

3.5 LIFE HISTORY STRATEGIES

The stages in the life cycle of many organisms are very well defined, marked by different body forms and ways of life. A toad begins as an egg, continues as an aquatic tadpole and ends as a terrestrial adult. Seeds sprout wiry saplings adapted to life on the forest floor, but they will grow into massive trees if a gap opens in the canopy.

The life history strategy of a species will have evolved to ensure the survival of as many offspring as possible. During their life cycle some organisms radically change their habits and physiology to exploit different niches. For example, the larval stage of some insects is given over totally to feeding and growth, while the adult, whose role is to disperse and reproduce, may not feed at all.

What are the benefits of partitioning the life cycle in this way? Most likely, making the best use of the resources available. The fact that different stages feed in different habitats or on different diets reduces the competition between age classes. So adult dragonflies preying on flying insects above the pond are not competing with their larvae feeding on invertebrates in the water below.

Other features of an organism's life history are adaptations to maximize survival. Large numbers of

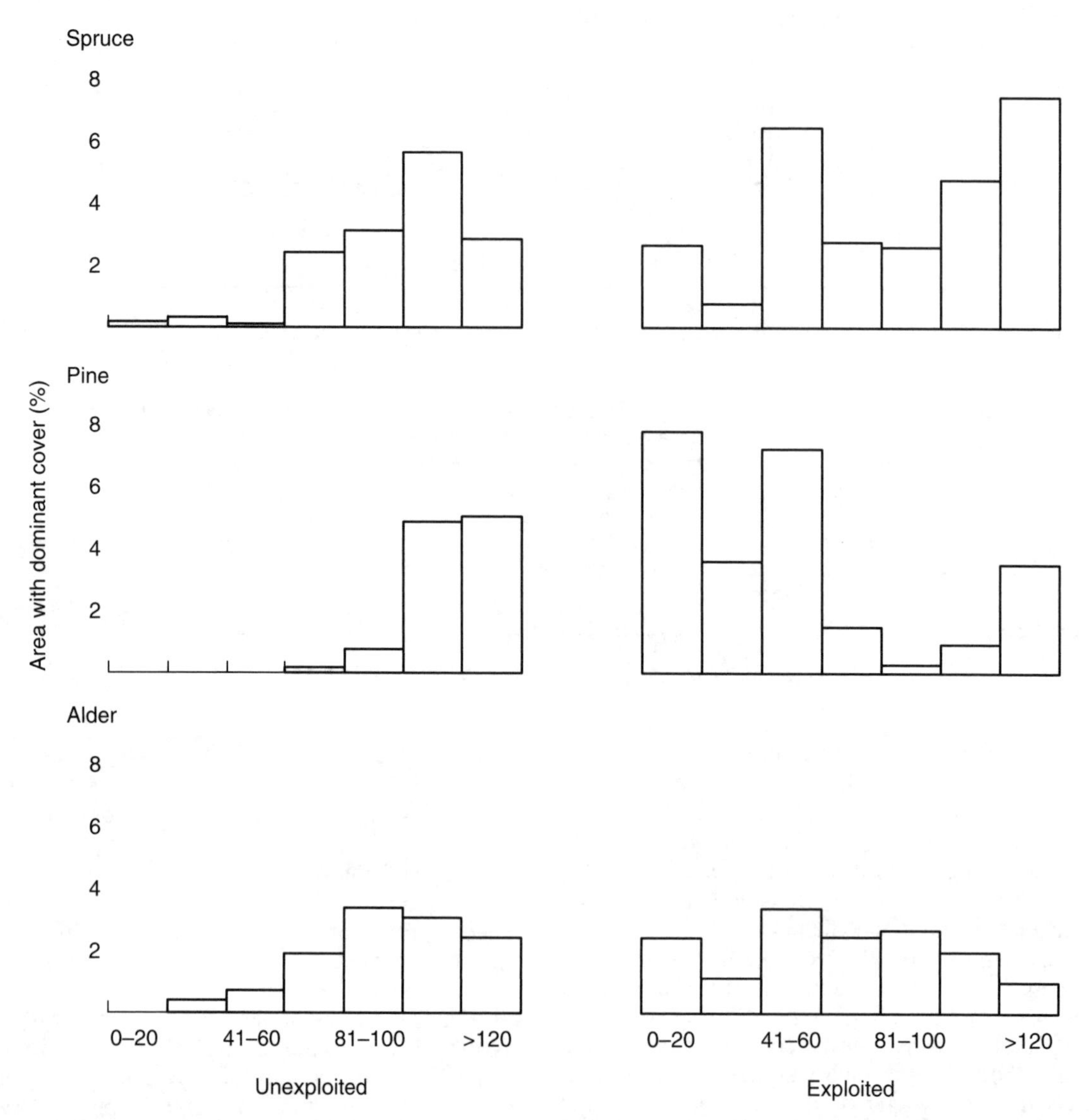

Figure 3.6 The age distribution of spruce, pine and alder trees in exploited and unexploited areas of the Bialowieza National Park in eastern Poland. The unexploited population is derived from a 'primeval' forest that has remained largely undisturbed since the fifteenth century, a community close to the natural forest that once covered most of lowland Europe. The exploited stands have been harvested since 1915; notice that the age distribution includes a greater proportion of younger trees, established here largely through active replanting.

eggs anticipate large losses. The massive wastage in most years is offset by the odd occasion when favourable conditions allow many to grow into adults. The alternative strategy is for parents to invest time and energy in protecting and supporting the few offspring they have produced.

A female cod can produce four million eggs at a single spawning, but devotes little attention to each egg or to its aftercare. On average, perhaps three or four will reach sexual maturity. An elephant produces a single calf, which it has fed in the womb, will suckle at the breast, and will nurture for several years thereafter. During this time, the young elephant benefits not only from the resources its mother provides, but also from her protection and her knowledge of the environment.

We should not think the elephant is more successful than the cod: indeed, there are certainly more cod than elephants in the world. They represent two different strategies, each adapted to their way to life in their particular habitat.

3.5.1 The spectrum from *r*-selected to *K*-selected strategies

A life history strategy consists of several elements: the average length of a life, the proportion of time spent in each stage, the number of reproductive events and the number born on each occasion. Together, these determine the speed of population growth and consequently have considerable adaptive significance.

Consistent patterns appear when we compare the strategies adopted by various organisms. In contrast to those that mate only once, species that reproduce repeatedly start later and produce a small number of offspring each time. Their eggs are usually larger too.

Life history traits are an adaptation to the permanence of the habitat. Where conditions are transient and opportunities for population growth are short-lived, natural selection will favour species that reproduce rapidly. This implies a rapid completion of the life cycle and a short generation time. Having a small body and large numbers of offspring aids dispersal, which is necessary if a species is colonizing transient habitats. At least some may then survive to reproduce elsewhere.

Their short life and prolific reproduction mean such organisms have a high *r* value. These ***r*-selected** species are small, short-lived and highly mobile. They also tolerate a wide range of conditions, due in part to their genetic adaptability (itself a result of their fast reproduction rate).

In a more predictable habitat there is little advantage in being adaptable. Nor is there any advantage in being small. Now the favoured strategy is to stand and fight, to crowd out competitors and invaders. The advantage lies with highly specialized and closely adapted species able to secure the scant resources for themselves.

Long-standing habitats are dominated by large species with long generation times. These ***K*-selected** species are adapted to compete in a habitat close to its carrying capacity, where resources are scarce. Typically they postpone reproduction until conditions are favourable, but they can reproduce many times. Their few offspring take a relatively long time to mature, but because of the competition for resources, mortality rates are highest among the younger age groups.

Compare the mouse, the elephant, the oak tree and the cod we met earlier. Many small mammals, insects and the annual plants found in disturbed soils show *r*-selected characteristics. Large mammals and long-lived trees are typically *K*-selected, growing slowly as individuals and as a population.

This is a very broad schema, with few organisms being readily classified as perfectly *r*- or *K*-selected. Many plants and animals show characteristics of each, while some, such as the dandelion, have races following different strategies in different habitats. Cod, along with many other fish, show *r*-characteristics in their reproductive strategy because of the uncertainty in recruitment from one season to the next. Even so, many of these will grow up to be absolute whoppers.

3.6 SURVIVAL AND EXTINCTION

Adult numbers of *r*-selected species fluctuate greatly as resources wax and wane. Yet for *K*-selected species it is juvenile abundance that varies most. This is because births exceed deaths only when there is spare capacity in the ecosystem. It is the young that suffer at other times.

These differences become important when we are trying to prevent a species from becoming extinct. We need to know how best to maximize reproductive success and reduce the death rate. Much of our conservation effort is directed toward large species whose reproductive rate is low.

However, conservation of all species means conserving their habitat. Life in a zoological or botanical garden may be the only option for a species whose natural habitat has disappeared, but we are then, in one sense, only conserving its genetic code. A habitat large enough to support a viable population is needed for long-term survival.

The numbers of a *K*-selected species, closely dependent upon particular conditions, will decline if its habitat shrinks and the carrying capacity is reduced; but *r*-selected species may be able to accommodate change or move to other locations. The poor dispersal of some *K*-selected species is one reason why they are often endangered. Their size is another, implying a low rate of population growth and a low population density.

K-selected species often play a critical role in shaping the community of which they are a part.

The browsing activity of elephant and giraffe in Africa is essential to keep the savanna grasslands free of thorn bushes and scrub, allowing other animals to graze (Plate 10). The loss of elephant can lead to major changes in the plant community so that many species, including some of the acacia trees upon which they feed, would disappear with them. Organisms with such critical roles are called **keystone species**.

We return to these ideas in Chapter 5. For the time being, it is worth remembering that single species conservation and habitat conservation are inseparable: encouraging the conservation of a large species with an equally large public profile often means that whole communities can be saved.

3.6.1 Persistence time and extinction

Extinction is a necessary part of the natural process of change. Many people are aware of the abrupt periods in the past when massive numbers of extinctions occurred on Earth. Fewer realize that we are going through such a period now. Although we are unsure about the causes of past extinction events, today we can be certain that the loss of natural habitats is primarily responsible. Today species are becoming extinct at between 1000 and 10 000 times the natural rate (Box 9.2).

As environments are degraded or destroyed the original habitat is broken up into fragments or patches, each with a relatively small carrying capacity. In the process, the distances between patches increases. Small, isolated habitats are much more vulnerable to change, and less likely to maintain the conditions required by a large or specialized species. A once continuous population is now divided into a fragmented metapopulation whose overall size has fallen (Box 3.1).

3.6.2 Fragmentation, extinction and colonization

Small and discrete populations can be severely threatened by relatively minor environmental changes. A small patch of habitat provides little protection when conditions deteriorate. For example, a flood will have a minimal impact if resident animals can escape to higher ground within the patch. But these refuges are scarce in small habitats and populations are more readily wiped out. Even without complete destruction, the carrying capacity of the patch has been reduced. Populations of all sizes may suffer in variable environments, but the implications are far more serious when numbers are low, and for species adapted to specific conditions.

Small populations face other difficulties. If there are few individuals, they have fewer opportunities for meeting a suitable mate. Even when they do meet, not every adult will be fertile or free from disease. Some will never produce offspring. For many populations the sex ratio is critical, especially when numbers are reduced. While a single male may inseminate many females, the reproductive capacity depends on the number of females, and in future years, on the number of females born. Together, the size, age and sex structure of a population govern its capacity to grow. Consequently the size of the breeding population (referred to as the **effective population size**) is always smaller than the actual population, often substantially so.

There are also genetic dangers in small populations. In any population, only some fraction of the total genetic variation finds its way into the next generation, as the eggs and sperm which produce offspring. Not all gametes will be successful and some of the code will be lost. Without migration into the population, or new mutations, the total genetic variation is thus eroded with each new generation. This erosion happens faster in small, isolated populations – the consequences of drawing from a small genetic pool (section 2.5).

As time passes the genetic differences between individuals become less marked, and the likelihood of mating with a partner sharing much of the same genetic code increases. With very small numbers, mating may only be possible with close relatives. The viability of any offspring may then be impaired by genetic inbreeding.

The principal danger here is the higher risk of each parent contributing the same recessive gene to their offspring. In this homozygous combination, the trait will be expressed in the offspring's phenotype. Many recessive traits are deleterious, often leading to weak, sometimes infertile offspring, and ultimately a depression of population growth. One way to restore vigour is to increase variation within the population, by introducing individuals from other populations.

Inbreeding depression is a common problem in domesticated plants and animals bred from a limited stock. It is also known in human populations where mating has been confined to small groups, isolated by geography or religion. Some people assume a status which divides them from the majority of the population and so choose from a range of limited partners.

Changing circumstances will challenge groups of genetically similar individuals. Their uniformity means there are fewer phenotypes for selection in a variable environment. With little difference between them, a selective pressure is likely to affect all individuals equally. Little significant genetic variation means future generations have little scope for further adaptation.

The fragmentation of the population exacerbates the problem, by reducing the exchange of genes between isolated populations. Thus while the size of the metapopulation may be large, the effective breeding population within any single patch can be very small. Alongside inbreeding depression, genetic isolation can produce locally adapted races, when some characters fit them for surviving in a specific habitat. As we saw in Chapter 2, such local adaptations are the first stages in speciation. Then reunited populations may not breed with each other. With isolation, physiological, anatomical and behavioural barriers develop between races to prevent successful mating. This is **outbreeding depression** and is sometimes a problem with *K*-selected species, particularly various trees that have evolved local races.

This combination of difficulties – a susceptibility to environmental variation, the problem of sustaining population growth with a small number of individuals, and the genetic consequences of a reduced breeding stock – can lead to a downward spiral in numbers, where one factor feeds on another, accelerating the species toward extinction. This has been called the extinction vortex (Figure 3.7).

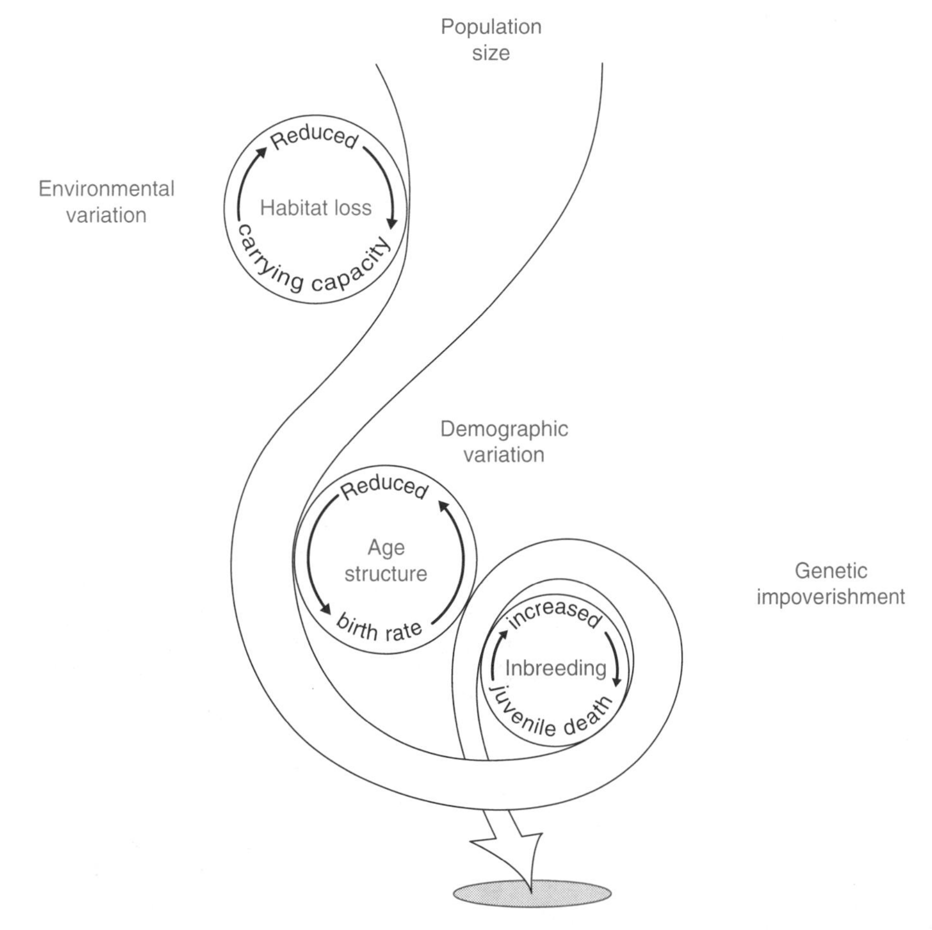

Figure 3.7 The extinction vortex. Habitat loss reduces the carrying capacity for the population; demographic factors reduce their birth rate and genetic factors can lead to increased juvenile mortality. Because all of these tend to reduce population size, and smaller populations exacerbate demographic and genetic problems, an endangered population may quickly race to oblivion.

3.6.3 The elephant and the rhinoceros

The creation of the Chitwan National Park in Nepal has saved the greater one-horned rhino (*Rhinoceros unicornis*) from near-certain extinction. This impressive species was confined to the margins of lowland rivers in the Himalaya, though until recently it had a population that ran into tens of thousands. Hunting and habitat destruction during the last 100 years had reduced it to only two groups of about 100 individuals by the early 1960s.

In Chitwan, this amounted to an effective breeding population of just 21–28 individuals (Figure 3.8). Today, with substantial protection, the total popluation has increased to more than 400 animals, while Kaziranga National Park in India now has 1500 individuals. Much of the credit must go to the rhinos themselves. Their spectacular recovery after the ending of poaching has been possible because they have retained a very high genetic variation.

We would normally expect genetic variation to be low in a population reduced to such a small size. Ironically, the rhino appear to have held on to much of their variation because their decline was so rapid. Two demographic factors contributed to this: the rhino have a long generation time (12 years) and females produce only one calf every four years. Thus the extent of inbreeding was limited over the very short period of their decline.

The poaching of rhino has devastated all five rhino species, particularly the two species in Africa (Figure 3.9). Despite legislation and a policy in some countries to shoot poachers on sight, the high price of the horn ensures that the slaughter will continue. Because of their masculine associations, the horns are prized as dagger handles and, when ground to a powder, as an aphrodisiac. While it probably makes the dagger unusable, it certainly has no physiological effect on the libido of the human male.

Nevertheless, the high price fetched by the horn is a major temptation to local peoples with meagre incomes. The dramatic decline of the African rhino species follows from intense poaching rather than habitat loss. Of the hundreds of thousands of black rhino (*Diceros bicornis*) that ranged over sub-Saharan Africa in historical times, just 65 000 were alive in 1970 (Figure 3.10). In 1990 there were

Figure 3.8 Indian rhino (greater one-horned rhino) in Chitwan National Park, Nepal.

Figure 3.9 Black rhinoceros (*Diceros bicornis*).

3800 distributed among 75 populations. Only 10 of these populations had more than 50 individuals.

The total population of the black rhino has been separated into relatively isolated pockets for some time. Based on their horn shape and body size, seven subspecies have been identified. One is now extinct. Present evidence suggests that these subspecies have become genetically differentiated only recently, and that breeding between them could be a viable conservation strategy.

The picture is less encouraging for one race of the white rhino. Although the southern race (*Ceratotherium simum simum*) has been through a major reduction in numbers, it has recovered to around 3000 individuals on various reserves in southern Africa. In contrast, just 22 of the northern race (*C. s. cottoni*) were left in the wild in 1990, and its future is much more bleak. Moreover, the white rhino shows very distinct genetic differences between the two races, probably because their geographical ranges have not overlapped in recent times. Consequently, conservation biologists have decided not to interbreed them.

In Asia, the security of the rhino populations and their habitats are more closely interlinked, as forest clearance and agriculture have led to habitat fragmentation. With only 50 Javan rhinos (*Rhinoceros sondaicus*) left, there have been fierce arguments whether to leave them all in the wild, possibly combining subpopulations on a single reserve, or to transfer some to zoos for captive breeding.

As for the horn, so for the tusk: the international trade in ivory has led to the slaughter of large numbers of African and Asian elephants. Laos and Cambodia once had 40 000 elephant but now have fewer than 4000. Much of the decline in the Asiatic elephant is associated with habitat loss, whereas poaching has been much more serious in Africa. Unlike their Asian counterparts, both male and female African elephants bear tusks and so both are slaughtered. Their dramatic decline in Kenya, from 130 000 in 1973 to 16 000 in 1989, led to a government ban on the ivory trade and a very public ceremony in which the President, Daniel arap Moi, set fire to a large pile of impounded tusks.

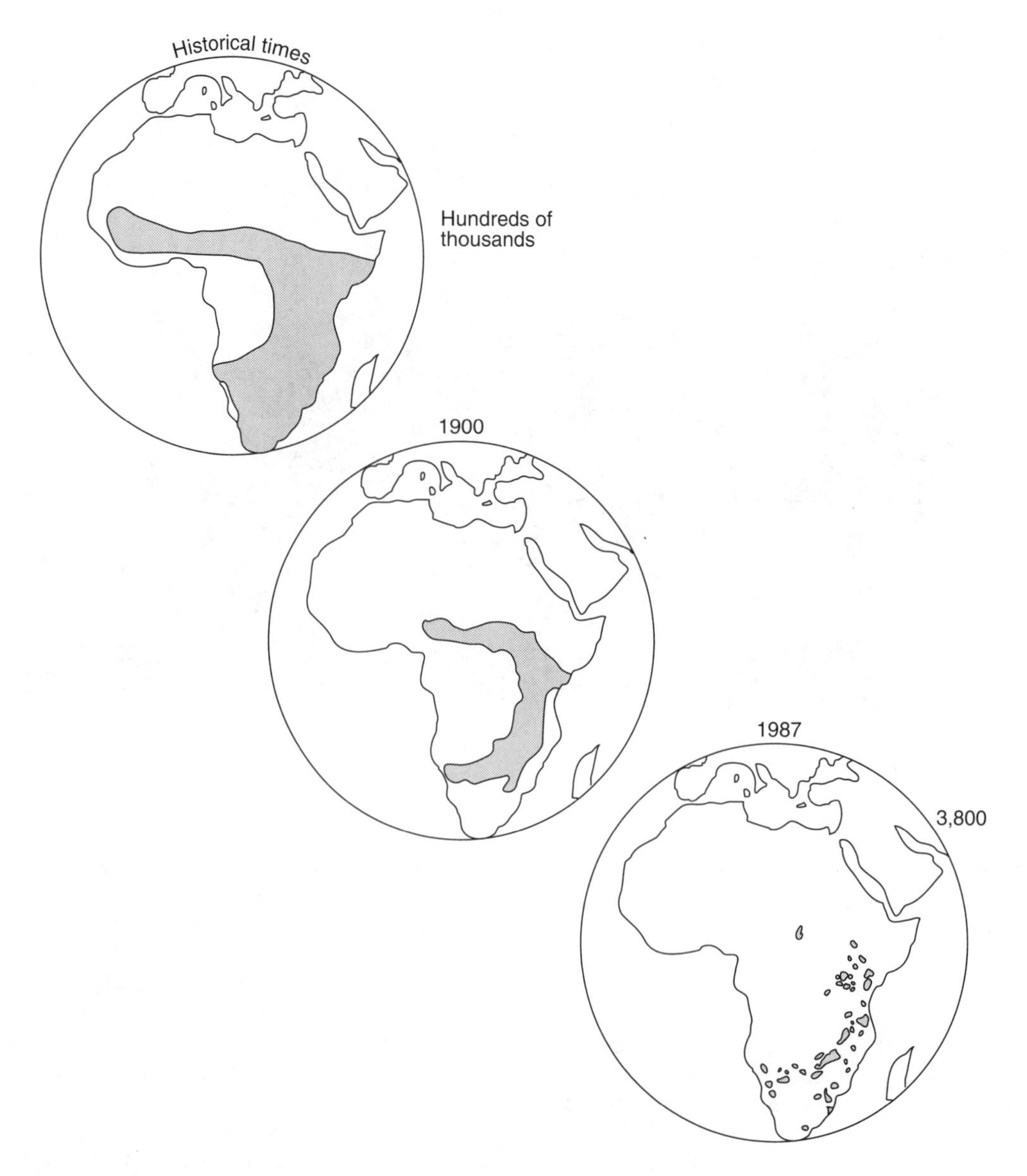

Figure 3.10 The decline of the black rhino in Africa.

Large tusks attract a larger price, so poachers favour the larger adults, especially the males. An elephant might live between 40 and 60 years, though today most adults are under 30 years old. Unfortunately, males do not begin breeding until they are around 30. A 1988 survey of the Mkomasi Reserve in Tanzania could find no adult males at all.

Changing the age structure of these populations not only affects their future growth, it also has social implications for the herds. Herds are normally led by elderly females, the matriarchs, whose knowledge of waterholes and foraging routes is passed on to the younger members of the group. Today most herds in Africa are led by females in their early 20s and teens.

The loss of the elephant also has important implications for the African landscape. Their browsing on shrubs and trees is crucial to maintain an open savanna. Grasses can then thrive, providing food for grazers such as gazelle, wildebeest and zebra. Less obvious are the many smaller vertebrates and invertebrates, who directly or indirectly rely on an open bush. Elephant and giraffe are keystone species, maintaining the grassland community of the dry savanna and an open scrub of the wetter parts of Africa.

However, there are complications when the elephant population is close to its carrying capacity. Regeneration of the shrubs and trees can then be severely reduced (Plate 11). Elephant will push over trees to browse, and their trampling damages grassland regeneration, particularly around waterholes. This is a problem in some southern African reserves where poaching is effectively controlled. The regeneration of the vegetation is being suppressed and habitat quality is threatened for future generations. Here herds have been systematically culled to reduce the pressure on the vegetation. In Zimbabwe, areas have been designated by the government where native agriculture runs side by side with a game management and conservation strategy, as part of the effort to control poaching. Government marksmen carry out organized culls, taking selected elephants, and this helps to maintain the integrity of the elephant herd by managing its age structure. Meat from the kill is distributed among the local villagers, while the skin and tusks have, in the past, been sold on. The proceeds are being used to fund conservation and to provide compensation to farmers in the form of additional food. Local people are thus being encouraged to tolerate and, indeed, value their wildlife resources.

Box 3.3

LEARNING FROM THE PAST

Professor Gary Haynes
University of Nevada at Reno

The crisis-clock is ticking for African elephants. Most of them will be able to survive only in crowded refuges, which offer dwindling resources and ever-shrinking carrying capacity. Eventually, the pressures of so many elephants in so few places may reduce biodiversity in their ranges to the point where their own long-term survival is unlikely.

This seems to be how their ancient relatives, the mammoths and mastodonts, died out too. The mammoths of Eurasia and America – open-land grazers and mixed feeders – disappeared forever at the end of the last ice age, about 11 000 years ago, and so did the mastodonts, woodland browsers. Their bone accumulations now provide us with important lessons about extinction, especially when they are compared with the bone heaps left from starved African elephants. In North America, a rapid cycling of climatic conditions from cold and dry to warmer and stormier led to major changes in the plant communities used by mammoths and mastodonts. As the extent of their feeding ranges shrank, many died out in locales undergoing profound ecological change with the climatic fluctuations. Other populations must have clustered at favoured resource points, such as water-holes surrounded by the kinds of open ground and marshy habitats that supported the animals' main source of forage. The competition for food and water weakened the mammoths and mastodonts, and young animals died in great numbers. Older females also perished. The social cohesion of normal herds broke down, and groups fragmented as each mother wandered away in company with only her youngest offspring to find food and to avoid competition. Fights between adult females and males must have been more common under these conditions.

The signs of stress in mammoth populations are present in many fossil bone collections that date to the end of the ice age. Dense accumulations of bones, often badly trampled and broken, attest to the large numbers of surviving animals crowded around water-holes where carcasses lay thick; at many sites, the presence of broken tusks suggests violent fighting over access to water. The skeletons of very young animals bear witness to a very selective process of mortality – the process of starvation.

The starving animals would have stuck closely to certain water-holes or other resource points and, as a result, much of the vegetation surrounding the points suffered from overfeeding pressures and trampling damage. Soils were compacted, microinvertebrates were destroyed. Many plants and key soil invertebrates were lost because mammoths or mastodonts crowded into the refuge areas. As female and young proboscideans died off, the potential for renewal in the next generation also disappeared. The last to die, as seen in modern studies of closely related wild mammals, would have been the mature males. Most of them died somewhere else, away from the carcasses of females and calves.

This is a grim scenario, but it is based on a decade and a half of studies of free-roaming African elephants. I personally have seen them die in great numbers. When death comes for elephants they are perhaps some of the most dignified creatures in the wild. If attacked and shot, they cluster together and try to protect their young; some will make their own defensive attacks. If starvation claims their lives, they often die alone and silent, after suffering the acute pains of gut cramps, diarrhoea or intestinal blockage. Most lie down in midday shade to rest, lowering their natural guard and inviting predators to an easy kill. At the end of the last ice age, America's prehistoric hunters would have found mammoths in similar weakened states, perhaps whole herds of starving animals, and opportunistically killed them whenever possible.

Abrupt climatic changes brought the mammoths and mastodonts to the brink of extinction, but human hunting may have toppled them over it for good. Whatever the ultimate cause of their dying out, mammoths and mastodonts undoubtedly suffered greatly in their final days. As their food sources proved inadequate, the young ones starved and died first, and their usual tenderness towards each other was replaced by desperate competition for the last water and the last mouthfuls of food.

Although there is an agreed world-wide ban on trade in new ivory, some countries would like to sell the ivory from culls. They point out that the proceeds will help to support their wildlife conservation programmes. One argument is that big game is a resource like any other, and they are harvesting elephant in the same way that other countries catch fish. Unfortunately, allowing an international market in ivory provides a route through which poached tusks can be sold, where they can be 'lost' in the regulated trade. All African countries, some reluctantly, ratified the ban again in 1994, but agencies are trying to find ways in which ivory from a legitimate source might be identified. One promising scheme is to create a databank on the genetic identity of the different herds to allow the origin of a tusk to be traced.

SUMMARY

Simple analytical population models help us to understand the key features of population growth, but lack sufficient detail for us to make realistic predictions about real world populations. We can identify the maximum sustainable yield as about half the carrying capacity of a population, but this tells us little about the weight of fish to be caught. For this detail, we need more information on rates of growth and recruitment, and to match our fishing effort with rates of replacement of both tissues and individuals.

For both fisheries and single species conservation, we have to consider the detail of age and sex structure within a population, and the life history and reproductive strategy of a species. Conservation efforts are often concentrated on the large, slow-growing *K*-selected species closely adapted to their habitat. These are most likely to suffer when there is significant habitat loss. They are also more susceptible to the negative effects of breeding from a small gene pool. On the other hand, *r*-selected species are adaptable, and more able to move between habitat fragments. Nevertheless, the scale of habitat loss today means that these species, too, become extinct with depressing regularity.

FURTHER READING

Beeby, A.N. (1993) *Applying Ecology*, Chapman & Hall, London.
Primack, R.B. (1993) *Essentials of Conservation Biology*, Sinauer.

EXERCISES

1. What simple measurements might you make in a woodland to determine how old it was? Justify your reasons for each measurement.
2. Use the equation in Figure 3.2 and the following data to derive the maximum rate of increase for a starting population of 50 when:

(a) $K = 1000$, $r = 0.36$
(b) $K = 1000$, $r = 0.18$
(c) $K = 500$, $r = 0.36$

You will probably find it useful to create a table like that in Figure 3.2b for each calculation.

What do you notice about the effect of *r* and *K* across these examples? What is a quick way to calculate the maximum rate of increase in these simple models?

3. When a population has reached its carrying capacity, which of the following statements are true and which are false?

(a) The birth rate is greater than the death rate.
(b) There are no further births.
(c) The birth rate and death rate are matched.
(d) The death rate now exceeds the birth rate.

4. How might the characteristics of a fish population change over many generations if overfishing leads to too many immature fish being harvested?
5. Which are more likely to become pests: *r*- or *K*-selected species? Which are more likely to become endangered: *r*- or *K*-selected species? Give your reasons in each case. Now find exceptions to each one!

Tutorial or seminar question

6. Should we attempt to conserve single species? Discuss the social, political, economic and ecological implications of setting aside reserves and protecting a species such as the white rhinoceros.

Acquiring and allocating resources – cooperating to maximize reproductive success – competition between individuals and between species – plant–animal interactions – plant defences against herbivory – predator–prey interactions – parasitism – using natural enemies in controlling pests.

4 Interactions

They made us many promises, more than I can remember, but they never kept but one; they promised to take our land and they took it.

Chief Red Cloud of the Oglala Teton Sioux

This is intraspecific competition in action – individuals of the same species competing for the available resources. We might call it war, because individuals had grouped themselves together and acted cooperatively to acquire or defend a resource. To Jacob Bronowski war was not a human instinct but, rather, a 'highly organized and cooperative form of theft'.

Humanity is not alone in cooperating in this way. Many animals, from ants to chimpanzees, will form themselves into bands to improve their chances of winning a competitive battle. In a world of shortages, individuals and societies need strategies to secure their needs, for this generation and for generations to come.

Of course, it is too simplistic to describe war as a contest for scarce resources. It may be a prime cause, but conflicts also begin when societies seek to promote or preserve their structures and their traditions. Individually and collectively, people make moral and ethical judgements to fight and to sacrifice resources to defend an idea. For both humans and animals, defending a society often means defending the system by which resources are partitioned amongst its members.

The problem for the individual is to make a living; that is, to acquire the resources that allow it to survive and to thrive and to ensure that its genes make it into the next generation. This means acquiring the energy and nutrients necessary to produce gametes and to nourish its offspring. Societies are often organizations to facilitate this process. At one extreme, an ant society is highly integrated with most individuals sacrificing their own reproductive effort to promote that of a single queen. Mammalian societies are much looser arrangements, but increase their efficiency by dividing effort and reducing wastage by partitioning resources amongst their members.

Seen in these stark terms, life appears to be very brutal. Yet the great variety of life on the planet, and the beauty of its plants and animals, has risen from this struggle for resources. Even so, cooperation may be beneficial, especially among related individuals sharing part of their genetic code, or even between very different species. Cooperative interactions are essential for the survival of whole groups of organisms. Sometimes the benefits are more one-sided, when one species is used by another as a resource. In all cases, these interactions evolve and develop by natural selection and help to tie communities of species together.

This chapter looks at the range interactions, considering their costs and benefits to each partner. Each individual seeks to maximize the proportion of its genes in the next generation, to maximize its reproductive success, whether this means preying on or cooperating with another species (section 1.2). In some cases, the trade-offs made by a species may not be immediately obvious and only by unpicking the detail do we find how it benefits from the association.

We begin by exploring some cooperative associations, and then go on to look at the competitive battles between individuals and between species. Some, more direct conflicts mean that one species loses out because another exploits it – say between a predator and its prey, or a parasite and its host. As we shall see towards the end, these are associations that we have used to our own advantage in our struggle with pests and disease.

4.1 ACQUIRING RESOURCES

There are some needs which are common to all living organisms. The two most fundamental are food and water: water, because it is the medium in which the chemical reactions of life take place; food, because it supplies the materials for these reactions and also the energy that powers them.

Food and water are often in short supply, either periodically or continually, and an organism needs strategies to acquire and to conserve supplies of both. All species require space in which to secure their resources. Plants need a volume both of soil and of air in which to capture nutrients, water and sunlight. Animals too need space – for example, sea anemones will jostle each other for room, pushing (albeit very slowly) against each other for a prime location. More obviously, dragonflies dart up and down their stretch of river bank, defending a territory in which they feed and mate. Even a partner can be regarded as a resource, necessary to produce the offspring that will carry some of their parents' genetic information into the next generation. 'Resources' thus refers to all of those elements of its habitat that an organism needs to survive and to procreate.

Many resources are so abundant that a species can afford to be profligate, making no special effort to conserve supplies. Oxygen is a prime example. In contrast, energy supplies are limited for most species, and they have to allocate the use of energy to best effect (Box 4.1), to decide how much to invest in reproduction, growth or further energy acquisition. Like the different reproductive strategies we have already examined (section 3.5), these acquisition strategies help to define the ecological niche of a species.

4.2 COOPERATION

One strategy is to share some of the costs in finding a resource with the neighbours. A group of female lions combine their strength to overpower a wildebeest. On their own they have little chance of bringing down a healthy adult; together, not only are they successful but they also share the costs (and risks) of securing their food. This works because the

Box 4.1

ALLOCATING RESOURCES

For any resource in short supply, be it energy or some key nutrient, an organism needs a strategy to make efficient use of what it can acquire. Since its prime objective is to reproduce and ensure the survival of its genetic code, this means allocating its resources to maximize its reproductive success.

The dilemma is often whether to play the short game or the long game. Resources could be used to produce offspring in the short term whilst the resources are in hand. Alternatively, resources could be devoted to survival now, perhaps growing larger and delaying reproduction, so that several reproductive events are possible.

Natural selection will favour different strategies in different environments, according to how predictable they are (section 3.5). Acquiring the resource also requires a strategy that balances the costs of acquisition against the potential return. A predator has to evaluate the energy costs of capturing a prey against the benefit it will get from consuming it. Clearly, if the energy invested in searching for, handling and consuming a prey is less than that gained from eating it the predator will show a net loss and have less energy to devote to reproduction or the next chase. An active predator is therefore continually reviewing how easily food is being found and will tend to stay where its prey is abundant. This is an unconscious cost–benefit analysis, with the consumer moving on when search costs start to rise too high.

All that we say of animals we could say, in slightly different terms, about plants. Deciduous trees shed their leaves in winter because of the cost of respiration. Simply keeping the leaf alive during the short days outweighs the return from any photosynthesis (Box 5.1). Similarly, plants have different strategies for investing in root growth (to acquire nutrients) or timing their shoot growth according to a variety of factors (including other species) in their environment.

In fact, any organism has to meet a series of costs that it cannot avoid, simply to stay alive (Figure 4.1). Only a fraction of the energy in the food will actually be assimilated (*A*) by an animal and then only a small fraction of this is actually fixed in the tissues (Box 5.3). Energy has to be used in metabolism, simply to maintain the tissues, in movement and in growing new tissues. This is the energy of respiration (*R*) and is eventually lost as heat. The difference between *A* and *R* can be used to produce either gametes (*Pr*) or new tissues (*Pg*). The balance is thus:

$$Pr + Pg = A - R$$

When food is in short supply and energy reserves are low, this allocation of resources becomes critical. Even if energy itself is not in short supply, a lack of some key nutrient, such as water, may limit its assimilation. Under these conditions an individual may only grow very slowly, often in competition with many others. Long-lived species able to match their growth against resource availability will be favoured in these habitats. Their strategy is to delay reproduction until its chances of success are most favourable. Alternatively, in an environment which is characteristically unpredictable, brief periods of plenty have to be exploited quickly, favouring short generation times that rapidly produce gametes.

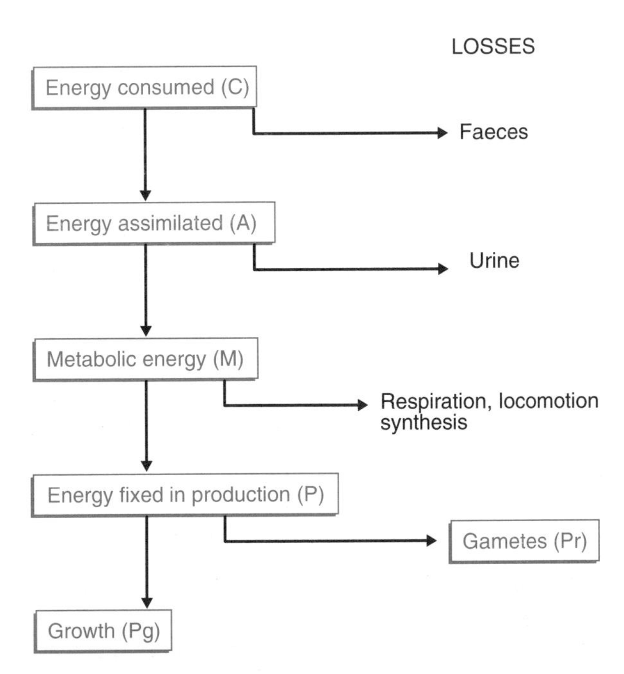

Figure 4.1 The allocation of energy to growth and reproduction in an animal. Note that the same diagram could be used to describe the allocation of any key nutrients, such as nitrogen or phosphorus.

wildebeest represents a meal for the whole pride and far more meat than a single female could eat in a single sitting. A range of animals cooperate in this way, from whales swimming in coordinated manoeuvres to confine a shoal of fish, to the more chaotic swarming of ants over a large beetle.

All of these examples describe cooperation between individuals of the same species, but we need to quantify the benefit to each member of the team. This is something we can best measure as an individual's reproductive success, rather than simply its satisfaction from eating a large meal. The fitness of its resource acquisition and allocation strategy is judged by the contribution an individual makes to the offspring of the next generation.

Sometimes, like the pride of lions, a group consists of close relatives, siblings and parents, which collectively ensure that their shared genetic code passes into the next generation. In other cases, genetic relatedness between individuals is much lower and the association is much looser. For these animals, the benefit comes from dividing the workload and improving the chances for each individual to feed itself. Living in a large group also affords protection and reduces costs. One individual in a shoal of a thousand fish has less chance of being eaten than when it is alone and the sole object of a predator's attention. Equally, the task of looking out for potential predators is shared between all the members of group, so each individual spends less time on the look-out (Plate 12).

Living in groups seems to have benefits for both predators and prey. The range and number of different animals that have developed cooperative social behaviour and the considerable investment most make to communicate with each other suggest this must be a powerful adaptive strategy. As we saw in Chapter 1, these same factors were a key feature of our own evolution.

Some of the more complex strategies for cooperation are best understood by looking at the genetic relationships between members of a group and the trade-offs that each is making. Working with a brother or sister can improve the prospects for the genes which are shared. Egg production in a colony of ants is delegated to one individual, the queen, and the rest of the females forego any reproductive activity (Figure 4.2). The individual worker is a sterile female, but she shares a large proportion of her genetic code with her sisters. The role of each worker is controlled by what they are fed. Particular chemicals, termed **pheromones**, govern the development of larvae and their subsequent behaviour. The prime source of these pheromones is the single fertile female – the queen – and these signals move through the nest by contact between workers and with the food. In this way, the numbers of each caste and their activity are regulated. Should the queen be lost, each female larva stands some chance of herself becoming queen, if fed a particular diet. Males are only produced by the queen when it is time to form a new nest. The queen controls the sex of her offspring by allowing the egg to be fertilized by sperm stored from her nuptial flight. A male is from an unfertilized egg and therefore carries only the female chromosomes.

Although the queen controls the sex of her offspring, it is wrong to assume that she controls the nest. In fact, the overall coordination derives from the combination of the chemical communication system and a large number of individuals doing a small range of predictable tasks. What may appear as chaotic behaviour from watching a small number of ants can make sense when viewing the colony as a whole. Order appears from the interaction of a large number of repetitious elements, something with which we shall find echoes in the final chapter (Box 9.5).

Overall, the collective effort of the females ensures that their shared genetic code has a greater chance of survival, better than if they wasted energy and resources by competing with each other. Interestingly, all of the sperm produced by a male are identical, so each worker has at least half of her genes in common with her sisters – those she inherited from her father.

An individual worker may be dwarfed by her monarch, but the sterile worker is not sacrificing herself for her queen. The genetic cost–benefit scales are tipped in favour of the code carried by the workers, rather than the queen. The workers, after all, are carrying code from their father which is not shared by their mother. A worker sacrifices herself for the code shared with her brothers and sisters who eventually leave the nest on a nuptial flight. It is they who carry her code into the next generation. This reproductive and resource utilization strategy is found in other insects, including bees, wasps and termites, and has produced some of the most elaborate and highly organized of all animal societies.

As you might expect, insect societies are also some of the largest. One ant society recently described from the Jura, in France, consists of 300 million individuals divided between 1200 ant hills,

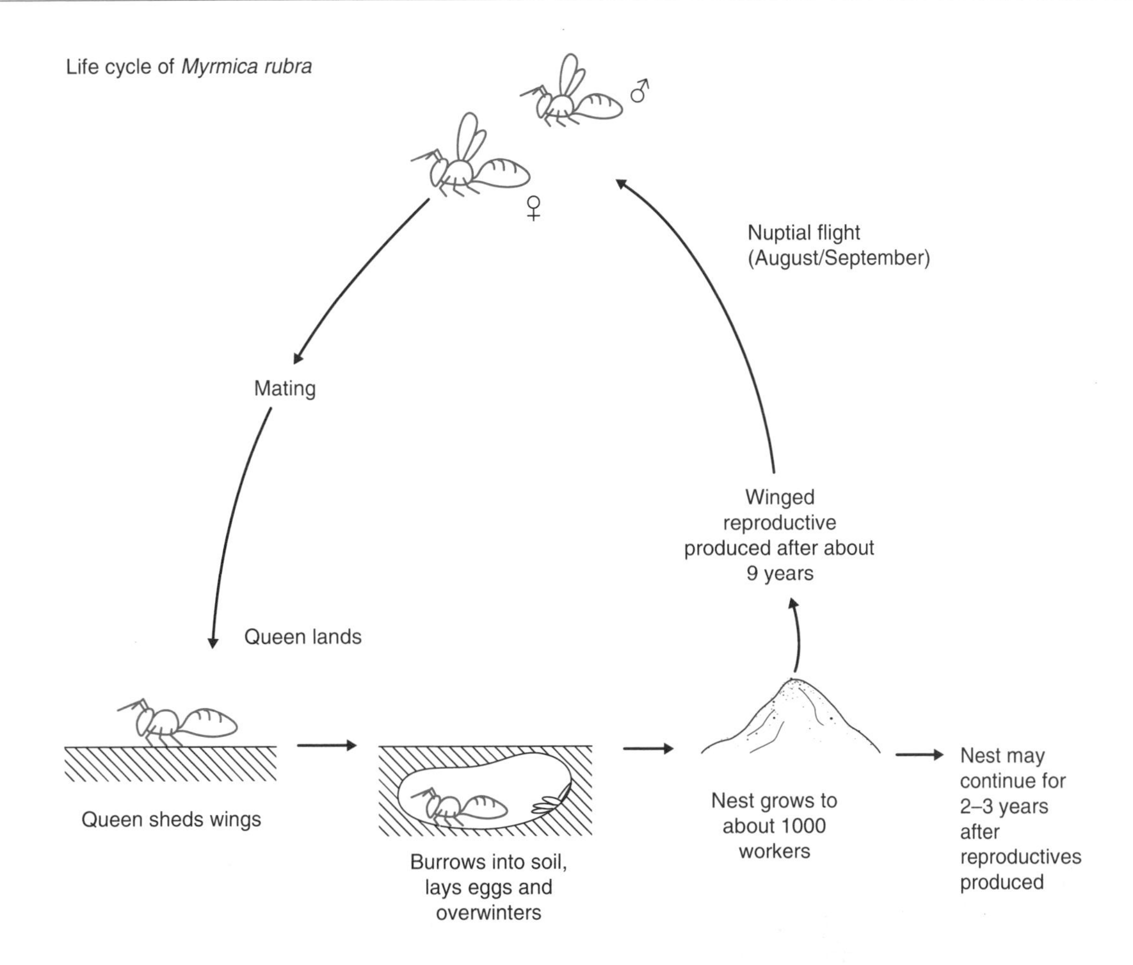

Figure 4.2 The life cycle of a temperate ant, *Myrmica rubra*, a species common in woodland and grassland areas where it can be found in loose soil or rotting logs.

covering around 70 hectares. The whole arrangement is connected by 100 km of tracks. Information, food and larvae move down these tracks, with the whole colony divided into sectors. There are around 15–20 main nests, and a number of smaller ones used to rear larvae in the summer months. Where food is short, resources in the form of food and labour are moved to give a more equitable distribution. We are still learning how this scale of coordination is achieved.

Large ant colonies are known throughout the world. Some species do not make permanent nests, but instead move *en masse*, continually foraging and setting up bivouacs for the night. Some raid the nests of other species and have castes adapted for fighting the battles. Slave-maker ants rifle the nest, removing larvae and pupae which they then rear in their own nest. The emerging workers then become slaves, keeping house for masters whose own workers are largely warriors. The costs of having specialist fighters is then offset by not having to produce eggs to make workers. Other species seize chemical control within a nest so that the original queen is replaced when her own offspring are induced to kill her. The parasitic queen is then protected and fed by these workers, who raise her offspring and their different genetic code. As Richard Dawkins points out, each individual parasitic queen has to make a small investment in subverting the chemical control system of the nest, yet gains immensely from the efforts of the duped hosts. It is an interesting question whether we should call this war, given that it is two different species fighting over resources.

4.2.1 Cooperating with different species

Cooperation between individuals of the same species may occur because they both require the same resources. With two different species, however, requirements differ and cooperative associations will only form if each species can derive some benefit from the other.

In **mutualistic** associations, both species benefit. (With commensalism, described at the end of this section, the arrangement is more one-sided, though the advantage to one is of no deteriment to the other.) Many insects share truly mutualistic associations with flowers. Flowers are adapted to attract insects to their reproductive organs, so that the insect may serve as a go-between, carrying the male gametes (pollen) to the female parts (style and stigma) of the next flower (Figure 4.3). In most cases, the flower itself is no more than an advertisement that food (nectar) is available, and many different species of insect may visit. Besides the costs of producing nectar (a relatively simple sugar solution), such flowers need to produce abundant pollen to dust the many insects that will visit them, perhaps only one of which will go on to visit another flower of the same species. Abundant pollen is also needed to make up for losses by those insects which consume it. Some plants actually produce a dummy pollen to reduce these costs.

Pollination can be made less haphazard by attracting particular insects to the flower. Under selective pressure, a plant which only allows insects of a certain size, or releases its pollen only to those with a certain configuration of mouthparts, may help to ensure that its pollen is passed on to the correct flower. In fact there is a range of strategies adopted by plants to reduce their pollen costs by concentrating on specific messengers – not only the configuration of their flower parts but also the scent used to attract the insects. Sometimes the coevolution of the two mutualists is so close that each depend completely upon the other.

The yucca (*Yucca filamentosa*) from the deserts of western North America is a dramatic example of this. The plant can only be pollinated by the female yucca moth (*Tegeticula yuccasella*), which does not

Figure 4.3 Insect pollination in action.

feed on the flower but instead collects its pollen and forms it into a ball. She then flies to another flower and lays her eggs in its seed buds. Thereafter she climbs up the stigma and applies some pollen, ensuring the flower is fertilized, before flying off to repeat the process in other flowers. In this way, her larvae will have an abundant supply of ripening seeds. Much of the seed is not consumed and so the plant still benefits from the care taken by the moth. Notice also that the moth spreads her risks by not laying all her eggs in one flower. This mutualism is complete: without the other, the yucca and its moth would become extinct.

We saw earlier how the coevolution of a number of species of orchid with bees and wasps had led to the flower mimicking the female insect (section 2.2). While female bees show no interest in the flower, the sight and smell of the flower is irresistible to a male, who will try to mate with the flower for some time – long enough for pollen to be dumped on his back. Ecologists would regard this as an example of commensalism since the benefits to the bee are rather more apparent than real.

Plants also enter mutualistic partnerships with animals to disperse their seed. Fruit is produced to attract consumers, but only when the seed is ripe and ready to be moved. At that stage, the fruit has produced the sugars to reward the carrier and advertises this by a change of colour. Until that time, the fruit remains inconspicuous and bitter.

Ants make very bad pollinators. Their subterranean lifestyle exposes them to pathogens and they secrete a range of protective substances, including antibiotics, which can affect the viability of pollen. However, they can make a very effective defence system, not least because of their high degree of cooperative behaviour. The bull's horn acacia (*Acacia cornigera*) of Mexico cultivates an association with the ant *Pseudomyrmex ferruginea*, by producing protein-rich nodules at the tips of its leaves. The ants collect these to supplement the protein content of their diet. The plant also provides carbohydrate from special nectaries located at the base of its leaves and, to ensure that the ants stay on the tree, it provides shelter in the form in large hollow thorns. Here the ants build nests within easy reach of food and from which they readily defend their home and larder. Few herbivores will feed where busy and aggressive ants are swarming. Both ant and acacia depend on each other – they have an obligate association since they each need their partner to survive.

Such close associations can be readily upset. This has happened in the South African fynbos (section 7.1) where one species of protea (*Mimetes cucullatus*) uses ants to disperse its seeds. The protea recruits the ants' services by providing seeds with little food parcels called elaiosomes attached to them. Once home with their 'take-away' meal the ants eat the elaiosomes and leave the seeds in the nest, or they are discarded with the waste from the colony. The seeds are, in effect, planted by the ants, and as a result are safeguarded from the periodic fires that surge through the low dry vegetation. The seeds are also provided with much-needed nutrients in what is otherwise a nutrient-poor environment.

This dispersal strategy could now be under threat. An Argentinian ant, *Iridomyrmex humilis*, has proved to be a formidable colonizer of South African shrubland since its accidental introduction at the turn of the century. When faced with protea seeds this species simply bites off the elaiosome and the seed drops to the ground, where it can be lost to fire or to seed-eaters. Since this ant nests beneath large rocks, any seed taken back to the nest has little chance of becoming established.

Many associations are not obligate, but one or other of the partners may fare less well without the interaction. One example is the association between many plants and fungi. Most higher plants readily form these associations, called mycorrhizae (section 6.2), in which the strands (hyphae) of particular soil fungi enter the root system of the plant. The plant benefits from the extension to its root system, improving its drought resistance and its capacity to absorb nutrients (particularly phosphates). The fungi benefit from the readily available sources of energy provided by the plant. Without a mycorrhizal association, a plant grows less well.

The association between photosynthetic plant and fungus is most highly developed in the lichens (Figure 4.4). This is clearly a successful combination as over 20 000 types of lichens are known. Again this is not an obligate association since both fungi and algae will grow in the absence of their partner. However, so close is this association that the lichen produces asexual reproductive structures (soredia) that contain both algal and fungal cells. The fungus grows to provide a superstructure (the fungal mycelium formed of strands of hyphae) which helps to support the photosynthetic partner, either green or blue-green algae (cyanobacteria). The fungus also produces complex organic acids which dissuade herbivores, help to protect against dessication and may mobilize minerals that the algae can absorb. The lichen benefits from the sugars produced by the algae's photosynthesis. In some cases, the fungus may 'cull' older algal cells to absorb their nutrients and some biologists have therefore argued that this mutualism may border on a controlled exploitation of the algae by the fungus.

The significance of mutualist associations and cooperation in evolution should not be underestimated. Lynn Margulis has established the key role that such associations have played in the evolution of the eukaryote cell and in the different lines which led to the main groups of organisms (section 1.6, Figure 1.8). They also play crucial roles in most of the major communities of the planet.

Commensalism is a much looser association. Very often a commensal is not dependent on a single species, but derives benefit from its presence or that of several species. Some dung beetles are relatively specific about the source of the dung in which to lay their eggs; others are not so fussy. Vultures have no capacity to overpower and kill a large game animal, but can clear up the mess after a pride of lions or other carnivores have picked over the carcass. Vultures will feed on any dead or dying animal if the opportunity presents itself; like many other scavengers (perhaps including ourselves in the past) they make use of the killing power of large predators, but without incurring any cost themselves.

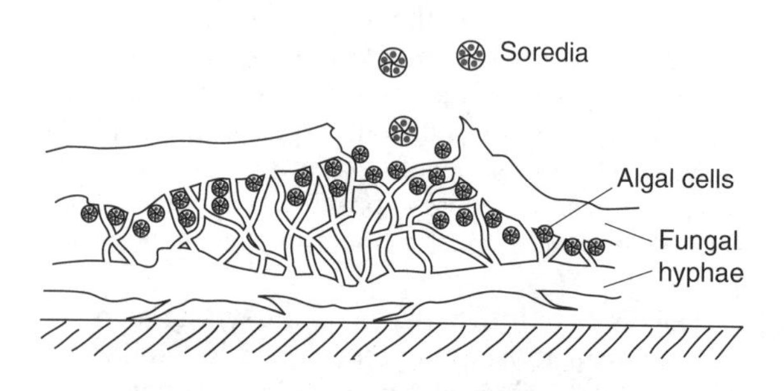

Figure 4.4 The typical structure of a crustose lichen. Lichens come in various forms from relatively large foliose species to the thin dry crusts that can be found colonizing bare rock. In all forms a fungal mycelium consisting of hyphal strands forms the main superstructure, with the symbiotic algal cells embedded within it. Soredia are shed periodically, consisting of both algal and fungal cells, ready to grow into a new lichen.

4.3 COMPETITION

When resources are limited, the best strategy may simply be to secure as large a share as possible. Competition is central to the theory of natural selection because the struggle for resources determines which individuals are able to reproduce. The fittest survive and have the resources to produce offspring.

This sort of competition, between individuals of the same species, is termed **intraspecific competition** (section 3.2). **Interspecific competition** is between two different species. In each case, competition can take different forms. Resource competition is the battle to secure part of a limited supply. Interference competition is where one individual or species prevents another from exploiting the resource by its activity or behaviour.

The nature of the competitive battle can also be classified according to the way it is fought (Table 4.1). With intraspecific competition, direct combat over a resource, be it a mate or food, is costly and carries considerable risk. Only when the stakes are high – perhaps when a harem of females is to be won – will competitors actually lock horns. Death rates amongst adult males can then be high. For this reason, many species have mechanisms to diffuse such confrontations, where size, coloration or ornamentation is used to establish the status of the owner or its position in the pecking order. Even so, the horns of many large mammals are not simply elaborate ornamentation but have primarily evolved to improve defence and attack (Plate 13).

In contrast, competition between two species for a resource rarely results in blood being spilt but is decided by which can usurp most of the supply or reproduce fastest. As we saw earlier (section 2.4), niche overlap may mean that two species competing for the same part of the resource spectrum leads to one of them being lost altogether.

Space is often limited, particularly in the best places. Plants and sessile animals compete by preemptive and overgrowth competition to swamp out

Table 4.1 Types of competition

Type	*Description*
Chemical competition	Production of a toxic or deterrent chemical to exclude competitors.
Consumptive competition	Competitive use of a renewable resource (e.g. food).
Encounter competition	Physical defence of a resource, possibly by aggression.
Overgrowth competition	Successful competitor overwhelms the opponent in size or number.
Pre-emptive competition	Rapid colonization of space when it is available.
Territorial competition	Defence of territory, breeding and feeding areas.

potential and existing competitors. One highly invasive weed that uses this form of interference competition is bracken (*Pteridium aquilinum*). Its dense foliage and deep litter layer forces out existing plants, preventing others from becoming established (Figure 4.5). Along with a series of chemical defences that dissuade most herbivores, this has made bracken a worldwide pest problem (section 4.5).

The chemical battle of plant against plant is known as **allelopathy** and may be quite widespread. We can see this readily for some species which show a widely dispersed distribution, with large distances between individuals. Plants of the creosote bush (*Larrea* species) are so regularly spaced that they appear to have been planted in ranks. This natural spacing may be because each plant has a clearly defined area in which it collects its resources, or it secretes an inhibitor which suppresses growth of other individuals near itself. Perhaps both effects act together. Certainly, the litter from species such as *Eucalyptus* has compounds

Figure 4.5 Bracken (*Pteridium aquilinum*), a major invasive weed found worldwide.

which inhibit the growth of competitors, both of their own and other species (section 7.1).

4.3.1 Intraspecific competition

We have already examined the effects of competition between individuals of the same species. In Chapter 3 we saw how limited resources lead to density-dependent population growth and how this would tend to stabilize the population around a carrying capacity (section 3.2). We have also seen how differences within a population might, with some degree of isolation, lead to character displacement and the formation of a new species. Specialization on a particular part of a resource spectrum can then result in two new niches being defined and two new species (section 2.4).

Now we go on to ask whether competition between two species for the same resource lowers the carrying capacity for each one. The question is whether their coexistence can continue indefinitely or if, over time, one displaces the other.

4.3.2 Interspecific competition

In a prolonged competition between two species for limited resources natural selection would produce one of two alternative outcomes:

- Each species becomes more highly adapted to its part of the resource spectrum and shows the character displacement of a specialist (section 2.4).
- One species extends the range over which it feeds, and makes more efficient use of that resource than its competitor. It may do this by interfering with the growth of its competitor or it may simply produce more offspring per unit of resource. It therefore out-competes its competitor, which must eventually be displaced.

The second alternative is the **competitive exclusion principle** (section 2.4). Simple mathematical models show that where there is strong competition between two species one must decline to extinction. There is now considerable evidence that competitive exclusion does indeed happen. A number of laboratory experiments using microorganisms, plants and cereal beetles, confirm that one species will be lost where there is significant overlap for a key resource, at least in these simple ecosystems (Figure 4.6).

Sometimes, these laboratory fights were resolved by the winner reproducing fastest (resource competition) or by interference competition, where the winner ate the opposition (or at least their young and pupae). The result of each battle was not always guaranteed, however, since changing the conditions of the experiment could sometimes mean reversing the result. Then the habitat more closely matched the niche requirements of the usual loser or the new circumstances more closely suited its competitive strategy.

We also have evidence of competitive exclusion from the real world, when two species with similar niches meet for first time. Many of these battles have been fought where humans have introduced two species that had never met before. Sometimes the introduced species have become pests and out-competed the native species, though more often the introduced species have lost out, presumably because they are less well adapted to their new habitat.

The common house gecko (*Hemidactylus frenatus*) is native to Asia but was introduced into Hawai'i in the 1940s. Geckos are lizards with highly flattened feet able to cling to vertical surfaces and are commonly found on the walls of tropical houses, where they feed on insects attracted to the lights. After the arrival of *Hemidactylus* three other species, previously introduced to Hawai'i from various locations around Polynesia, all suffered major population reductions in their suburban habitats. However, the common house gecko has not dominated the whole of Hawai'i. Another Polynesian species, the mourning gecko (*Lepidodactylus lugubris*), survives well in drier habitats and seems to be holding on, even though its numbers have been reduced. The eventual outcome may depend on interference competition again, since *Hemidactylus* is known to eat *Lepidodactylus* juveniles, at least in laboratory experiments.

Elsewhere, species of *Hemidactylus* seem to winning most of the battles in the Pacific. Its numbers have risen on Suva (Fiji) at the expense of other species and although it was unknown on Vanuatu in 1971 it is rapidly becoming the most common species here in some areas. Ted Case and his colleagues recorded the arrival of *Hemidactylus* in Tahiti in 1989 and within two years it had spread beyond the docks in a 10 km radius.

Although the confrontations between *Hemidactylus* and its rivals are far from resolved, its

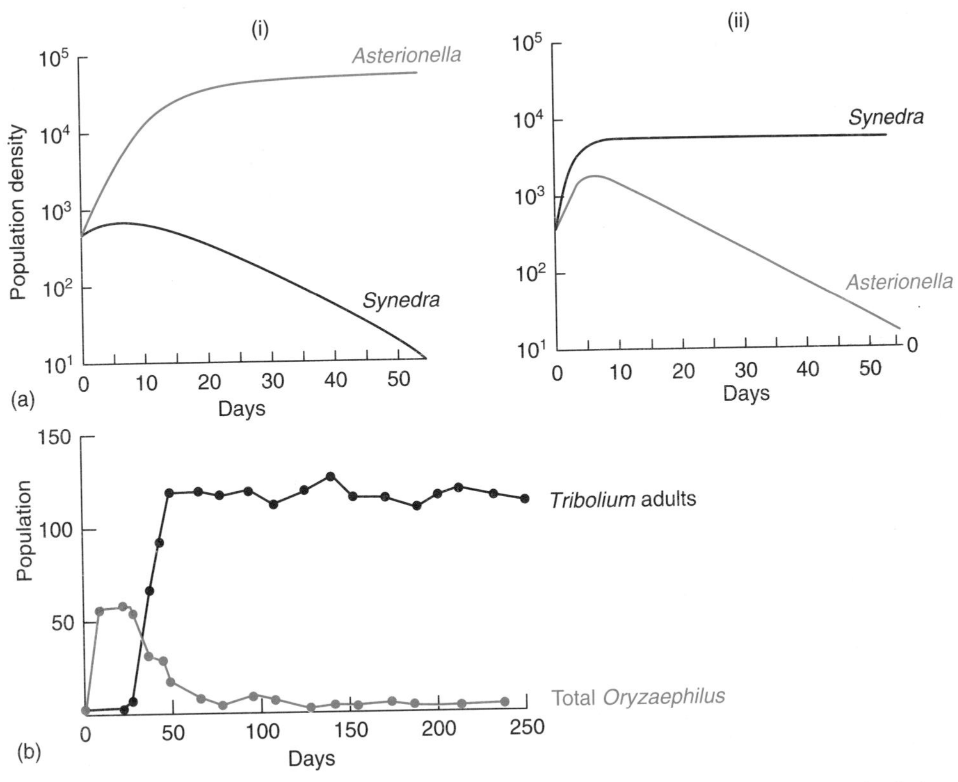

Figure 4.6 The results of a series of laboratory trials looking at the competitive battles between two species that occupy similar niches.

(a) Between two diatom species (unicellular algae with a siliceous case), *Asterionella* and *Synedra*, in cultures where silica is limiting. (i) In most trials between the two species, *Asterionella* won, whether *Synedra* had a higher intial population or not. (ii) However, when the temperature was raised to 24°C, *Synedra* won.

(b) Between two cereal beetles, *Tribolium* and *Oryzaephilus*. These species predate each other's larvae and pupae when placed together in a simple flour ecosystem. *Tribolium* won most battles because it ate more of the opposition. Coexistence was possible if *Orzaephilus* larvae had a refuge from adult *Tribolium*. In this example, both species start with the same population size; the growth in the *Tribolium* population is matched by the decline in the *Oryzaephilus*.

successes and occasional failures, along with the evidence from more controlled experiments, do show that competitive exclusion is not the only possible outcome of interspecific competition. In a variable environment, when conditions might change or where there is scope for avoiding competition, coexistence might be possible.

Unfortunately, the success of *Hemidactylus* over much of the Pacific also demonstrates how readily species can be lost when two competitors meet for the first time. These extinctions are entirely natural but with our propensity to travel we have speeded the process. The brown snakes and rats we have unwittingly distributed throughout the Pacific islands have decimated many native vertebrate faunas, especially birds and small mammals.

4.3.3 Competitive types

Highly invasive species, such as the rat, often share a particular set of characteristics that fit them to their way of life. A rat approximates to the *r*-selective type described in section 3.5 – as an opportunisitic species capable of rapid reproduction. Although many introduced pests share similar characteristics, there are plenty of examples of alien invaders who are closer to the *K* end of the spectrum – species built for more long-term competitive battles.

You may remember that species can be classified along the *r–K* continuum primarily according to their reproductive strategy in habitats of varying predictability. Although simplistic, it does distinguish the two main alternative strategies. At one extreme (true *r*-selection) opportunist species grow rapidly and reproduce prolifically when conditions are favourable. At the other, *K*-selected species are competitive and are found in stable habitats and among populations close to their carrying capacity. Here resources are limited and natural selection favours those that use them most efficiently. These life history strategies imply very different allocation of resources. The true *r*-selected species invests considerably in each reproductive event, whereas a *K*-selected species invests relatively little, but can do so repeatedly.

However, Richard Southwood pointed out that life history strategies were selected according to not only the predictabilty of the environment and the availability of resources, but also the presence of stressors – shortages of nutrients, for example, or the presence of toxins, both of which might limit rates of growth. Southwood called this adversity selection. Philip Grime has developed a three-way classification of plant strategies, noting that some are highly adapted to living in poor, highly stressed environments (Figure 4.7). Where resources are limiting, or there are other stresses (such as toxins or extreme soil pH), a plant may only be able to grow slowly. Since few other species can withstand the adverse conditions, species growing here, termed **stress-tolerators,** will have few competitors, and slow growth is not a disadvantage. Equally, fast-growing species would not be able to secure sufficent resources in these habitats: as **competitors**, their strategy is to outgrow or crowd out the opposition where resources are abundant.

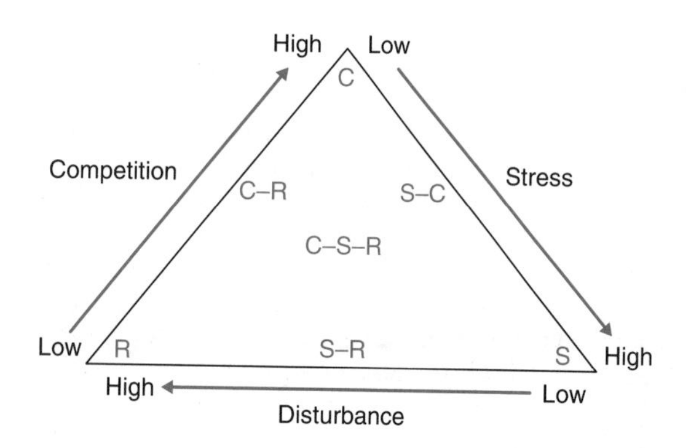

Figure 4.7 Grime's classification of resource acquistion strategies in plants. Ruderals (**R**) are roughly equivalent to *r*-selective weeds (e.g. *Stellaria media*). They are found in habitats that are regularly disturbed and they are typically fast growing and short-lived. Competitors (**C**) are found in undisturbed habitats where resources are abundant. Their strategy is to try to crowd out other plants by rapid growth (e.g. stinging nettle, *Urtica dioica*). Stress-tolerators (**S**) are slow-growing species that can tolerate shortage or other forms of stress in the soil; they prevent many other species from establishing themselves (e.g. sheep's fescue, *Festuca ovina*). Between these are various intermediate types.

The final category is equivalent to the weed-type, *r*-selective species previously described. **Ruderals** live in disturbed and unpredictable environments but where resources may be occasionally abundant and available.

There are intermediate forms between the three main types, but this scheme acknowledges that a plant must have a strategy for resource allocation that matches the availability of resources, the frequency of disturbance and any adverse features of the habitat. In fact, we also need to recognize that these demands will often change during the organism's life cycle, and the conditions under which a seedling flourishes may be very different from those demanded by a mature tree (section 3.5).

Often, it is only by taking these changes into account that we can understand the outcomes of some competitive battles. For many adult plants the worst of the competitive fight has been fought. In a garden lawn (Plate 14), for example, species of grasses and broadleaved plants form a tight-knit community which is a far from ideal place for young seedlings. There is fierce competition for water and minerals, and there are other constraints on space and access to light. Even when gaps appear, seedlings have to be tolerant of trampling, soil compaction and the regular grazing of the lawn mower.

Gaps with different conditions favour different species. Dry compacted gaps favour plantains (*Plantago* species), moderately damp areas daisies (*Bellis perennis*) and the wetter areas buttercups (*Ranunculus* species). Occasionally, in some lawns at least, the gaps will actually be colonized by grass, but these species vary too according to the nature of the gap. Fine-leaved grasses such as bents (*Agrostis* species) and fescues (*Festuca* species) fare better than annual meadowgrass (*Poa annua*) when water and nutrients are in short supply.

Yet the resources demanded by each plant are relatively similar and many are in direct competition with their neighbours for whatever is available. This presents us with something of a paradox: why should there be such a vast variety of plant species, even within a small area, when they all require more or less the same resources from their environment? Furthermore, how do they manage to coexist without one or two species becoming dominant and excluding the rest?

David Tilman offers one possible solution. He suggests that what is crucial is the amount of a resource needed by each species relative to its availability in the environment. Coexistence is only possible if each species is prevented from dominating a nutrient supply completely. That would happen if the growth of each competitor is checked by a shortage of another resource.

Let us say that the outcome of a competitive battle between two plant species depends on two key resources – for example, water and a nutrient. For each resource, either species would survive if its minimum requirement was met, but it cannot dominate the supply because its growth is checked by a shortage of the second factor. According to Tilman, coexistence is possible if each species is limited by a different factor, but each gets most of that which it finds most limiting. Both can then sustain population growth. Because they are limited by different resources, each will persist while leaving enough of each nutrient for its competitor.

This might help, in part, to explain the remarkable diversity of plants in the tropical forests, on soils which are relatively infertile (section 9.3). There is no reason why this principle should not also extend to animal communities.

Much of the diversity of a plant community derives from coevolution between the plants and their associated animals. As we have already seen, the evolution of higher plants has been intimately linked with that of terrestrial animals, as pollinators, seed dispersers and herbivores. Dependence on animal pollinators or seed dispersers may itself lead to the plant being only locally abundant or declining at particular times. The pressure of being grazed or browsed may well be as important in preventing some competitive exclusions.

The same is true of animals: the attentions of a predator or parasite may well prevent competitive battles reaching a definitive conclusion. Overall, the abundance of any species will respond to a range of key factors, including other species, all of which are continually changing. This is one reason why communities are dynamic – variable conditions favour different species at different times. Some species are abundant because of past success, others are beginning to enjoy their moment. We should see the forest or the lawn as an ongoing competitive battle, with no forseeable end and no winner.

4.4 CONSUMERISM

Consumers are those animals that consume other living organisms. Herbivores and carnivores are therefore both consumers, distinguished by their choice of diet. Herbivory is the consumption of photosynthesizers, and predators are flesh-eating animals that kill their prey. Herbivory might be regarded as a form of predation (one organism eating another) but often this does not kill the plant so here we shall keep a distinction to avoid confusion. We might also consider the action of parasites or even pathogens as predation, but again the host may not be killed. As ever, we are faced with variations on a simple theme where the demarcations between one category and another are sometimes arbitrarily drawn.

4.4.1 Herbivory

Being confined to one place, plants have limited strategic options to ensure their reproductive success in the face of herbivory. All plants seek to maintain some sort of balance between their photosynthetic tissues and their roots: the leaves are net consumers of nutrients and net producers of energy-rich sugars. The reverse is true of the roots. If the balance between the roots and the shoots is not maintained then either respiration exceeds photosynthate production (leaves are lost) or photosynthesis is inhibited by a lack of key nutrients (roots are lost: section 5.1). When they are attacked, plants will alter their growth pattern to restore this balance, so leaves or roots are replaced as necessary.

One strategy is simply to grow quickly and produce seed as rapidly as possible, perhaps before being noticed by herbivores. This is the strategy of ruderals, the weeds common in habitats where there is a relatively high frequency of disturbance. An example is chickweed, *Stellaria media* (Plate 15). Rarely found in the closed community of the lawn,

Stellaria is certainly common in the cultivated borders around it.

The alternative for a plant is to make a more substantial investment in leaves, roots and shoots, perhaps as a **biennial** (living for two years) or a **perennial** (living for several years) and perhaps seeking to reproduce more than once. This long-term strategy may also allow both sexual and asexual reproduction. The interest of herbivores cannot be avoided, but it might be dissuaded if the plant seeks to protect its investment. Plant defences include physical, chemical and biological methods.

Spikes, thorns and stings may be effective against larger herbivores and unsuspecting people, but they are less effective against many invertebrates (Figure 4.8). Finer hairs, sometimes with sticky secretions, can prevent smaller insects from getting close to the leaf surface. A woody trunk protects against slug and fungal attacks at the soil surface. Tough silicate deposits on the leaves of some grasses limit the range of animals prepared to tackle them. Others simply toughen the leaves and deposit distasteful chemicals, such as tannins and phenols, in them.

Plants contain some of the nastiest chemicals known (Table 4.2). The origins of these defences are still debated. Many of these chemicals, known as secondary plant metabolites, probably started out as being waste products shunted into leaves, to be excreted when the leaf was shed. However, if herbivores find such leaves unpalatable then there is some selective advantage in concentrating these compounds in living leaves. Deciduous oaks accumulate tannins and phenols in their leaves during the year, gradually making them less palatable and less nutritious. Because plants seem to allocate their chemical defences according to their growth pattern and because they will also lower their production when no longer browsed, they appear to be avoiding the costs of production wherever possible. Thus, even if these chemicals started out as metabolic waste products, they have since been refined by natural selection and now are produced on demand to protect the plant.

This would also explain the vast range of chemicals synthesized. A large variety of alkaloids, terpenes, phenolics and others are responsible for a myriad of effects on herbivores. Not only does their taste seem to dissuade some herbivores they also act as anti-feedants, nerve poisons, carcinogens and (in human beings at least) hallucinogens. Some of the clover and rose family release deadly hydrogen cyanide when chewed. Phenolic compounds and tannins bind to digestive enzymes, reducing protein assimilation. Animals grazing such vegetation lose condition and, once weakened, begin to lose their own competitive battles. However, while this deters many herbivores, others have hijacked these defence mechanisms for their own purpose (Box 4.2).

The coevolution of defences between plants and their herbivores is as impressive as that between plants and their pollinators. For example, ecdysone, the hormone that initiates moulting in insects, is produced by bracken, upsetting an insect's growth rate and maturation. Several plants produce precursors of hormones able to disrupt the reproductive cycles of the animals that feed on them. We have used this for our own purpose: the first contraceptive pill was based on naturally produced progesterone from Mexican yams. This is just one example of the many thousands of pharmacologically active compounds we have derived from plants. Traditional medicines based on plant extracts are used throughout the world and are still the primary treatment for many peoples.

Plants that make these compounds are long lived and are seeking to reduce the costs of replacing leaves. Many such plants come from relatively undisturbed habitats, and typically are competitive or stress-tolerant species. A good example is the plants of the Mediterranean maquis (section 5.2). These are stress-tolerant species, adapted for surving a long summer drought. Typically low bushes, most keep their leaves throughout the year and these become very tough and leathery with age. They are also responsible for the characteristic perfume of the Mediterranean. Thyme, rosemary and lavender produce a range of secondary metabolites which dissuade most insect herbivores but which we find attractive as flavourings or scents.

Some plants are far from passive in their interactions with animals. The carnivorous plants, more properly the insectivorous plants, are typical of environments where nitrogen is in short supply – often very damp places. Yet nitrogen is abundant in the insects that visit their flowers to feed and, using various mechanims, some plants have found

Figure 4.8 Rose thorns, a deterrent to would-be herbivores.

ways of exploiting this nutrient source (section 6.2).

4.4.2 Predation

The most direct effect of a predator species on its prey is to reduce the latter's numbers. But if the prey species is the only food source for the predator then the abundance of the prey will itself govern the abundance of the predator. The question is: who is in control, the predator or the prey?

A series of mathematical models, again based on the logistic equations (section 3.2), has been used to examine predator–prey relationships (Figure 4.9). In their simplest form, these models assume that the predator population can only grow if there are enough prey. When prey numbers are high, the predator has abundant resources from which it produces offspring. That allows the predator to increase its population size, but these hungry mouths now represent a further demand on the prey population. If the reproduction of the prey species cannot match the

Table 4.2 Secondary plant products

Class	Number of compounds	Occurence	Physiological effect
Nitrogen compounds			
Alkaloids	5500	Found in many flowering plants in roots, leaves and fruits	Toxic and unpleasant tasting
Amines	100	Found in many flowering plants, principally in flowers	Unpleasant smelling, some hallucinogenic
Amino-acids (non-protein)	400	Found in seeds of many flowering plants – particularly legumes	Toxic
Cyanogenic glycosides	30	Occasional (e.g. clover, Rosaceae rose family)	Toxic (hydrogen cyanide)
Glucosinolates	75	Cruciferae (cabbage family) and 10 other plant families	Bitter tasting, acrid smell.
Terpenoids			
Monoterpenes	1000	Found in many flowering plants – principally as essential oils	Strong smelling (not unpleasant)
Sesquiterpene lactones	600	Mainly in Compositae (daisy family) and other groups of plants	Toxic, bitter, allergenic
Diterpenoids	1000	Found in many flowering plants – mainly in latex and resins	Toxic, sticky
Saponins	70	Widely found (in 70 plant families)	Haemolytic – damage blood cells
Limnoids	100	Mainly in Rutaceae – citrus, Meliaceae – mahogany and Simaroubaceae – quassia families	Bitter-tasting
Cucurbitacins	50	Cucurbitaceae – cucumber family	Bitter-tasting and toxic
Cardenolides	150	Mainly in Aponcynaceae – periwinkle, Asclepiadaceae – milkweed and Scrophulariaceae – figwort families	Bitter and toxic
Carotenoids	350	Widespread in fruits and flowers	Coloured pigment
Phenolics			
Simple phenols	200	Widespread in plant tissues	Antimicrobial
Flavenoids	1000	Widespread in higher and lower plants	Coloured
Quinones	500	Widespread (particularly in Rhamnaceae – buckthorn family)	Coloured pigments
Other			
Polyacetylenes	650	Mainly in Compositeae – daisy and Umbelliferae – umbellifer families	Some toxic

losses to the predator, the prey population must decline. In these models there is a level of predation where prey deaths exceeds prey births and prey numbers fall.

This, in turn, means that food now becomes limiting for the predator. Its population growth is checked, and as the prey population declines further, so predator numbers start to fall. Once again, some point will be reached when predation is no longer limiting prey population growth and their numbers start to grow. So, in turn does the predator population. The cycle repeats itself, and the two populations stay locked in their deadly waltz.

In effect we are running two population models together, where the capacity for growth in one is determined by the population size of the other, and vice versa. In these simple terms, we can see that the answer to our question is that prey and predator control each other. Because these populations begin out of step with each other, they continue to oscillate around each other, never settling down to constant population levels. If we impose other checks

Box 4.2

OBJECTIONABLE BEHAVIOUR

Caterpillars are soft-bodied, often fat, always slow-moving eating machines which, at first, might seem to be an easy target for predators. Not surprisingly, this vulnerability has led to them evolving various ways to avoid being eaten. Some produce a mass of irritant hairs which defy all but the most persistent predator. Others swell their bodies or produce sticky strands to make them awkward to swallow, or flash false eyes and appear to be part of something larger and much more able to defend itself. While some use cryptic coloration to camouflage themselves as leaves or twigs, others use warning coloration to advertise the threat that they pose to potential predators (Table 4.3).

Bright and stark coloration will signal that the owner is distasteful, can sting or will poison. Once experienced, a predator will readily associate those markings with an animal to be avoided. Yellow and black stripes are used by both vertebrates and invertebrates as such a warning. One example is the striped caterpillar of the cinnabar moth (*Tyria jacobaeae*) which feeds on ragwort (*Senecio jacobaea*) and incorporates the plant's poisonous alkaloids into its own tissues; the adult moth is a striking red and black. Similarly the yellow–black–white stripes of the monarch caterpillar (*Danaus plexippus*) warn that it contains an alkaloid concentrated from its diet of milkweeds; again, the adult has stark orange and black wings that will be readily disgorged by any bird that tries to eat it. It shares this wing pattern with another butterfly, the viceroy (*Limenitis archippus*), which is also distasteful to birds. Together the monarch and the viceroy represent an example of **Mullerian mimicry**, where the shared pattern means that predators learn more quickly to avoid any butterfly with these markings, so that each species benefits from the matching pattern.

In contrast some species bear the warning coloration but avoid the costs of producing or concentrating any poison in their tissues. This is termed **Batesian mimicry**. Hoverflies which sport black and yellow livery are often mistaken for wasps though they have no sting. Edible species copying distasteful species may save resources in the short term, but they also risk lessening the impact of the signal on a predator. If their numbers are large, and a predator has few or no encounters with the original distasteful species, an aversion to the warning coloration may not develop. Given long enough, natural selection would then favour predators not fooled by Batesian mimics and which are more discerning about what they eat.

When stripes and false eyes fail, the defensive behaviour of caterpillars can be equally colourful. Some will vomit on their attacker. As if being pelted with the half-digested content of its gut was not bad enough, then what was on the prey's menu can be even worse. For example, one species feeding on the cocoa plant will shower assailants with a cocaine-laced vomit.

Stephen Peterson studied similar behaviour in eastern tent caterpillars (*Malacosoma americanum*). These seek out young leaves of black cherry (*Prunus serotina*) and then carefully mark out pheromone trails leading to the leaves with the richest supply of a compound called prunosin. This will generate cyanide. When attacked by ants, caterpillars simply regurgitate the part-digested leaves, which release hydrogen cyanide, dissuading the ants from continuing their attack.

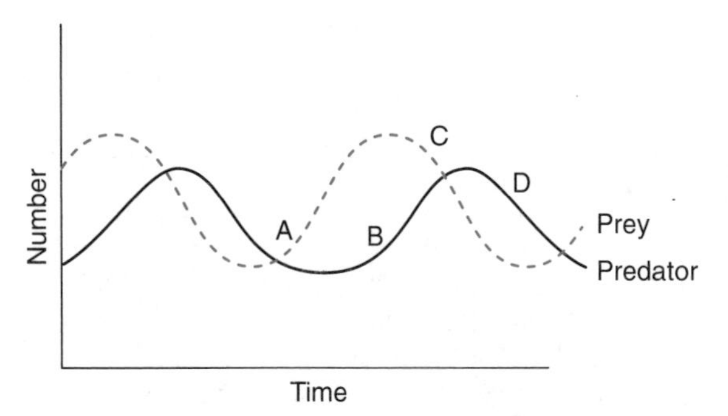

Figure 4.9 Predator–prey relationships. Changes in the population of a predator and its prey based on mathematical models where the predator exploits one prey species only. At point A the prey can increase because of low predator numbers. Predator numbers begin to rise (B) as their prey become more abundant. Eventually predation checks the population growth of the prey (C). The decline in the prey is followed by a decline in the predators (D) and the interaction begins to cycle again.

on the population growth of either species, then the oscillations begin to disappear.

In fact, it is hard to show that such cycles exist in nature and even more difficult to create them in the laboratory. Simple bench-scale ecosystems invariably lead to the predator eating all the prey and then itself starving to death. Adding refuges for the prey where some can escape predation allows for these oscillations to become established but they rarely last more than a few generations without further interference in the experiment.

Consequently, we need to ask what it is about real ecosystems that allows predator and prey to coexist. Complexity is part of the answer – adding detail to the models helps to defuse the simple relationship between predator and prey numbers:

- First, we should not assume that each predator or each prey is equivalent. Some predators are better killers than others and some prey are more likely to get eaten. Field observations show that predators often take the weak and the infirm, including older prey which may have already ceased reproducing.
- Second, many predators do not confine themselves to one prey species. Predators are often opportunists and will take what is easily available to reduce their costs in finding and acquiring their food. Then a predator population can maintain itself when one prey species is in short supply. By the same token, a predator will compete with several other species to utilize a prey resource. Together, this complex of interactions can mean the abundance of either the predator or the prey are not closely coupled to each other but reflect changes in the larger community.
- Third, a predator does not only respond numerically to an increase in prey numbers. Each individual may consume more (termed a **functional response**) and only later may it produce more offspring (the **numerical response**). If the predator only shows the functional response, then prey numbers will decline without any change in predator abundance. Most often, the predator will respond both functionally and numerically, but each type of response will have associated delays. It may take time for the predator to increase its feeding rate (and perhaps switch its attention from an alternative prey).
- How quickly a glut of food translates into new offspring depends on the generation time of the predator, or whether there are young alive whose survival depends on the food supply. The significance of such time-lags is often seen with major insect outbreaks in temperate forests, where considerable damage may be done before birds begin to feed exclusively on the pest. The prey abundance may not translate into an increase in the adult population of the birds for several months, or even a year.
- Finally, real ecosystems are not uniform in space but have patches where prey may accumulate or where they may escape predation. To reduce its search costs, a predator will stay where prey are abundant so some prey will go undiscovered. These individuals will be important for maintaining the prey population. Escapees are crucial if both the prey and the predator populations are to continue from one generation to another.

Taken together, these different factors mean predator and prey numbers may not be closely linked and that we should not be surprised to discover that the simple oscillations are rare in nature. Nevertheless there are plenty of real-world examples where predator abundance is the best explanation of variations in prey numbers and, conversely, some where predator abundance follows that of one of its prey (Figure 4.10).

We should also recognize that prey rarely sit around waiting to be picked off by a predator. An aphid feeding may seem to be oblivious to the ladybird consuming its sister a centimetre away but its

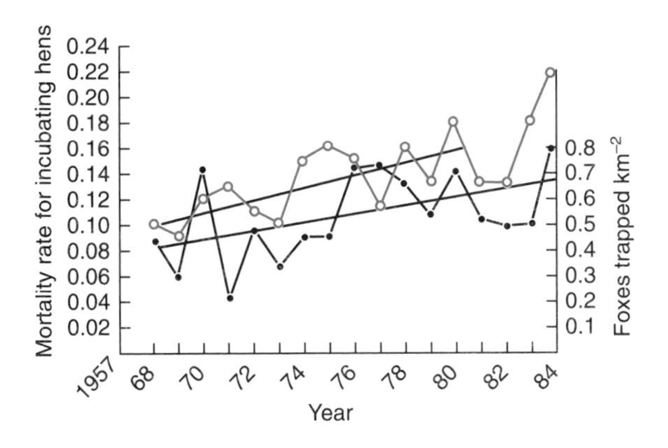

Figure 4.10 Changes with time in the number of foxes trapped in five farms in Sussex and its correspondence with the number of grey partridge hens lost due to predation by foxes. The level of predation is closely linked to the number of foxes, but this study showed that the size of the prey population was primarily regulated by two predators: the foxes, and crows taking eggs. However, when the bird population is growing, abundance also depends on their food supply, primarily insects, and its effect on chick mortality. Blue circles = mortality rate for incubating hens, primarily due to fox predation; black circles = number of foxes trapped within the study area.

species has not given up the battle, nor is the aphid itself ready to give up the cares of this world. Like the ants, the individual's fate is almost immaterial since its genetic code is likely to be passed on by one of their sisters, mothers or daughters.

Other species have adopted different strategies.

4.4.3 In self-defence

One other misleading impression we might get from these simple models is that the stakes are the same for both predator and prey. This is not so, for while a predator may give up on a chase and only be short of breath, the prey cannot afford to do this. Richard Dawkins and John Krebs call this the 'life–dinner principle': for one, it is a matter of life or death; for the other, it is a matter of dinner or no dinner. Although a predator cannot keep failing indefinitely, it need not win every chase. The prey, however, cannot afford to lose even one chase. Natural selection favours those prey that escape, as well as those predators that succeed most often. This pressure is so direct, so intense, that we see a wide variety of dramatic defence and attack strategies (Box 4.2, Table 4.3).

A predator that has little impact on adult prey numbers may exert a relatively weak selective pressure.

Table 4.3 Animal defence strategies

Type	*Strategy*
Aggression	Threat behaviour, intimidation displays and overt aggression are used to frighten off potential predators or competitors.
Aposematic coloration	Warning colours signal danger. These tend to be stark colours and patterns which are used for predators to associate with stings, poisons or being distasteful, for example black and yellow stripes.
Chemical defences	This can include passive defences such as the secretion or accumulation of poisonous and distasteful compounds within or on the body, or active in the case of venomous or unpleasant bites, stings and sprays.
Crypsis (camouflage)	Merging into the background to avoid attention of predators. *Catalepsis* (frozen posture) adds to the effectiveness of the defence.
Masting	A strategy which relies on strength in numbers, by producing numerous progeny which either overwhelm, confuse or swamp out predators.
Mimicry	A strategy in which animals mimic living and non-living things. *Batesian mimicry* involves innocuous prey impersonating a dangerous or unpalatable organism. *Mullerian mimicry* reinforces aposematic coloration. Dangerous and distasteful organisms independently evolve similar forms of patterns and coloration.
Polymorphism	Groups of populations within a prey species avoid being eaten by looking sufficiently different from most of their species to go unrecognized by predators.

Instead, the prey species may be responding primarily to its other interactions or some abiotic factor in its environment. The pressure for improvement is likely to be greatest for predators and prey whose populations are closely coupled. Here, some evolutionary change in one species will have a direct effect on the reproductive success of the other species. Internal parasites particularly have to be highly adapted to the environment inside their hosts.

4.4.4 Parasites and pathogens

Parasites have special strategies that avoid some of the risks of wiping out the species on which they depend. Most only consume a small part of their host. They can do this because they are often much smaller than their host. Classically, parasitism has been viewed in terms of the weak attacking the strong.

In fact, the range of parasitic strategies and the intricacies of parasite life cycles is remarkable. They range from highly specific associations, where a parasite is locked into a single host, to much more opportunist species with hosts belonging to a large group, such as the mammals (Box 4.3). Some of the most specialized parasites are the **parasitoids** – insects from the Hymenoptera (bees, ants and wasps) and the Diptera (the true flies) that practise a form of parasitism bordering on predation. These lay a single egg inside the host insect, often a pupal or larval form such as a caterpillar. As the larva grows, so does its parasite and all the effort of the caterpillar will ultimately benefit its passenger in a form of controlled predation. The parasitoid then pupates inside the husk of its host, using this as protection until it can emerge as an adult.

The effects of most other parasites on their host are far less dramatic, as long as their numbers do not rise too high. In a particular form of intraspecific competition, an existing parasite may prevent other individuals becoming established in their host. The host is then protected from too severe an infection and the parasite thereby safeguards its resource.

Unlike predators, which devour their prey and quickly move to the next victim, parasites tend to establish long-term associations whether they live on the outside of their host (ectoparasite) or inside it (endoparasite). During this time they will reproduce frequently and produce eggs or larvae that attempt to migrate to another host (Box 4.3). A parasite will typically exploit the host's associations with other species to pass between hosts (Box 4.3).

Box 4.3

THE HUMAN ECOSYSTEM

Any quick review of the variety of organisms that live within each of us makes it plain that we represent a valuable resource to other species. Over 100 animals are parasitic on human beings. In addition, a range of fungi and bacteria make their home in or on us, with our bodies offering a variety of niches to be exploited.

Some of these associations are benign, with little overall effect on our well-being. Others are positively beneficial – many of the microorganisms inhabiting our gut and other tracts are important for maintaining their internal environment. Indeed, they provide some protection against invasive pathogenic species. Such associations are mutualistic, benefiting both ourselves and the resident flora. Other organisms are true parasites, exacting a cost to the host. Occasionally, an organism may change from one to the other – from a benign form to a life-threatening infection. Amoebic dysentery, for example, occurs when *Entamoeba histolytica*, normally a well-behaved resident, switches to a form that attacks the wall of the large intestine, in a transformation we do not fully understand.

Table 4.4 Some animal parasites of humans*

Species/group	*Other hosts*	*Vectors*	*Principal sites in host*	*Distribution*
Protozoa				
Trypanosoma (sleeping sickness)	Mammals	Flies, bugs	Blood	Tropical Africa, S. America
Leishmania (leishmaniasis)	Mammals	Sandflies	Blood	Tropics
Entamoeba (amoebic dysentery)	–	House fly helps to transmit cysts	Large intestine, various organs	Tropical and warm temperate areas
Toxoplasma	Mammals	Domestic animals	All organs, especially brain	Global
Plasmodium (malaria)	Mammals	Mosquitoes	Blood/liver	Tropics, warm temperate
Flatworms				
Fasciolopsis (large intestinal-fluke)	–	Resistant external stages. Aquatic snails	Intestine	S.E. Asia
Clonorchis	Dogs, cats	Resistant external stages. Aquatic snails, fish	Liver	S.E. Asia
Paragonimus (lung fluke)	Mammals	Aquatic snails, crabs, crayfish	Lung	Asia, Americas
Schistosoma (bilharzia)	–	Resistant external stages. Aquatic snails	Blood vessels, liver	Africa
Dyphyllobothrium	Dogs, cats	Resistant external stages. Aquatic snails, fish	Small intestine	Most of Northern Hemisphere
Hymenolepsis	Dogs, rodents	Resistant external stages. Fleas	Small intestine	Global, warm areas
Taenia	Pig	Cow, pig	Small intestine	Global
Roundworms				
Trichnella (Trichinosis)	Mammals	Other mammals used as food	Small intestine and migratory	Global
Ancylostoma, Necator (hookworms)	–	Resistant external stages	Migrates through different organs during development. Adult in small intestine	Tropics
Ascaris (Intestinal roundworm)	–	Resistant external stages	Small intestine. Migrates through organs during development	Global
Wucheria, Loa (Elephantiasis)	–	Various biting flies	Connective tissue, blood. Migrates during life cycles	Tropics
Onchocerca	–	Black flies (gnats)	Connective tissue beneath skin	Africa
Dracunculus (Guinea worm)	–	Aquatic crustacea	Deep connective tissue	Tropical Africa, Middle East

* Table 4.4 gives only those animals whose association with humans is semi-permanent and for which humans are a definitive host. It excludes those who may feed on humans, but for whom the association may be intermittent and so omits specific external parasites such as the human headlouse, bedbugs, ticks and mites, as well as fleas, leeches, and vampire bats that show less and less dependence upon humans as their main host. You may notice that this list gradually strays into the grey area between parasite and predator.

The range of animals that live in us is truly impressive, even if we confine the list to parasites alone (Table 4.4). Different species attack different parts of our bodies in different parts of the world. Which and where depends not only on the local climate, but also on the possible routes for infection. Local habits and hygiene determine our parasite burden, as do our associations with domesticated animals, soil and water.

Improvements in our understanding of these sources have enabled us to prevent infection, interrupting the parasites' life cycles and in some cases leading to their eradication. Several viral diseases have been so controlled. Smallpox now exists only in laboratory cultures and poliomyelitis may have the same fate in the near future. Others have not succumbed so easily – malaria is the greatest killer of humanity, as it has been throughout our history, despite the recent chemical war waged against the malarial mosquitoes that transfer the protozoan parasite (*Plasmodium* spp.) from one individual to another (Figure 4.11).

Breaking the cycle of reproduction and transmission is the first step in control. All parasites eventually face the problem of moving from one host to another. As an ecosystem we are relatively short-lived and have a limited carrying capacity. Besides our behavioural or technological defences, our biology also has a highly adaptive immune system that the parasite has to overcome to establish itself. Many parasites have sophisticated mechanisms for 'fooling' or suppressing the immune system.

Most animal parasites are from ancient animal groups with a long evolutionary history. They have relatively simple body plans which enable them to retain a large capacity for asexual reproduction and regeneration. This means they can produce large numbers of infective stages at various phases in their life cycle, and this improves their chances of finding the correct host.

Another strategy is to infect a number of different species and reproduce in a variety of hosts sharing the same physiology. For example, trypanosomes (sleeping sickness, Chaga's disease) and *Leishmania* (leishmaniasis) will attack most mammals and the final or **definitive host** (where the adult develops) depends on the species of blood-sucking insect (the **vector**) that carries the parasite from one host to another. Usually the vector, or intermediate host, transfers a larval form that will only mature in the definitive host; some parasites also have resistant stages that can survive in the outside world. Human beings are the sole definitive host for the tapeworms *Taenia solium* and *Taeniarhynchus saginatus*, though the first arrives in pork and the second with beef. The beef tapeworm may produce 600 million eggs during its life, which is necessary because this is the resistant external stage that has to be consumed by cattle if the life cycle is to be completed.

Most internal parasites use the stages of their life cycle to adapt to the different environments they encounter as they move between hosts. Life cycles so dependent on such transfers could only evolve where close associations between hosts are maintained. Many of our parasites use our close association with domesticated animals to move between hosts (Table 4.4). This implies a rapid evolution on the part the parasite, within the 1.5 million years of *Homo* or the 10 000 years of *Homo sapiens*, the pastoralist.

More generally, parasites may play a fundamental role in the evolution of many species. One suggestion why so many species reproduce sexually, rather than using less costly and less risky asexual reproduction, is that it helps to avoid the attentions of parasites. Within a population, the argument goes, individuals with the most common genotype are the ones most likely to be attacked and to which most parasites will be adapted. Variation can mean a parasite is not adapted to the environment produced by the new genotype. Novelty has a selective advantage and sex is the prime means of generating it.

While parasitoids blur the distinction between predator and parasites, so do some of the protozoan parasites blur the line between parasite and pathogen. A **pathogen** is any organism which causes a disease, and at high levels of infection this would include many multicelluar parasites. Pathogenic disease may lead to the death of the host, and a parasite can only afford to adopt this strategy if it can transfer readily from one host to another. This is possible in many unicellular parasites because of their short generation times and rapid population growth. Malaria is caused by a series of sporozoan parasites that attack a wide range of animals and it has been the biggest killer of humanity throughout our history (Figure 4.11).

There is, of course, a massive range of bacterial and fungal pathogens that infect us, as well as the viral infections that perhaps make the most direct attack on cellular life by hijacking its gene replication machinery.

At the other extreme, some parasites have sought to reduce the cost of passing their genes into the next generation by usurping the reproductive effort of other species. The cuckoo (*Cuculus canorus*) is a **brood-parasite**, manipulating the behaviour patterns of other bird species to improve its reproductive potential. Not only are the eggs, laid in the nest of a host bird, reared by the foster parents; but also the young cuckoo automatically practises interspecific competition, easing out the chicks of the nest's owner. Meanwhile, freed from the responsibility of rearing her young, the female cuckoo can lay eggs in other nests.

Similarly, the queen of one ant species (*Lasius reginae*) will kill the queen of another species (*Lasius alienus*) so that its workers care for her larvae. Eventually the *L. alienus* workers die, but by that time the *L. reginae* workers are in control of the nest. Several wasp species have pararsitic strains of queens who will lay their eggs in the nest of another.

Several duck species also have a form of intraspecific brood parasitism, where eggs are laid in the nests of neighbours. All the birds benefit from the protection that communal living affords against predators, but there is always the risk of raising somebody's offspring. Even so, this is offset to some extent because many in the group are closely related.

Plant parasites are divided into two main types. **Holoparasites**, such as broomrapes (*Orobanche*) and their relatives (Plate 16), have abandoned photosynthesis and must therefore acquire all of their resources from their host plant.

Hemiparasites, on the other hand, hedge their bets and will photosynthesize, to supplement their energy needs or keep them going in the absence of a suitable host. Some, such as yellow rattle (*Rhinanthus minor*, Figure 4.12), are able to exploit a range of species (mainly grasses), slowing down their host's growth. This helps to prevent vigorous grasses from out-competing other plants in a grassland community and this is why yellow rattle is sometimes included in 'wildflower' seed mixes used to produce low-maintenance grassland.

The most famous parasitic plant is undoubtedly mistletoe (*Viscum album*). A hemiparasite, it taps into the living vascular tissue of trees, but photosynthesizes itself. Bad infestations can reduce the growth of a tree by almost 20%. Despite this, *V. album* is not a major economic pest, but there are more troublesome members of the mistletoe family (Loranthaceae). *Scurrula cordifolia* has spread from the Himalaya into northern India, where it

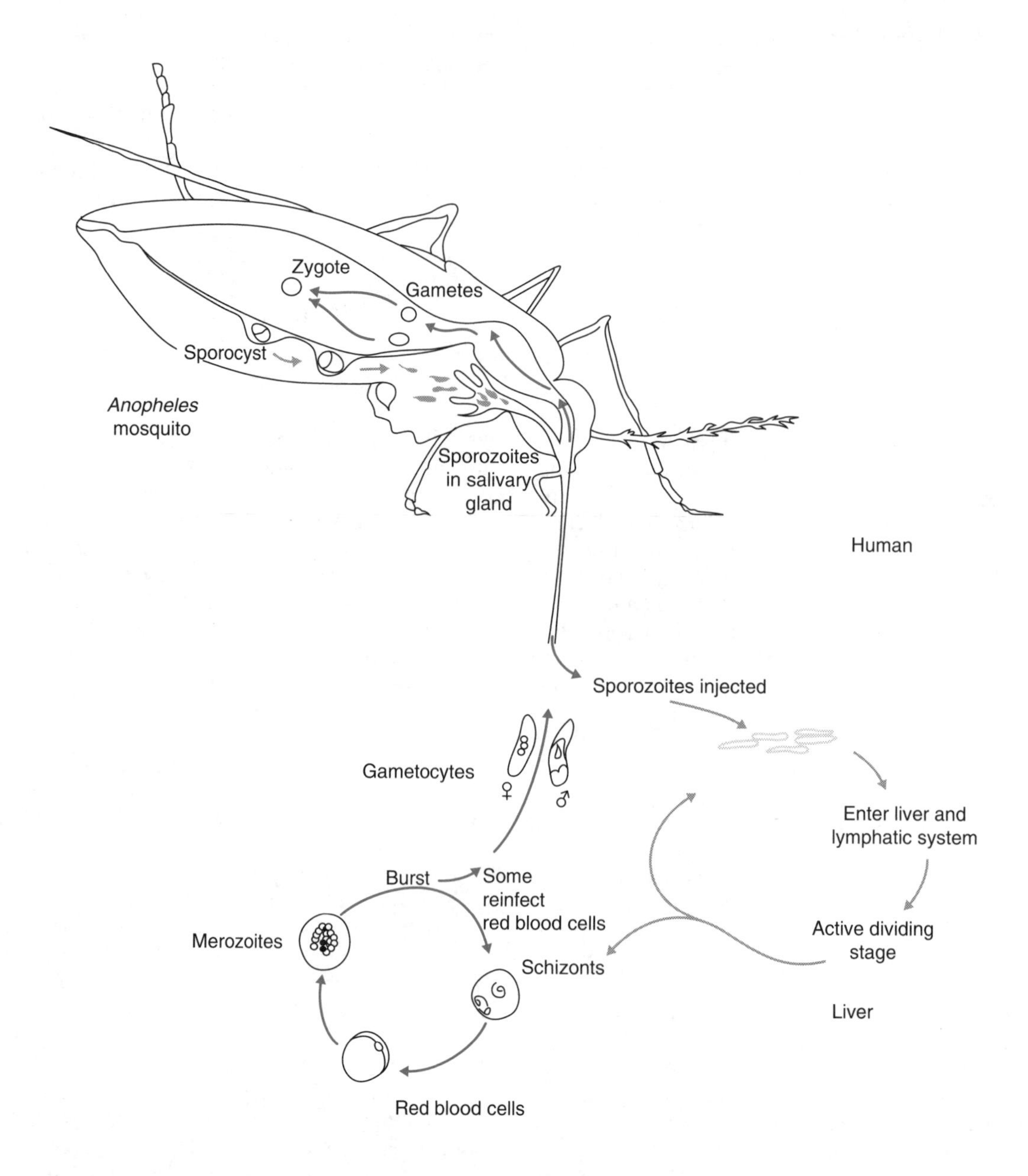

Figure 4.11 The life cyle of *Plasmodium*, the genus of protozoan parasite that causes malaria in humans and other mammals. The timing of its life cycle depends upon the species, but the stages are the same. **Sporozoites**, injected when the mosquito takes a blood meal, enter the liver and undergo rapid division. After they burst from the host's liver cells, some re-enter liver cells but others enter red blood cells. Here they become **schizonts** and undergo further asexual multiplication. When the **merozoites** are released from these cells, some reinfect other red blood cells, and some form sexual cells. If these **gametocytes** are then taken up by a mosquito they form gametes within its gut and fertilization produces a **zygote**. This encysts in the insect gut wall, eventually to release sporozoites that invade the insect's salivary glands, to be injected at the next feed.

The insect itself is acting as an ectoparasite by taking the blood meal (or at least the female is, since only she feeds on blood). *Plasmodium* is parasitic on both the mammal and the insect: all of its sexual stages take place in the mosquito and it is entirely asexual in the mammal, so we might be regarded as vectors, with the mosquito as the definitive host.

Figure 4.12 Yellow-rattle (*Rhinanthus minor*), a hemiparasite.

attacks commercial orchard and forest trees. Control is difficult, because herbicides could harm the trees as much as the mistletoe itself. However, *Scurrula* has an enemy of its own: another mistletoe, *Viscum loranthi*. In the wild *V. loranthi*, unable to live directly on trees, **hyperparasitizes** *Scurrula*, weakening and sometimes killing it. Now introduced into plantations and orchards, *V. loranthi* has proved to be a safe and efficient means of control because it has the most desirable property of the ideal biological control agent: it only attacks the pest.

4.5 CONTROLLING PESTS

Understanding the close association between species and the details of their interactions enables us to manipulate these to our own advantage. Sometimes, this means correcting past mistakes, where our interference has led to species escaping the checks imposed by their natural enemies and competitors. Released from these constraints they have become pests.

A species is usually designated a pest by the cost it incurs to human health, wealth or well-being. The term pest has no biological meaning, but is a label that we apply to a species that causes us to spend money. Increasingly, this includes expenditure to safeguard threatened species. As we saw earlier, some introduced aliens are so effective as competitors or predators that they have devastated the native species. Sometimes, in our attempt to reverse our mistakes we make matters worse. The Indian mongoose was released on several Caribbean islands to control the brown rat but it, in turn, also had a disastrous effect on the islands' bird, reptile and amphibian fauna.

Not all introduced species become pests. Many fail to establish themselves in the new habitat. This can be because they are poorly adapted to the conditions or fail to out-compete the better-adapted flora or fauna. However, there are numerous examples of introduced species, often adapted to harsher or more limiting environments, flourishing in their new homes. Sycamore (*Acer pseudoplatanus*) and *Rhododendron ponticum* are non-native species which have slowly invaded large areas of native vegetation in Britain and their competitive staying power has made them difficult to control.

Many weeds are *r*-selected opportunists, which grow rapidly where agriculture creates opportunities for them. A range of insect pests, such as aphids, also follow this strategy, and are able to outgrow the functional and numerical response of their natural enemies.

Pest control requires difficult choices and hard work. Often the fastest (and cheapest) way to control a pest is to use chemical methods. Insecticides, herbicides, molluscicides and so on can rapidly bring a pest outbreak under control. In the past, many of the chemicals used were general-purpose poisons that killed non-target as well as the target, pest species. Many have also been highly persistent – they go on killing or impairing species long after they were applied (Box 5.4). This is one reason why several pests have developed resistance to some of these compounds: DDT resistance in malarial mosquitoes is one good example and an impressive demonstration of natural selection in action over the last 40 years. Nowadays, a more detailed study of the biology of pest species is undertaken to find out how a poison works and the best way of using it to control a pest. Increasingly we are able to produce highly specific pesticides which have low persistence.

While these may be part of future answers to pest outbreaks, they will not be the final answer. Modern pesticides are expensive to develop and add considerably to the farmer's costs. Their use also implies some ecological costs: even if the evolution of resistance can be avoided (and the experience from many case histories is that it cannot), there is a host of other impacts from chemical control methods – ranging from the risks to workers to subtle shifts in species interactions.

Biological control can simply amount to reuniting a pest with its natural enemy (Box 4.4). Even so, a considerable amount of research is needed to find a natural enemy and to ensure that it will indeed control the pest. Often this means finding where the pest came from and whether any species regulate its numbers in its home range. Even if we can identify a natural enemy, we need to be sure that we can safely release it.

Research begins with a detailed study of the ecology of the potential natural enemy and its chances of success in the new habitat. In particular, we need to establish whether it will compete with the native fauna and if it can maintain a population without itself becoming a pest. Ideally, it will only consume the pest species, and that usually implies a specialist predator or parasitoid. To avoid the need for reintroduction, the natural enemy should not kill all of the pests; it should allow a low though economically unimportant population to survive. In that way both species remain at low numbers, with the natural enemy present as some safeguard against future outbreaks (Box 4.4). Thereafter, our prime concern is that the natural enemy should not affect any other species in the community nor become a nuisance or create other problems. Ideally, it will be cheap and easy to rear and to release (Table 4.5).

Table 4.5 Ten desirable attributes of a predator or parasitoid as an effective biological control agent

1. Rapid functional and numerical response to a rise in pest abundance.
2. Able to find pest quickly (high searching efficiency).
3. Able to maintain a low pest population and sustain itself over the long term.
4. Able to survive competition with native predators and parasitoids.
5. High prey specificity with minimum impact on non-target species.
6. Its activity and its life cycle show a close match to that of the pest.
7. Easy to culture in the laboratory.
8. Easy to release in the field.
9. Cheap to use with rapid results to inspire confidence.
10. No social nuisance.

We are not restricted to manipulating predator–prey interactions in a biological control programme. We can also use competitive interactions. An ingenious use of intraspecific competition is central to the control of a very destructive parasite, the screw-worm *Cochliomyia hominivorax*. This fly from central America lays its eggs inside the open wounds of mammals, including humans. The fly feeds on the flesh, enlarging the wound and preventing it from healing. The stench from the wound attracts other adults, who add their eggs to the seething mass. With a very high level of infestation the host may die. Today the screw-worm is under control in most of central America and a outbreak in north Africa in 1988 was controlled within two years.

The technique attacks a weak point in the life cycle of the screw-worm, which is that the female mates only once. The pest is bred in vast numbers at a factory in Mexico, where males are separated and irradiated just as their testes are forming, sterilizing them. They are shipped and dispersed as pupae to emerge as properly formed adults. The female

Box 4.4

PRICKLY PROBLEMS

The prickly pear is the classic case of an intentionally introduced species becoming a pest in its new habitat. It also a classic case of biological control, where a single natural enemy was found that provided rapid control and still keeps the plant in check.

The prickly pear cactus was introduced into Australia from the Americas in the latter half of the nineteenth century. Settlers imported two species (*Opuntia inermis* and *O. stricta*) for use as stock-proof hedges. These not only thrived in their new location but began to spread, invading the land they were meant to surround. Within a few decades vast tracts of eastern Australia were dominated by the cacti, turning natural landscapes and agricultural land into spiny thickets were no mammal could graze.

Efforts to control the weed began before the First World War, but searches in the 1920s identified around 50 potential control species which were sent to Australia for testing. The first group of species, including a moth borer (*Olycella*), a plant-sucking bug (Chelinidea) and the cochineal beetle (*Dactylopius*), failed. A second attempt fared better, since one particular insect, *Cactoblastis cactorum*, was found to be especially effective. The caterpillar bored into the cactus pads, not only destroying its internal tissues but also allowing fungal infections into the plant. Following a two-year breeding programme, large numbers of its eggs were released in eastern Australia. The initial success here led to further releases over the next few years covering a much wider area. The effect was dramatic. Large areas were cleared of *Opuntia* within three or four years, so much so that the *Cactoblastis* population itself crashed in the early 1930s.

Although *Opuntia* showed signs of recovery shortly after, within a year or so the *Cactoblastis* population had responded and the cactus was prevented from becoming a problem again. Today, *Opuntia* remains in Australia, confined to small patches that form a metapopulation (Box 3.1). *Opuntia* can maintain a low population because the moth lays its eggs in clumps and some patches may take some time to be colonized. When that patch dies out, so does its population of *Cactoblastis*.

This demonstrates several of the ideals of a biological control programme (Table 4.5), not least the capacity of the natural enemy to maintain a viable population that can keep the pest in check over the long term.

cannot distinguish sterile males from wild males, so swamping an area with millions of factory-bred individuals means that most copulate to no purpose. The reproductive rate plummets and the population crashes. This programme has removed screw-worm from the southern United States and most of Central America. The drive now is to eradicate it entirely from the Americas.

We may also check the growth of a pest by changing our cultivation techniques such as the timing of our crops. Some cotton farmers grow a 'trap crop' to catch overwintering boll-weevils, which they then destroy. Another method is to grow strips of alfalfa to attract other pests away from the cotton.

The sequence of crops grown on an area of land can also help to avoid pest outbreaks where a pest is closely tied to one food plant. Hedgerows and patches of wild vegetation are important for their role in harbouring generalist predators that can check pest growth before it takes off in the crop. In Britain, trials of specially planted grass banks in large fields have shown this to be a useful control measure. In effect, increasing the patchiness of the habitat decreases the distance between the predator and the pests at important stages in their life cycle.

The same technique can help specialist predators or parasitoids. In the vineyards of California control of an introduced pest, the grape leafhopper *Erythroneura elegantula*, is achieved by using a native parasitoid, *Anagrus epos*. This lays its eggs inside the leafhopper egg, but the parasitoid overwinters as an adult feeding as a predator on a native leafhopper that lives on bramble. By leaving bramble around the edges of the vineyards, the natural enemy can sustain itself over winter, and then control the pest in the new growing season.

Using biological, cultural and chemical methods together in a control programme is termed **integrated pest management** (IPM). This uses the principal advantages of each method to reduce costs and to give effective long-term pest control. Biological and cultural methods need a full understanding of the ecology of the pest or the natural enemy but will, ideally, give a sustained, low-cost control that holds a pest in check. However, the inherent delays in the response of a natural enemy may mean we need to resort to chemical methods in the short term to prevent a pest outbreak. Chemicals can also be used to allow a natural enemy to survive, perhaps by reducing a predator or a competitor, or even to attract control agents to a pest outbreak. Some of these techniques have become possible as we have developed pesticides that are increasingly specific to a target species. As a result, IPM is being used as an economic and viable alternative to the simple application of general-purpose poisons.

The interactions we have described between individuals and between species serve to structure natural communities. Our problem is to unpick their threads to help our understanding and our attempts to control them. In doing so we recognize how complex some of these arrangements are. At the level of the individual the underlying principle is that each is attempting to ensure its own reproductive success. Evolution has explored a range of strategies. The variety and subtlety of some of these should not surprise us – competition and predation are powerful selective forces and are therefore likely to favour the novel.

SUMMARY

Organisms interact in various ways to secure resources essential for their survival and reproductive success. These takes various forms. In mutualism different species interact to each other's benefit. In commensalism, one benefits by its association with another species, but without any effect on the other species.

Some highly complex and specialized relationships have evolved from these partnerships in a process called coevolution. Many higher plants rely on animal pollinators or seed dispersers. Lichens represent a highly integrated association between algae and fungi.

Competition is an interaction where individuals of the same or different species compete for a limited resource or inhibit each other. In simple laboratory experiments, competition for the same resource frequently leads to one species excluding the other, but often the interactions in the real world are far more complex, and competitive battles do not lead to a clear resolution. Some introduced species have caused the loss of native animals but there are also many cases where the native fauna have be able to exclude the invader. Similarly, predator–prey dynamics are complicated by the interactions of each species with the rest of the community and also on their abiotic environment.

Predation, herbivory and parasitism are interactions in which one organism depends upon another for a resource, to the detriment of the other. In all cases, interacting species have evolved elaborate means of reducing the costs of the interactions (through adaptations to avoid predation, for example) and of maximizing their advantage (say, by becoming a more effective predator).

These associations are used in biological control programmes to limit pests. Species often become pests because they have escaped the checks imposed by the interactions within their natural community. Biological control uses a range of techniques to re-establish these interactions and to limit pest damage.

FURTHER READING

Beeby, A.N. (1993) *Applying Ecology*, Chapman & Hall, London.

Horn, D.J. (1988) *Ecological Approach to Pest Management*, Guilford Press, New York.

Krebs, C.J. (1994) *Ecology* (4th edn), Harper Collins, London.

EXERCISES

1. Classify each of the following interactions as mutualism, commensalism, intra- or interspecific competition, predation or parasitism. (Remember that some of the distinctions between these categories are not sharply drawn!)

   (a) Nitrogen-fixing bacteria in specialized root nodules of a leguminous plant.
   (b) Egrets feeding off insects disturbed by cattle.
   (c) Algae that live on the fur of sloths.
   (d) Lampreys that attach themselves to the sides of fish.
   (e) Slave-maker ants collecting workers from another species' nest.
   (f) A young male lion attempting to take over an established pride.
   (g) Aphids tapping into the phloem of a plant.
   (h) A vampire bat feeding on a range of mammals.
   (i) A female wasp that lays its eggs in the nest of another of the same species.
   (j) Bracken invading grassland and heathland communities.
   (k) House geckos invading Pacific islands.

2. Complete the following paragraphs by inserting the appropriate words, using the list below (some words may be used more than once, some not used at all).

   A predator may respond in one of two ways when a prey species increases its abundance. It may show a functional response in which its __________ increases or it may show a __________ response where its population increases. Predator numbers can rise because of reduced __________ mortality and by an increase in __________ effort. Both have __________ associated with them, but the response will be fastest with improved __________ survival.

   Predators may __________ from another prey species to reduce their __________, but again there will be a delay. Generally predators tend to keep to __________ where prey numbers are __________ and that means some prey may go __________. Some __________ species, such as aphids, outgrow their natural enemies by simply having a much higher rate of increase.

   juvenile, adult, switch, lost, feeding rate, delays, reproductive, patches, prey, undiscovered, low, metabolic costs, high, birth rate, search costs

3. Write a paragraph describing the circumstances under which a generalist predator might be used in a pest control programme.

4. Look at the table of characteristics (Table 4.5) of the ideal pest control agent. For each of the ten attributes, identify, as far as possible, how far *Cactoblastis cactorum* meets each of these criteria.

Tutorial or seminar topic

5. What are the dangers of comparing the organization of our societies with those of the social insects? In what ways do they differ, both in terms of organization and in the role of the individual? In what ways are they similar?

Primary producers and photosynthesis – gross and net primary productivity – metabolic costs – limits to productivity – secondary productivity – food chains and food webs – the role of decomposers – agricultural systems.

5 Systems

All flesh is grass.

Isaiah 40:6

On this point we have to differ. Although Isaiah is correct in the sentiment, he is wrong in the detail. Certainly, most of the animals of the planet ultimately depend on photosynthetic plants for their nourishment, even if they are not actually grasses. However, it seems there was a time when there were no plants but there were consumers – simple single-celled animals that survived by scavenging organic molecules or eating other cells. At its outset, life on the planet depended on energy captured from chemical reactions.

Indeed, we can still find communities that depend on such energy today. At the bottom of deep and dark oceanic trenches, where volcanic vents supply both heat and reduced sulphur compounds, entire ecosystems have been developed around very primitive life-forms which use this chemical energy to build long-chained organic moleclues.

Lynn Margulis has investigated the possible origins of primitive life, trying to describe the conditions under which the modern groups became established. Something similar to these oxygen-poor conditions must have prevailed on the young Earth. The first organisms able to use radiant energy did not appear until much later, around 3 billion years ago (Figure 1.8). Later still, some groups entered into symbiotic associations with cells that lacked their own energy-fixing capacity and the eukaryotic, self-feeding precursor of all major photosynthetic organisms had evolved.

The success of this association was to have profound consequences for the planet. As we saw in Chapter 1, it led to a major shift in the chemistry of our atmosphere. It also meant that the abundant energy in sunlight could be used to fuel the synthesis of the building blocks of cells, more so than the meagre supply coming from the earth.

With energy fixed by photosynthesis, complex communities developed, adding consumers that used the energy locked in the tissues of other organisms – herbivores, carnivores, and also decomposers eating the non-living remains. In the last chapter we saw how species' interactions could be interpreted as a trade-off in terms of costs and benefits for an individual. In this chapter we look at energy flow through the whole community and how it helps to determine community structure. Community processes – nutrient flow, population dynamics and interactions – are driven by energy capture and transference, and by analysing how the community functions in relation to its abiotic environment we describe its properties as an ecosystem (Box 1.1).

We begin this chapter by examining energy fixation and the evolutionary innovation of photosynthesis. We then consider how energy moves from one species to another, along food chains, and in food webs. The details of food web construction seem to be repeated in very different ecosystems, and we consider what this tells us about community organization. Finally, we look at the oldest profession, agriculture, as a means of securing energy supplies for ourselves.

5.1 ECOLOGICAL ENERGETICS

Energy is the common currency of the sciences. The physical, chemical and biological worlds are ultimately linked by the limits and rules that govern energy transformations and movement. As we saw earlier, the harnessing of energy in molecular replication was key to the development of life on Earth. That meant storing energy in chemical bonds and releasing it slowly to drive the reactions needed to construct more organic compounds. Even in these systems, we see a pattern that is true of the most complicated of living systems – energy is either stored in chemical compounds or used in the maintenance of life, or metabolism (Figure 4.1).

Energy is formally described as the capacity to do work or, more loosely, the means to bring about change, and is measured in various units. The internationally accepted unit is the **joule** (J) – the amount of energy in a force moving 1 kilogram through 1 metre – though you may be more familiar with the calorie (cal), which is the amount of heat needed to raise the temperature of one gram of water by one degree centigrade. Despite measuring different kinds of work (raising a weight or the temperature of water) these units are entirely interconvertible (1 cal = 4.2 J).

More important than its units are the first and second **laws of thermodynamics**, which describe the properties of energy. These state that energy may never be created or destroyed and that energy transformations always lead to a reduction in usable energy. When energy is transformed from one type (say, movement) to another (say, electrical energy) some is dissipated as heat. Any system, living or otherwise, operates through a series of energy transformations, so without further inputs these heat losses mean that usable energy must eventually run down, or run out.

Energy at work is **kinetic energy**. Stored energy is termed **potential energy**. Living systems have means of storing energy in chemical bonds. Besides synthesizing structural molecules to build new cells or gametes, much of the activity of the cell is about adding to these stores and drawing upon them to drive metabolic processes. We detect this as the heat lost through respiration. We can also measure the energy fixed in its tissues, termed its **productivity**.

5.2 THE PRODUCERS

Energy enters the biosphere as sunlight, with each square metre of the Earth's surface receiving an average of 48 million kilojoules (kJ) of radiant energy per year. Without photosynthetic organisms, this energy, after heating the surface, would simply be radiated back to space.

Energy enters the living community through the photosynthetic capacity of green plants, termed **primary producers**. Because they can fix their own energy they are known as **autotrophs** (literally, 'self-nourishers') or, acknowledging their primary energy source, **photoautotrophs**. The most primitive photoautotrophs are specialist bacteria, including the cyanobacteria which have a rudimentary photosynthetic apparatus. Before these evolved, living communities relied on a very different source of energy: the potential energy in reduced chemical compounds. These **chemoautotrophs** can still be found in extreme environments where oxygen is absent.

Photosynthesis involves a two-stage process of light capture followed by storage as chemical energy. The end-products are sugars which can subsequently be built into more complex molecules (Box 5.1). However, not all of the light hitting a plant gets used – most passes straight through the leaf and never encounters the energy-absorbing pigment, **chlorophyll**. Just 44% is at a wavelength that can be absorbed by chlorophyll (Figures 5.2. and 5.3). Overall, only 2% of the radiant energy falling on a leaf ends up fixed in sugars by photosynthesis (though the warming effect of the rest helps to increase photosynthetic activity). Despite what seems to be a very low efficiency, the productivity of autotrophs makes up 99% of the global biomass, and it is upon this that the rest of the living planet is built.

The productivity of a primary producer is the rate at which it converts sunlight into living material or **biomass**. This can be expressed either in terms of the amount of matter within a given area over time (kg/m^2 per annum) or its energy content (kJ/m^2 per annum). The full photosynthetic output of a primary producer is known as its **gross primary production** (**GPP**), but not all of this energy becomes fixed in the tissues. The largest proportion, perhaps more than 60%, is used in metabolism, eventually to be lost as heat. Additionally, photosynthesis itself has energetic costs, one of which, photorespiration, is a consequence of the nature of the enzymes used in the process (Box 5.1). Only the energy remaining can be devoted to growth or the production of gametes, and this is termed **net primary productivity (NPP)**:

$$\text{net primary production (NPP)} = \text{gross primary production (GPP)} - \text{respiration (R)}$$

Box 5.1

TRAPPING THE LIGHT FANTASTIC

Plants capture light by the process of photosynthesis, in which the biochemical machinery within their leaves uses radiant energy to build carbon dioxide and water into glucose (or other simple sugars):

$$6CO_2 + 12H_2O \longrightarrow C_6H_{12}O_6 + 6O_2 + 6H_2O$$

carbon dioxide | water | radiant energy | glucose | oxygen | water

Whilst this tells us what happens, it does not tell us how it happens. Light is absorbed by chlorophyll, a group of green pigments common to all plants (Figure 5.1) and contained in chloroplasts (Figure 5.2). Here, chlorophyll is arranged within a mass of tightly folded membranes, called thylakoids, to pack as much pigment into as small a volume as possible.

Chlorophyll appears green because it absorbs light at the red and blue ends of the spectrum and reflects back green (Figure 5.3). At these wavelengths electrons within the chlorophyll molecule become energized. Leaves that are not green still have chlorophyll but also have a range of other pigments. These accessory pigments trap light of different wavelengths and feed the energy back to chlorophyll. Accessory pigments are used by plants which grow in poor light conditions, in the shade of the forest floor or deeper coastal waters. We see some of these in the autumn when the leaves turn red: then the chlorophyll is degraded to reveal the pigments it once masked.

There are two stages in photosynthesis. The first involves the conversion of light energy into a usable form and the second locks this energy into energy-rich compounds (sugars) which can be stored. These are called the **light reaction** and **dark reaction**, respectively (Figure 5.4).

The light reaction consists of two systems, photosystem I and photosystem II, operating side by side. Photosystem I was the first to evolve and is found in many primitive autotrophic bacteria. Photosystem II is found in the cyanobacteria and all other photosynthetic plants and is used to aid energy capture by photosystem I. This is the source of the high oxygen concentration of our atmosphere.

In photosystem II, energy captured by chlorophyll is used to break off an atom of oxygen from a water molecule to generate two electrons. These are used to replace electrons used in photosystem I. Here the electrons, along with a hydrogen nucleus, reduce a molecule of NADP to NADPH, which can then be used as electron carrier. In addition, photosystem I operates a special electron roundabout (called cyclic phosphorylation) shunting electrons round and round, driven by light, to produce adenosine triphosphate (ATP), the molecule used to store energy. Both NADPH and ATP feed into the dark reaction to fuel the process of sugar manufacture in the dark reaction.

At the heart of the dark reaction is the Calvin–Benson cycle, a biochemical production line driven by the products of the light reaction and which produces sugars (Figure 5.4). Outside the chloroplast, carbon dioxide enters the leaf through the stomata, dissolving into water that lines the surrounding cells. An enzyme, ribulose bisphophate carboxylase (rubisco for short), splits off the carbon from the carbonated water and attaches it to a 5-carbon sugar (ribulose bisphosphate – RuBP) to make a 6-carbon sugar. This then splits into two 3-carbon sugars. These enter the chloroplast, where ATP and NADPH from the light reaction are

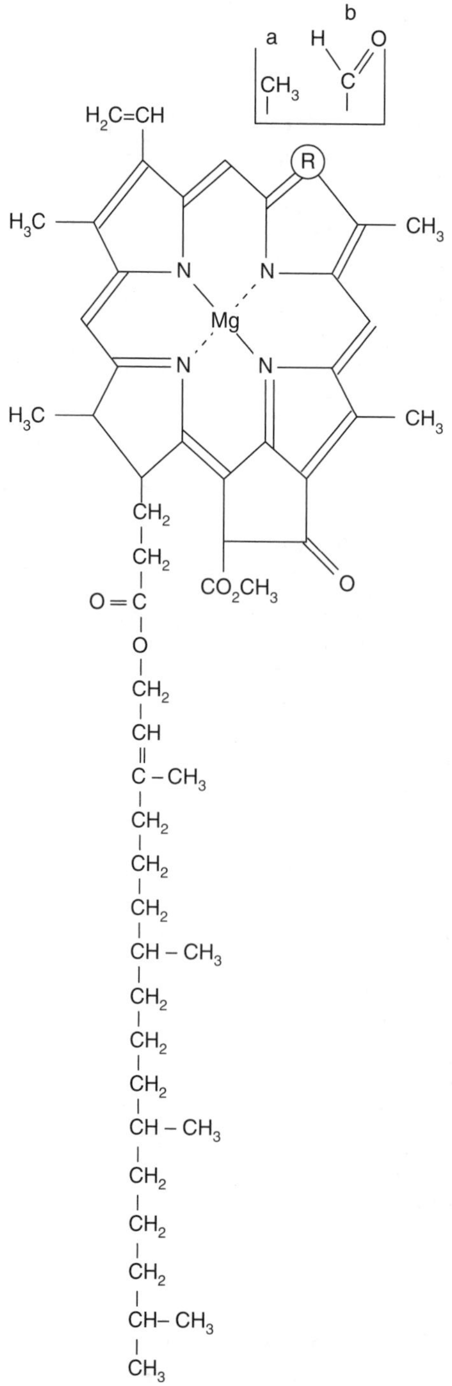

Figure 5.1 Structure of chlorophylls *a* and *b*. Chlorophyll is an organic molecule with an atom of magnesium (Mg) at its centre and is therefore known as an organo-metallic compound. The structure shown is the basic form – variations for chlorophyll *a* and chlorophyll *b* are shown in the inset and are attached to the molecule at point R.

used to combine them and form glucose or fructose and to restore RuBP. At this stage some rubisco may react with oxygen to be lost in a process termed **photorespiration**. Maintaining their rubisco represents an important energy cost for the plant and one reason why photosynthetic surfaces – leaves – are expensive to maintain.

Because the first product of the carbon fixation is two 3-carbon sugars this is called a **C_3 pathway**.

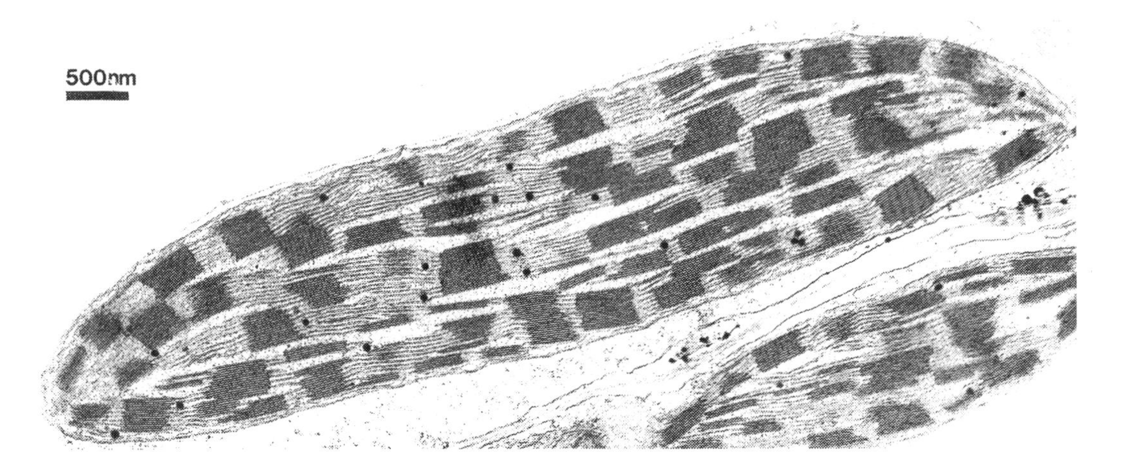

Figure 5.2 Electron micrograph of the chloroplast, a cell organelle responsible for photosynthesis. It consists of a series of membranes (thylakoids) on which chlorophyll and its associated proteins are situated. The non-membranous part (the stroma) is where the dark reaction takes place. Chloroplasts have their own DNA and this is thought to be evidence of their evolutionary past as free-living organisms which took up symbiotic residence within the other cells.

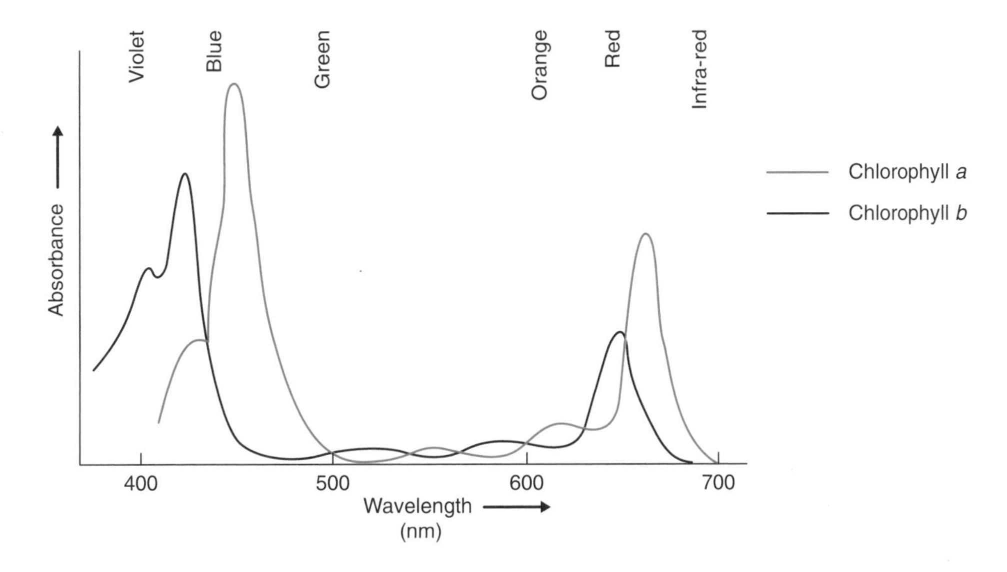

Figure 5.3 Absorption spectra of chlorophyll *a* and *b*. The graph shows the pattern of light absorption of chlorophyll. Note that the absorbance peaks at the red and blue ends of the spectrum: these are the wavelengths that are best at exciting electrons within the chlorophyll molecule. Green light is not absorbed and is reflected back.

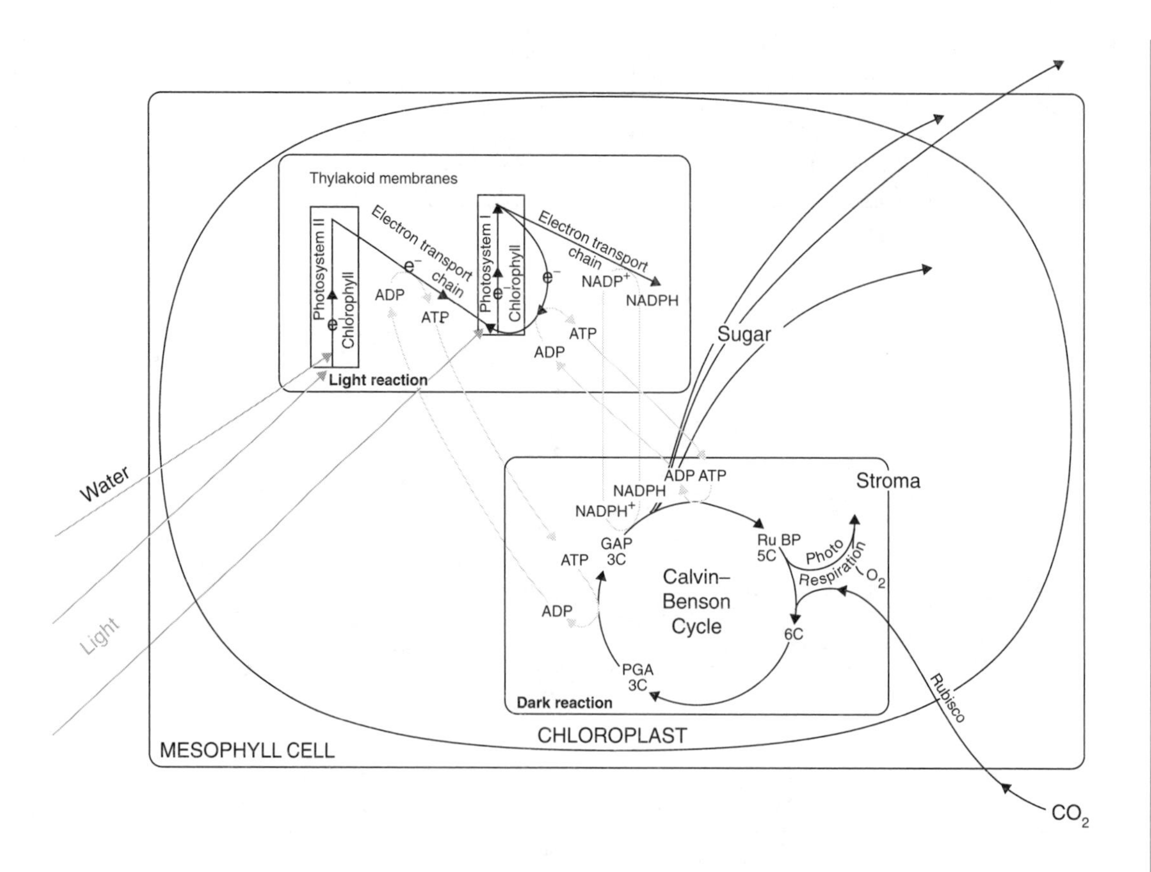

Figure 5.4 An overview of photosynthesis. Energy capture, the **light reaction**, takes place on the thylakoid membranes. Chlorophyll groups together with its other related pigments and proteins to form photosystems of which there are two types, I and II. Light hitting photosystem I excites an electron, which then gives off some of this energy as it cascades down an electron transport chain and this is captured in the chemical bonds of adenosine triphosphate (ATP). Light also powers the splitting of water in photosystem II and the electrons generated are passed to photosystem I to reduce the electron carrier NADP (nicotinamide adenine dinucleotide phosphate) to NADPH.

ATP and NADPH then power the **dark reaction**, which takes place in the stroma. Here the enzyme ribulose bisphosphate carboxylase (rubisco) fixes carbon dioxide and attaches the carbon to a 5-carbon compound, ribulose bisphosphate (RuBP). The newly formed 6-carbon splits into two 3-carbon compounds, phosphoglyceric acid (PGA). This is recycled using ATP and NADPH from the light reaction to produce sugars which can be used by the growing plant. RuBP is recycled but some is lost to photorespiration.

Biomass represents the energy fixed in the tissues – the NPP accumulated in an area over a period of time. This is also termed the **standing crop**, and it differs widely both between species of plant and between ecosystems (Table 5.1).

As we saw in the last chapter, resources can be allocated in different ways according to the strategy adopted by a species and its life history. Some plants have very short life spans: thale cress (*Arabidopsis thaliana*) completes its life cycle in little more than a month; while one bristlecone pine (*Pinus aristata*) nicknamed Methuselah, has an estimated age of around 5000 years. The energy a plant devotes to growth rather than reproduction depends very much on its longevity.

Annual plants, as the name suggests, complete their life cycle within a year. Early on, energy is used to produce as many leaves as possible and the plants only shift their resources to reproduction towards the end of their life. Then, as much as 90% of their

Table 5.1 Global primary production

Ecosystem type	*Mean net primary productivity (g/m² per annum)*	*Mean biomass (kg/m²)*
Continental		
Tropical rainforest	2200	45.0
Tropical seasonal forest	1600	35.0
Temperate evergreen forest	1300	35.0
Temperate deciduous forest	1200	30.0
Boreal forest	800	20.0
Woodland and shrubland	700	6.0
Savannah	900	4.0
Temperate grassland	600	1.60
Tundra and alpine	140	0.60
Desert and semidesert scrub	90	0.70
Extreme desert, rock, sand, ice	3	0.02
Cultivated land	650	1.00
Swamp and marsh	2000	15.00
Lakes and streams	250	0.02
Mean continental	773	12.3
Marine		
Open ocean	125	0.003
Upwelling zones	500	0.02
Continental shelf	360	0.01
Algal beds and reefs	2500	2.0
Estuaries	1500	1.0
Mean marine	152	0.01
Grand total	333	3.6

energy is invested in flowers and seeds. Such plants are characteristically *r*-selected species, fast-growing opportunists (section 3.5). Longer-lived **perennials**, which survive a number of seasons, have an allocation strategy that invests energy in structural tissues or in energy storage that allows rapid growth at the beginning of a new season. Energy reserves often take the form of tubers, roots or swollen underground stems, rich in starch. Perennials may delay reproduction until they have accumulated sufficient energy to fuel the process but they may do this repeatedly. You may recognize this as a *K*-selected strategy, adopted by competitive species that dominate stable and predictable environments.

Trees and shrubs go a stage further. As they grow, much of their energy goes into producing large amounts of woody tissue. When they are saplings, more than half of their biomass may be in the form of leaves, but this changes as they grow. A mature tree may consist of 95% non-living woody tissue, with the live biomass comprising the shoots, roots, leaves and a thin layer of living tissues beneath the bark, wrapped around a trunk of dead wood. The wood serves as a scaffold to support the light-gathering surfaces of the crown and the nutrient gathering network in the roots.

A large tree represents a considerable accumulation of energy and this becomes obvious when it is released as heat as we burn wood. The plant's investment in this water-conducting and structural tissue pays if water is abundant and there is severe competition for light and nutrients. Forests dominate the wetter parts of the world (section 8.1) and represent the most productive ecosystems on Earth (Table 5.1).

Even under optimum conditions, different plants have different capacities to fix radiant energy. Some differ according to the biochemical machinery they use in their photosynthetic process (Box 5.2). Generally, plants adapted to habitats which have abundant nutrients have the highest photosynthetic efficiencies and these include many of our agricultural crops. Photosynthesis also rises with temperature, though each species will have a range to which it is adapted and in which it is most productive.

Light is rarely limiting to photosynthesis in terrestrial ecosystems. Most plants have their highest efficiency at relatively low light levels and bright light can inhibit photosynthesis. Light is quickly extinguished with depth of water, both by absorption and due to reflection by suspended matter, and only certain wavelengths can penetrate any distance. Plants living in deep water therefore have to be able to collect energy from whatever light is available. Large kelps that live in deep coastal waters have a range of pigments that absorb energy over these parts of the spectrum. For the same reason, many terrestrial plants growing under deep shade or dim conditions may also have accessory pigments to collect energy over these residual wavelengths.

Plants can only invest in new tissues when the energy fixed in photosynthesis is greater than that used in respiration. If the two are in balance the plant is said to be at its **compensation point**. This translates into a physical reality in aquatic systems, being the depth where there is insufficient light to exceed the demands of respiration. Photoautotrophs can only persist above this depth. Similarly, some species cannot survive the deep shade of a forest.

Even so, many plants have to survive extended periods below their compensation point when they are burning more energy than they can fix. Most plants are able to go into debt for a while, living off energy reserves in the form of starch, but they cannot live beyond their means indefinitely. One strategy is to reduce repiration costs by shedding photosynthetic surfaces that are costly to maintain. This happens annually in the temperate regions. Here, autumnal leaf-fall is triggered by a decline in temperature, light quality and day length which, together, are reliable signals of the forthcoming winter.

This is not the only possible response to low light and temperatures. Evergreens continue to photosynthesize through the winter, albeit at a low level. Compared with deciduous trees, which produce new leaves each spring, evergreens are relatively inefficient, but because they collect light over an extended season, they can match or beat the shorter period of productivity of deciduous trees (Figure 5.5).

Leaf-shedding and dormancy will happen under other stressful conditions. Many plants of mediterranean and arid areas shed their leaves with drought stress (section 7.1). Another strategy is to adapt the photosynthetic process, to make its water-use more efficient (Box 5.2). Photosynthesis demands a supply of carbon dioxide from an atmosphere with a very low concentration, just 0.03%. This means keeping the gas-exchange pores (the stomata) open during the day when photosynthesis is under way. Because of this, water vapour is lost readily from the interior of the leaf and, in dry conditions, this can place a considerable strain on its transpiration stream from the roots. For this reason productivity is closely related to water availability and explains why the most productive ecosystems are those with abundant moisture (Table 5.1).

The productivity of the forests can also be attributed to their complex structure. Unlike grasslands or communities of low scrub, forests are multilayered with a tree canopy, shrub layer, herb layer and forest floor. Leaves do not need bright sunlight to be effective and light filtered or reflected from upper layers is sufficient for plants adapted to shade. This adaptation takes various forms – from large flat leaves to a range of additional light-absorbing pigments that help chlorophyll do its work (Box 5.1). A multilayered community can develop if there is abundant moisture to allow all of these photosynthetic surfaces to work.

We compare the complexity of the primary producer community of different ecosystems using the **leaf area index (LAI)**; simply, the leaf area as a proportion of the ground area it covers:

$$\text{leaf area index} = \text{total leaf area } (m^2)/\text{area of ground } (m^2)$$

	Beech *Fagus sylvatica*	Norway Spruce *Picea abies*
Lifespan	89 years	100 years
Leaf shape	Broad	Needle
Annual leaf production	High	Low
Photosynthesis	High	Low
Length of growing season	176 days	260 days
Primary productivity (C fixed)	8.6 t ha^{-1}	14.9 t ha^{-1}

Figure 5.5 Evergreen species sacrifice short-term productivity for the ability to photosynthesize for longer. The Norway spruce (*Picea abies*) photosynthesizes for almost half as long again as the beech tree (*Fagus sylvatica*) and, despite its low annual leaf production and photosynthetic rate, it is more effective over the long term.

Box 5.2

ALTERNATIVE ARRANGEMENTS

Photosynthesizing in hot, dry conditions can be extremely risky. Plants must open their stomata to absorb carbon dioxide and, in the process, lose large amounts of water. Being able to reduce the time these pores have to remain open would clearly be advantageous. Some plants in these situations (and also places where they may suffer from physiological drought, such as the salty water of coastal marshes) invest some of their energy in a photosynthetic pathway that helps to keep water loss to a minimum. Among them are sorghum (*Sorghum bicolor*) and maize (*Zea mays*), both staple crops of semi-arid regions.

This pathway was first noticed in sugar cane (*Saccharum officinale*) which was shown to make 4-carbon compounds in its dark reaction (Box 5.1). By fixing an extra carbon to 3-carbon phosphoenolpyruvate (PEP) it makes a series of 4-carbon acids such as oxaloacetic, malic and acetic acids (Figure 5.6). The **C_4 pathway** then feeds these into the Calvin–Benson cycle as in the normal dark reaction.

Inside the chloroplast, the 4-carbon acids are converted into pyruvate and

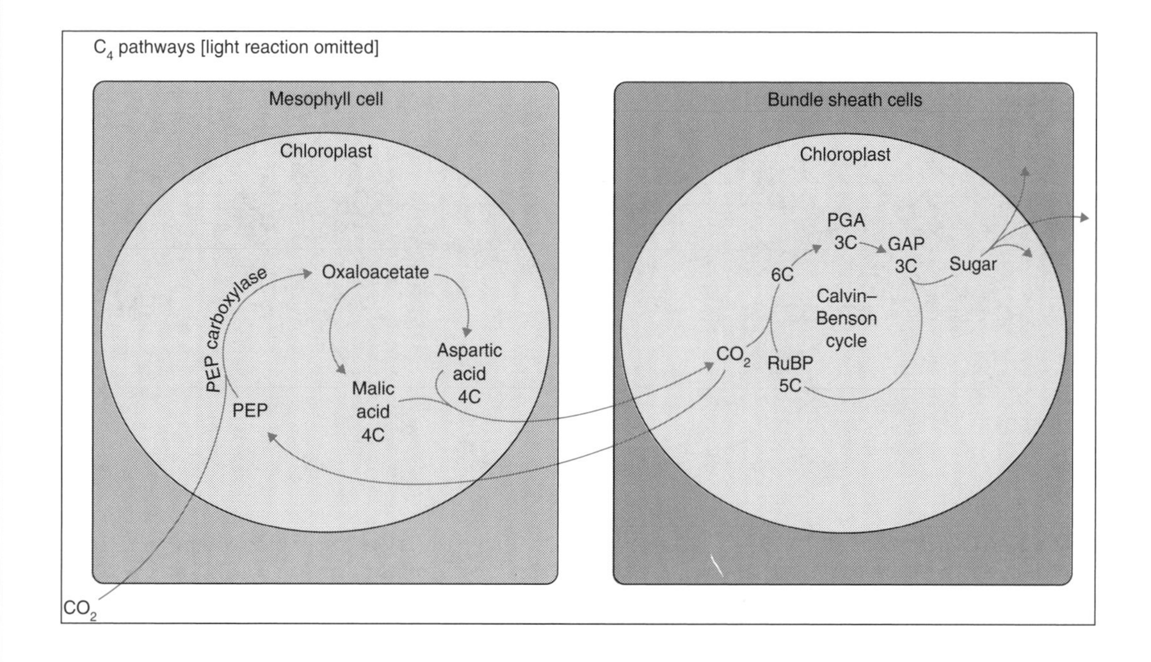

Figure 5.6 An overview of the C_4 metabolic pathway. In C_4 plants, carbon dioxide is fixed into phosphoenolpyruvate (PEP) by chloroplasts within the mesophyll cells. The 4-carbon compounds – oxaloacetate, malic acid and aspartic acid – then pass into the special chloroplasts of the bundle sheath cells where the carbon is transferred into the Calvin–Benson cycle.

release carbon dioxide, to be captured by rubisco and fed into the normal C_3 pathway (Box 5.1). The advantage lies in the enzyme responsible for building the 4-carbon acids and its ability to scavenge carbon dioxide at minuscule concentrations. As we have seen, rubisco is not a good carbon fixer and its efficiency plummets outside the chloroplast, where it can combine with oxygen. The C_4 pathway enables carbon to be brought into the chloroplast by intermediates so that rubisco can operate with relative efficiency, protected by the low oxygen levels inside. Note that these anaerobic conditions (in both C_3 and C_4 plants) echo the environment in which the chloroplast first evolved (section 1.6).

C_4 plants have a distinctive set of chloroplasts, surrounding the vascular cells of the leaf (Figure 5.7). Some C_4 plants, such as the prickly pear (*Opuntia* spp., Box 4.4), have gone a stage further. They have the anatomy of C_4 plants but only open their stomata in the cool of night, again fixing carbon dioxide as a series of 4-carbon acids to be stored until sunrise. They can then photosynthesize without having to open their stomata in the heat of the day. Then the acids lose carbon dioxide to the normal C_3 pathway. This is termed **crassulacean acid metabolism (CAM)** and is a particular feature of succulent plants, whose leaves contain cells with large fluid-filled vacuoles in which the C_4 acids can be stored. Such plants grow in extremely arid places and although shunting carbon dioxide around has energetic costs, it does reduce water loss. A further group of CAM plants are able to hedge their bets and save energy by reverting to C_3 metabolism on a rainy day, returning to CAM when the heat is on.

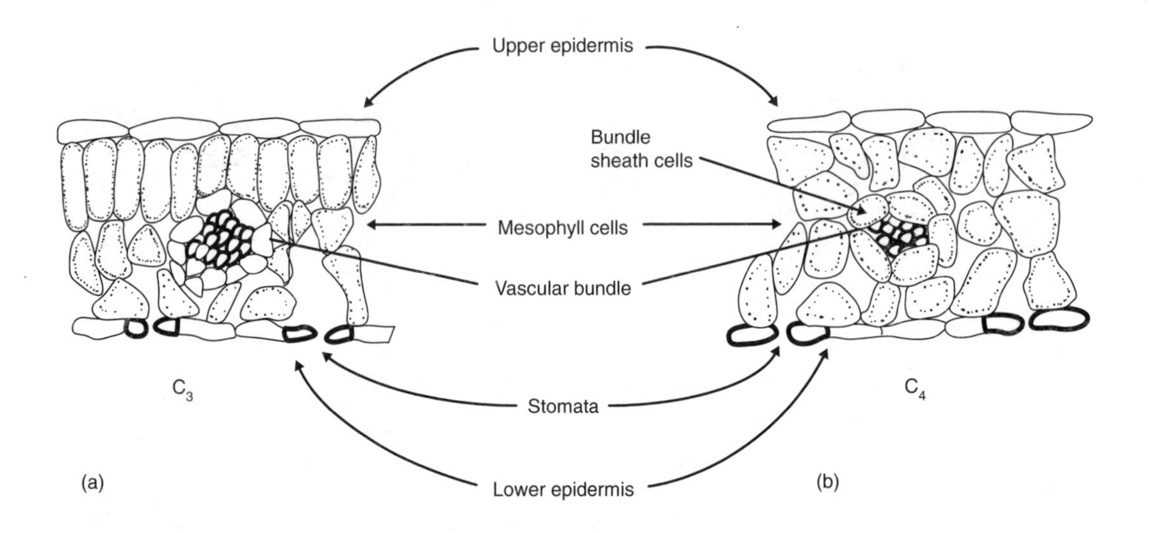

Figure 5.7 A comparison of (a) C_3 and (b) C_4 plants. Whilst C_3 plants have their photosynthetic cells in the palisade layer, C_4 plants have a special set of cells, known as bundle sheath cells, forming a ring around the vascular tissue supplying water and nutrients to the leaf. This is known as kranze anatomy (after the German word for halo).

Forest ecosystems can have LAIs as high as 9, which means that light passes through nine layers of leaves before reaching the ground. Arid and semi-arid areas may have an LAI of less than one.

Algal beds and reefs have the highest known rates of primary productivity (Table 5.1). With an annual NPP of 2500 /m² they can out-perform even tropical rain forest. They too have a multilayered structure, with plants of different sizes, though these are primarily adaptations to different depths of water – once again using accessory pigments, obvious in the red and brown seaweeds. Swamps and marshes too are highly productive, being both shallow and with an abundance of nutrients arriving with the river and marine sediments.

Grasslands tend to have LAIs somewhere between forests and desert, and with an intermediate NPP of 600–900 g/m^2 each year. They have none of the vertical construction of a forest and this is reflected in their standing crop (1.6–4.0 kg/m^2 compared with 20–45 kg/m^2). These differences follow from the allocation strategies of the grassland plants, adapted to a habitat where rainfall is seasonal. Grasses are perennials which invest little in structural tissues, but instead devote most of their resources to leaves that will function as long as water is abundant. The characteristic dry season of grasslands brings primary production to a close and the grass withdraws resources from its photosynthetic surfaces. Although the plant does try to protect its leaves from grazers, it protects its living tissues by keeping them are close to the ground. From here new leaves will sprout when water is next available. This is a highly successful strategy, one which also works well in moister areas, if there is intense grazing pressure. This strategy also led to much of the temperate forests of the northern hemisphere, once felled, being replaced by grazed pasture.

Cultivated land appears to have a low net productivity, (650 g/m^2 per annum), but this merely reflects the highly seasonal nature of agricultural activity. Often this land may support only one or two crops during the year, otherwise being in preparation or waiting for the next growing season. However, when they are actively growing, crops are among the most efficient primary producers (Table 5.2). Photosynthetic efficiencies close to 10% are achieved in some cases, primarily because we remove some of the checks on photosynthesis, by adding water and nutrients. Notice, too, that many crops are C_4 plants (Box 5.2) adapted to drier conditions.

5.3 LINKS IN THE CHAIN

A **food chain** describes one route by which energy passes through a community and the feeding relations between some of its species. Primary producers represent the first stage in fixing the sun's energy. When they are consumed, this energy passes to other organisms, which are known as **heterotrophs** since they are quite literally 'nourished by others', be they plants or animals. Consumers are collectively called **secondary producers** and ultimately depend on the productivity of plants. Secondary producers incorporate energy gained from photosynthesis into their own tissues, much as Isaiah suggested.

A simple food chain, and its terminology, is shown in Figure 5.8. The heterotrophs are divided into a series of **trophic levels** or feeding positions according to what they feed upon. Primary consumers are **herbivores**, feeding on plants. Although vegetation is abundant, we should not assume that herbivores are well supplied with energy. Plant tissues are generally low in nitrogen but rich in cellulose and this is not easily broken down. Plant material often represents a poor quality food which requires a considerable investment to digest it, including a long digestive tract. This means the assimilation efficiencies of herbivores are generally low (Box 5.3). Some herbivores have concentrated on parts of the plant which are more nutritious or more readily digested than others, such as new shoots or buds. Seed-eaters consume a food rich in stored carbohydrates and oils. They benefit from an

Table 5.2 Photosynthetic efficiencies and growth rates of crops

Crop	*Species*	*Country*	*Crop growth rate (g/m^2) per day)*	*Total radiation (J/cm^2 per day)*	*Light conversion efficiency (%)*
Maize	*Zea mays**	USA	52	2090	9.8
Millet	*Pennisetum typhoides**	Australia	54	2134	9.5
Sugar beet	*Beta vulgaris*	UK	31	1230	9.5
Millet	*Pennisetum purpureum**	El Salvador	39	1674	9.3
Sugar cane	*Saccharinum* spp.*	Hawai'i	37	1678	8.4
Tall fescue	*Festuca arundinacea*	UK	43	2201	7.8

* C_4 plants

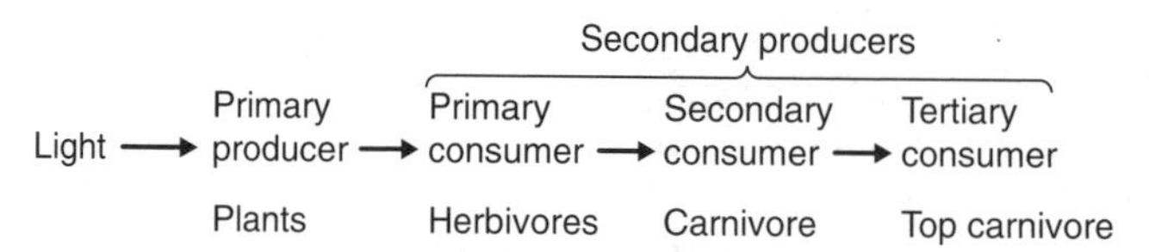

Figure 5.8 A generalized food chain – the pathway of energy through a simple food chain. The organisms can be divided up into primary producers which fix radiant energy and secondary producers which consume this energy second-, third- and fourth-hand.

easily digested food with high levels of nitrogen which otherwise would have fuelled the germination of the seed. Other herbivores have found ways of unlocking the energy contained within the indigestible biomass (Box 5.5).

Carnivores have a different set of challenges. As secondary or tertiary consumers, they live on the energy fixed in the tissues of herbivores or other animals. Flesh is primarily protein and fat, compounds which are more readily degraded, with a high energy content and, in the case of protein, rich in key nutrients. All this means that carnivores have a shorter digestive tract and can digest their food relatively rapidly. However, they still have to meet the costs of catching and killing their prey – costs which are minimal for a herbivore. A variety of adaptations have arisen, from the spider's web to the claw of a cheetah, to make predation efficient (section 4.4).

Some feeding strategies do not sit conveniently on one trophic level. **Omnivores** straddle different trophic levels and may feed on both vegetation and meat, or on herbivores and carnivores. As a result, they do not fit neatly into a linear food chain and energy flow does not follow a simple sequence of trophic levels. These patterns become even more complex when we realize that some omnivores may also scavenge dead animals or plants.

We ourselves are omnivores and have been accused of being scavengers in our past (section 1.1). A scavenger diverts energy heading for a decomposer food chain back to one based primarily on herbivores. We can distinguish two basic routes for energy moving through ecosystems: from herbivores to carnivores (termed a **grazing food chain**), and from decomposers to carnivores (a **decomposer food chain**). These two chains are connected every time a blackbird pulls a worm from the soil, or any other decomposer is eaten by a consumer from the grazing food chain. The amount of energy moving down each pathway depends upon the ecosystem. For most ecosystems a large proportion of the primary production passes into the decomposer route. In some aquatic systems, however, a sizeable fraction of the phytoplankton is actually grazed by herbivores. On average, only 10% of net primary productivity in terrestrial ecosystems finds its way into herbivorous consumers. The rest, all 103.5 billion tonnes of it globally, fuels the decomposer chain. The decomposer chain is therefore the most significant route by which energy passes into the rest of the system.

Box 5.3

ENERGY EFFICIENCY

Much of the energy of primary production never passes to herbivores or carnivores because it is never consumed. Some part of the production of each trophic level passes not to consumers but to decomposers. This is why the energy flowing through the decomposer food web is the larger component in most ecosystems.

Of that consumed, a sizeable proportion may not be digested. Amongst the most indigestible of tissues, at least for most animals, are the celluloses and lignins used in plant structures and support. Similarly, animals produce scales, hair, bone and shells which are often discarded or pass untouched through the consumer's digestive tract.

Because it involves a transformation, any movement of energy from one organism to another inevitably leads to some energy loss. Of the energy that is assimilated, much is used in the metabolic processes – the cost of building, maintaining and degrading the cell. From that assimilated (*A*), some energy goes to repiration (*R*) and some to production (*P*). Production can be divided into that used in growth (*Pg*) and in reproduction (*Pr* – Figure 4.1).

We can work out the proportion of energy in the diet taken up by a consumer simply as its assimilation efficiency:

$$\textbf{assimilation efficiency} = \frac{\text{energy assimilated } (A)}{\text{energy consumed } (C)}$$

(An equivalent equation derives photosynthetic efficiency by making the divisor the radiant energy received at the leaf surface.)

A simple extension of this allows us to work out how efficiently energy consumed is converted into new tissues:

$$\textbf{production efficiency} = \frac{\text{energy fixed in tissues } (P)}{\text{energy consumed } (C)}$$

The proportion of energy assimilated that is converted into tissues is the growth efficiency:

$$\textbf{growth efficiency} = \frac{\text{energy fixed in tissues } (P)}{\text{energy assimilated } (A)}$$

Ecological efficiencies vary amongst animals according to their metabolic costs. Warm-blooded animals (**endotherms**) such as birds and mammals have high metabolic costs and can spend over 90% of their energy income in maintaining their body temperature. **Ectotherms**, on the other hand, rely primarily on external heat sources and do not have these costs. They can devote more of their energy to production (Table 5.3).

In general, organisms further along a food chain have higher assimilation efficiencies, largely because of the quality of their diet. Herbivores may have a plentiful supply of plant material but up to a third of this may be cellulose, which they are unable to digest without the help of bacteria in their gut (Box 5.4). Their assimilation efficiencies are low – typically around 10% – and herbivores therefore need to consume large amounts of vegetation to meet their energy demands.

Carnivores receive their energy in more usable form, as proteins and fats which are both richer in energy and more readily digested. They have relatively high assimilation efficiencies, as much as 90% in some exceptional cases. This efficiency and the high nutritive value of their food mean that carnivores, compared with herbivores, need to eat less and to eat less often.

Production efficiency, which takes into account the energy used in respiration, also varies with trophic position (Table 5.3). Within a group sharing a similar metabolism, say the terrestrial invertebrates, herbivores again have the lowest efficiency (21–39%) and carnivores the highest (27–55%). Across groups, ectotherms have production efficiencies

Table 5.3 Production efficiencies of different animal groups

Group	*Production efficiency (%)*
Endotherms	
Insectivores	0.9
Birds	1.3
Small mammals	1.5
Other mammals	3.1
Ectotherms	
Fish and social insects	9.8
Non-insect invertebrates	25.0
Non-social insects	40.7
Non-insect invertebrates	
Herbivores	20.8
Carnivores	27.6
Detritivores	36.2
Non-social insects	
Herbivores	38.8
Detritivores	47.0
Carnivores	55.6

of between 10 and 55% and endotherms of just 1–3%.

These figures might seem to argue for being a carnivore and never being a herbivore. However, most of the biomass of the planet is vegetation and there is consequently more energy available to herbivores than carnivores. The figures also suggest that being an ectotherm is a better strategy when energy is in short supply, and this is probably true.

Ecological efficiencies also make it obvious why energy decreases so rapidly along a food chain. A carnivore three or four steps away from the primary producers of an ecosystem might have one ten-thousandth of the energy originally fixed by the plant. Figure 5.11 shows how little energy becomes fixed in each successive trophic level of a grassland food chain, so that a weasel fixes a mere 0.0026% of net primary productivity into its own tissues. That helps to explain why there are so few weasels and why everybody cannot be a carnivore.

5.3.1 The reducers

Decomposers are crucial for their role in nutrient cycling and for redirecting energy from dead organic matter back into the living system (Figure 5.9). A large proportion of primary production results in indigestible components that must first be processed by this trophic level before they can be used by the rest of the consumer community. To do this, consumers must rely on enzymes produced by fungi or bacteria, by culturing them either in their nests (in the case of termites and some ants) or in their gut (Box 5.5).

The most conspicuous reducers are detritivores such as earthworms. These can process **detritus** (fragmented plant and animal remains) into humus at a rate of between 50 and 170 tonnes per hectare each year. In terrestrial systems, the litter layer includes a variety of invertebrate decomposers, especially mites, woodlice, millipedes and nematode worms. Fungal threads (hyphae) are especially important in degrading organic matter with a low nitrogen content, and do so most of the year unseen. Only when their caps appear above the litter or sprout from a log do we get some idea of their variety (Figure 5.10). Even less conspicuous, though much more abundant in most soils, are the bacteria.

A gram of soil may contain as many as nine million bacteria, but these come in many different forms. They often appear in a distinct successional sequence after organic material is added to the soil. The sequence depends on their capacity to break down various components, with the detritus becoming increasingly intractable as its decomposition proceeds. Similarly, this massive variety of bacterial metabolisms is used by many invertebrate decomposers in their gut, much as herbivores do in the food chain above the soil (Box 5.5).

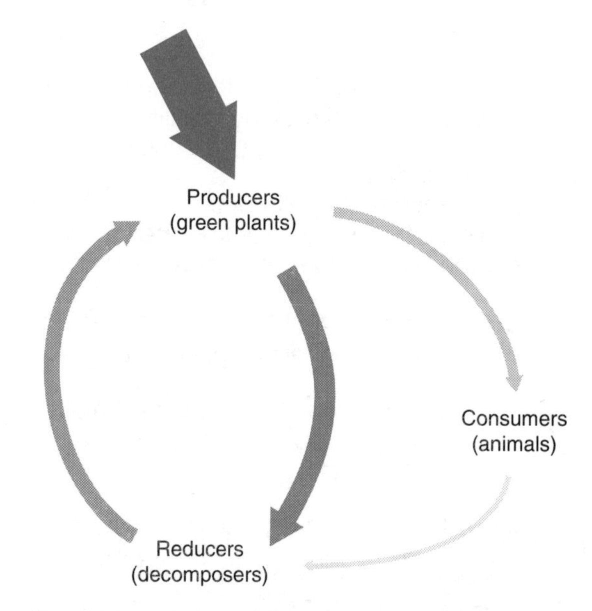

Figure 5.9 Decomposers. In most ecosystems the bulk of the energy fixed by primary producers pass to the decomposer community. This not only fuels the productivity of the decomposers and detritivores; it also drives the recycling of nutrients.

5.3.2 Chain length

Figure 5.11 shows a simple food chain for a grassland ecosystem and the amounts of energy passing

Figure 5.10 The familiar sight of the fruiting bodies of fungi. Underground is a large network of fungal hyphae which live by breaking down organic matter.

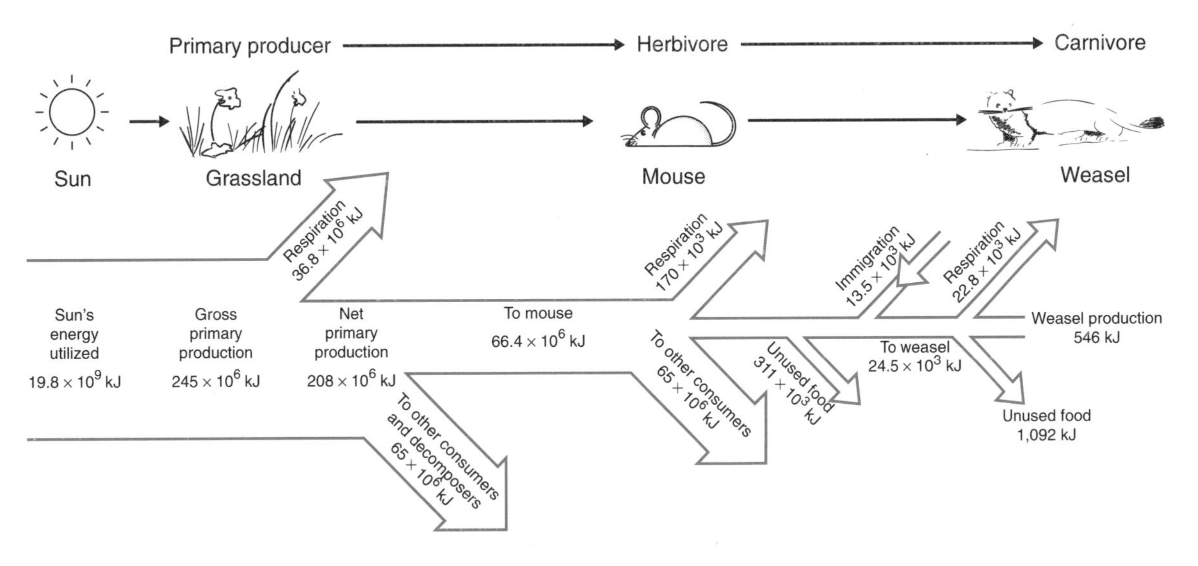

Figure 5.11 The loss of energy as it passes along a simple food chain (in this case a North American grassland). Energy values are in kilojoules (kJ).

through a herbivore (a mouse) to one of its predators (a weasel).

Notice how little energy actually becomes fixed in the tissues of the carnivore. The inefficiency of energy transfer (Box 5.3) and the inevitable losses with each transformation mean that little is available to the higher trophic levels. With every level, energy is expended in respiration and, of that consumed, only a fraction becomes fixed in the tissues. A plot of the energy content of the trophic levels in sequence produces a pyramid (Figures 5.12 and 5.13). In our example, the NPP of the primary producer (200 million kJ) is reduced to just 546 kJ for the weasel.

This chain has just three links. Long food chains are rare in nature and it is unusual for them to extend beyond five links. The dramatic decline in energy at each trophic level was once believed to account for the average chain length being just four links long. It would seem obvious that too little energy remains left to support a fourth or fifth trophic level. This suggests that chain length is limited by the energetic efficiencies of its constituent species. If so, we would expect food chains dominated by the more energy-efficient ectotherms to be longer than those with endotherms (Box 5.3). Surprisingly, the evidence from a number of studies is that this makes little difference to chain length.

Perhaps, then, chain length is controlled by the primary productivity of the ecosystem itself? If this was the case we should expect food chains to be longer in communities with higher primary productivity. Stuart Pimm, John Lawton and Joel Cohen compared food chains from highly productive biomes, such as tropical rain forest (Table 5.1.), with low productivity biomes, like tundra, and found that there were no significant differences in their number of trophic levels. The average remained around four. Pimm and Kitching also tried boosting the energy input (by adding leaf litter) to some simple communities but this led to no increase in chain length.

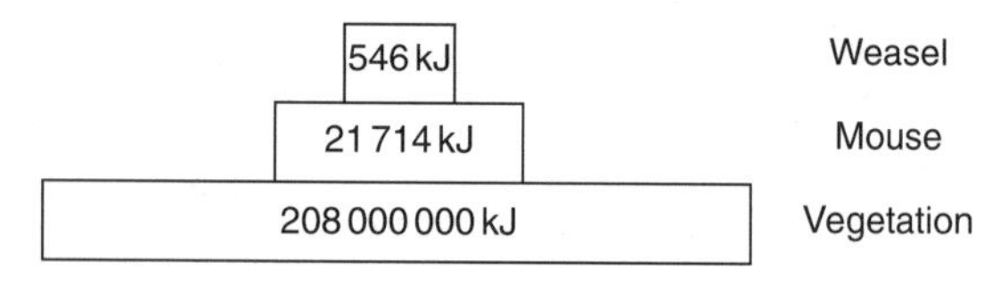

Figure 5.12 Pyramid of energy (not to scale), containing the productivity data from the food chain shown in Figure 5.11. Note the rapid decrease in the amount of energy available at the higher trophic levels.

It therefore seems that we have to look more closely at other features of food chains. Some patterns tend to recur, sometimes for fairly obvious reasons. One is the tendency for consumers to become larger and fewer with each link of the chain (Table 5.4). Large carnivores require larger territories from which to collect sufficient prey. Organisms at higher trophic levels tend be longer-lived, delaying reproduction until later in their life cycle.

Many of these characteristics result from an increase in size, and this could be why some food chains are not longer. As Paul Colinvaux has suggested, it may simply be that a predator would have to be so large to feed on the big carnivores at the end of most food chains that it is unlikely to secure enough energy to maintain itself or a population.

Predators will also be competing with each other for prey. Collectively, they may well hold herbivore numbers in check, so that despite the abundant vegetation in most ecosystems, herbivore numbers are capped by their consumers. In turn, their numbers limit their predators' numbers (section 4.4). In this case, it is the interaction between trophic levels which serves to limit the whole system.

These effects follow from the structure of the community. To understand them, we have go on to consider how food chains are connected in food webs.

5.4 THE WEB

If a food chain marks one possible route by which energy flows through an ecosystem, a food web maps all the possible routes. In them the significant food chains of a community are linked together as a network.

Figure 5.14. shows a food web for the community associated with oak (*Quercus robur*) trees in an Oxfordshire woodland. This web represents an aggregation of several food chains, some of which overlap. Most are only four links long, while others are even shorter. Note that the birds and mammals within this ecosystem have a tendency to be omnivorous. Titmice (birds of the family Paridae) feed at several trophic levels, and both blue and great tits feed on seeds (primary producers), winter moths (herbivores), beetles and spiders (carnivores). Mice and voles are also omnivorous, taking insects and also eating plant material.

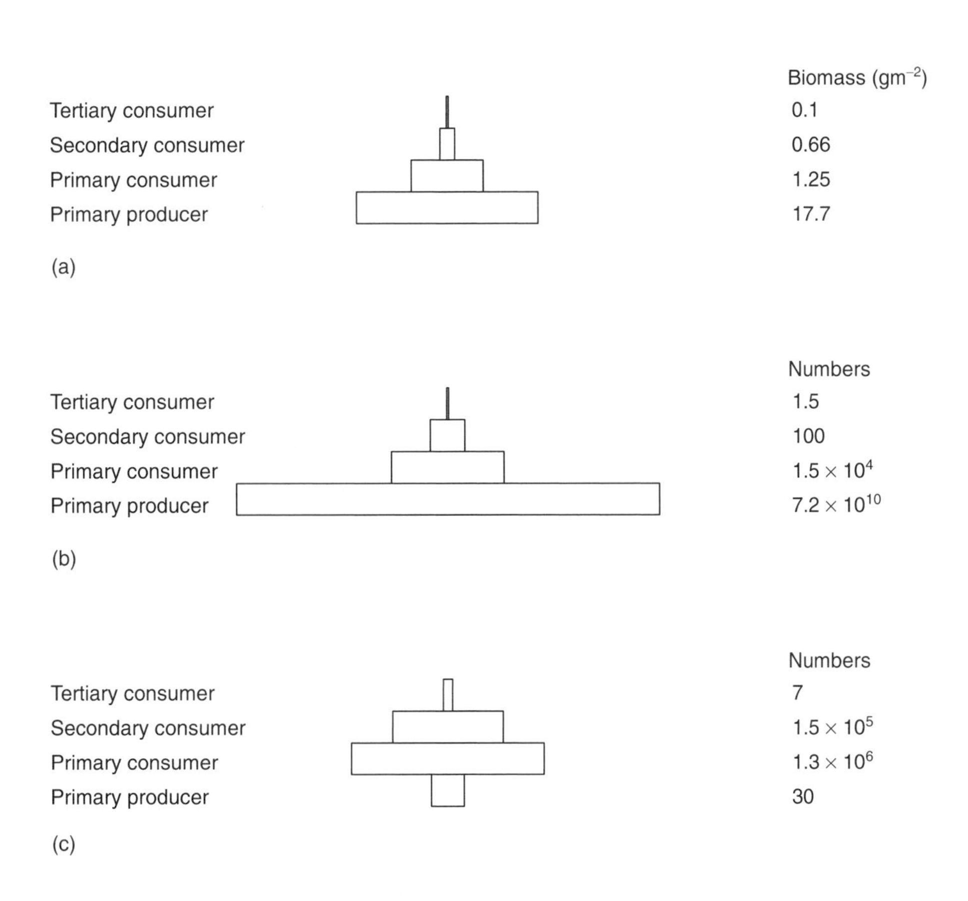

Figure 5.13 Some other ecological pyramids. Plots of (a) biomass or (b) numbers of individuals at each trophic level will also show a pyramidal shape for successive trophic levels within an ecosystem. However they can be misleading: (c) a single tree may support a large number of herbivores, so that a simple plot of number will produce an inverted pyramid.

Webs allow us to break way from the linear constraints of the food chain. In them we can include organisms that do not fit neatly into one trophic level. We can include detritivores and decomposers and so unite grazer and decomposer food chains. It is also possible to include parasites and hyperparasites, the tertiary consumers ignored in many food chains.

We also need to be aware of their limitations. Like any map, our representation of a web is necessarily a simplification of reality, with a somewhat arbitrary boundary around the system. Not all the trophic connections may be identified, if only because of the considerable amount of work needed to map every pathway. Many get left out, and sometimes species are simply lumped together into trophic or 'feeding' groups.

Webs are at best snapshots of nature and freeze a dynamic community into a particular structure. Species come and go with time, and some even change their feeding position during their life cycle. Most webs fail to give any measure of the strength of the trophic connections, and do not

Table 5.4 General trends along food chains

Trend	*Description*
Fewer species	Each trophic level of the food chain generally supports fewer species than the preceding one.
Larger body size	Moving along the food chain, species tend to display an increase in body size as predators are generally larger than their prey.
Lower population numbers	Higher energy costs of species nearer the top of the food chain mean that ecosystems can support relatively few of them.
Longer lived	Larger species tend to have longer lifespans (e.g. contrast a butterfly with an eagle).
Lower reproductive rates	Longer-lived species tend to mature more slowly, delaying reproduction until later on in their life history.
Increased home range	Larger organisms have to cover larger areas to meet their energy requirements in terms of prey. They therefore have bigger territories.
Higher powers of dispersal	In order to cover large territories an organism has to be able to move across a wide area.
Increased searching ability	Looking for food within a large area requires the ability to find and recognize food species.
Increased behavioural complexity	Large territories, long life span and lower densities mean that species near the top of food chains have more complex behaviour patterns with regard to feeding and reproduction.
Higher metabolic cost	Increased size brings with it an increased energy requirement for the maintenance, growth and reproduction of the organism.
Require food of higher calorific value	Food must have a higher energy value so that it can fuel the higher metabolic rates of species nearer the top of the food chain.
Reduced feeding specialization	At higher trophic levels there is a tendency for species to be more generalist in their feeding strategy; this also includes omnivory.
Greater assimilation efficiency	Species which are further along the food chain have increased efficiency in the use of the food they ingest.

distinguish between very rare linkages and very strong interactions.

Nevertheless, on comparing a large number of webs, Pimm, Lawton and Cohen found that some patterns recur. The trophic structure of a web – the changes in abundance and size along a food chain – appears to impose regularities, repeating patterns in different communities. These arise from the interaction between its elements, whether they are single species or species grouped together by feeding behaviour.

For example, there is evidence that the proportion of predator species to prey species is relatively constant. So too is the proportion of consumers on each trophic level, largely independent of the total number of species in the web. The number of linkages is typically close to twice the number of species and, as we have seen, food chain length appears to be independent of the web. Food chains, in fact, seem to be generally shorter in small habitat patches or in frequently disturbed habitats, or those with little vertical development (a grassland, say, compared with a forest), but these changes are relatively small.

What causes these patterns? Certainly, some follow from organizing species into a hierarchy and the consequences of position in a food chain, as we saw in the previous section. Predators need to be big, will get less energy and are likely to have a smaller population size.

Computer models of food webs suggest there may be limits on the number of species a web can contain. Once we take into account the population sizes and the intensity of the interactions between species, the models predict that an invader will find it more difficult to establish itself in webs with large numbers of species. Consider what happens if a predator seeks to invade a well-developed community. To gain access to a prey species it must compete with established predators. This means that interspecific competition must increase (section 4.3) and become more intense for more species at that trophic position, making life

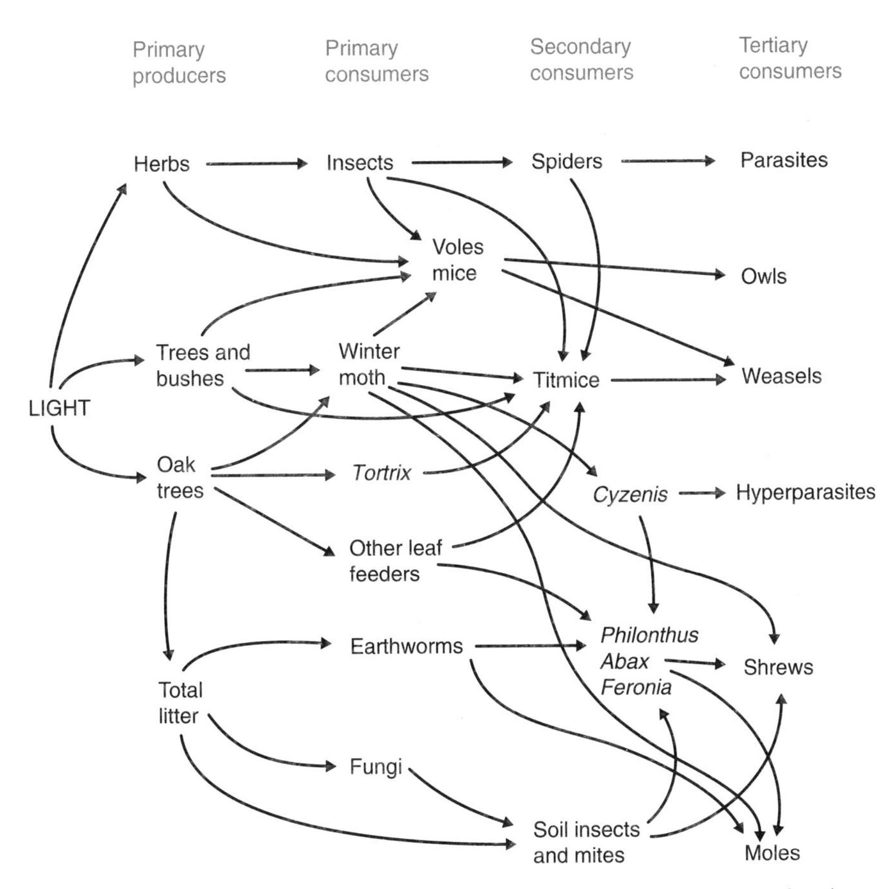

Figure 5.14 A food web as a series of interconnected food chains. In this case the web shows connections between animals associated with oak trees in Wytham Wood, Oxfordshire.

more difficult for everyone. Similarly, interactions between trophic levels will also serve to constrain webs: from their study of 62 published food webs, Frédéric Briand and Joel Cohen found evidence that the number of predator species was a better predictor of the number of prey species, rather than the other way around.

5.4.1 Are communities real?

It seems that these interactions constrain and define the web. They are perhaps **rules of assembly** which govern how a food web can be constructed. Looking at a simple food chain gives the misleading impression that its length is solely limited by energy, but that ignores all the competitive battles and other interactions a species will have with its neighbours. Whether they are only certain configurations of particular species, or of species carrying out particular roles, is another and larger question.

This is important. There may only be certain ways in which a food web can be put together. Perhaps some configurations are unstable. From what we know about ecological energetics, one rule of assembly seems to fix the proportions of primary producers, herbivores and carnivores. But is the community itself, and the species that compose it, somehow constrained – an inevitable outcome of its abiotic environment?

Clearly, the species composition of a community will depend on species being adapted to the abiotic conditions of temperature, moisture and so on, in addition to their trophic interactions with each other. If communities are very tightly linked, functioning as a unit, then we would expect to find some of the unit's species always associated with each other (section 7.2). One test of this idea is whether we could remove a species from a food web without the rest of the community collapsing. Certainly some species are more important to the whole trophic structure than others (section 9.2) but a highly integrated community may not allow too many bricks to be taken away before the pyramid collapses.

On the other hand, if a community is a loose assemblage of species sharing similar adaptations, brought together by history and circumstance, the trophic structure will be much more flexible. Then the community is largely a product of chance and we should expect its composition to change in both space and time. Unfortunately, static food webs are not the best tests of whether dynamic communities are tightly interlinked.

These questions will recur throughout the remainder of this book, and extend beyond the energetic relations between species.

As with many ecological concepts, food webs are artificial: they are attempts at describing a complex living world which has many possibilities. However, despite their drawbacks, webs have their uses. In particular, they can be used as ecological road maps by which we can track the route of pollutants, such as pesticide residues, heavy metals and radioactive isotopes, through the ecosystem (Box 5.4).

5.5 WORKING THE SYSTEM

Some of the shortest food chains are the ones we use ourselves. As omnivores, our agriculture allows us to feed as herbivores, consuming our crops directly, and also as carnivores, feeding on our herbivorous livestock (Figure 5.16).

From the earliest times, human beings responded to the patterns of productivity in the world around them, but a capacity to manipulate primary and secondary productivity marked the outset of civilization. Jacob Bronowski maintained that the change from nomad to village agriculture was the largest single step in the 'Ascent of Man'. Since then, our ability to work the system has been a key factor in our proliferation.

Hunter-gathering (Figure 5.17) is still practised by many peoples in tropical and subtropical areas. Amongst the cultures still practising hunter-gathering are the bushmen of the Kalahari and the Aboriginals of Australia. Although it never supports a high density of people, certainly in semi-arid areas, it does represent a viable and sustainable feeding strategy. Energetically, at least, it provides a relatively high return on the energy and effort invested – about 10 times as much, in fact. By

Box 5.4

BIOMAGNIFICATION, BIOACCUMULATION, BIOCONCENTRATION

For many people, it is almost axiomatic that a pollutant entering a food chain will poison the species at the end of the line. Food chains are seen to concentrate pollutants so that, inevitably, each successive trophic level is threatened by a larger dose.

The classic example, and the first pollutant that appeared to show this behaviour, is DDT (dichlorodiphenyltrichloroethane): after a few years of use as an insecticide, this chemical was found to be responsible for the population collapse of a number of predatory bird species. The realization that spraying insects killed birds two or three trophic levels away sparked off the environmental movement in the west and the tradition that pollutants were 'biomagnified' along food chains.

When we look at this process in more detail, we see the picture is not so simple. **Biomagnification** occurs if a pollutant's concentration is higher in a consumer (or a plant) than in its diet (or the soil). We can measure this as the **concentration factor**:

$$CF = \frac{\text{concentration of the pollutant in the consumer}}{\text{concentration of the pollutant in the diet}}$$

With CF > 1 we have biomagnification. If we include both the diet and water as a source for aquatic organisms, then we have **bioaccumulation**. **Bioconcentration** refers specifically to uptake from water alone.

This equation is equivalent to the assimilation efficiency used to measure energy transfer in Box 5.3. A trophically mobile pollutant will trace the energy pathways within a food web.

So what we are measuring? Firstly, all of these terms are time-dependent: levels in the diet will change continually and the same is likely to be true of the consumer. But a CF only has meaning when the two terms are in equilibrium with each other.

So, what should we measure: the whole-body concentration or just those tissues likely to become part of the diet of the next trophic level? In comparing trophic levels, we should only consider whole-body concentrations. Often, much of the pollutant burden will remain in the leftovers, in the bones, shell or hair, and not be passed on. Comparing concentrations in a particular tissue in different species, where it may represent different proportions of each animal's total mass, is questionable. So too are comparisons between a single prey species and a predator with a varied diet. Many consumers are omnivores and feed on more than one trophic level. Demonstrating bioconcentration means that the food web has to be fully described and the problem of placing a species in an appropriate trophic position can confound the situation still further.

Once we begin to address these questions (and consider whether they have been addressed in previous studies), we realize that biomagnification may not be so widespread as many believe. To be in equilibrium with its diet, a consumer must change in step with dietary concentrations and for a CF greater than 1 its rate of uptake must exceed its rate of loss. That may not always be the case. Another factor is body size: big animals tend to eat more and consume a greater mass of a pollutant, but lose it more slowly. The larger animals found at the end of food chains may have a higher concentration simply because they are bigger, rather than because of their trophic position (Table 5.4).

A further complication is the pollutant itself and its reaction with the living organism. Many organic compounds can be degraded by animals, so the level we measure may not represent that assimilated.

Even with pollutants that do not degrade, demonstrating biomagnification requires a large number of factors to be taken into account. Consider the data in Figure 5.15. This gives the concentration of two metals, lead and zinc, in various components of a grassland ecosystem established on old fluorspar waste ('tailings'). Since there is little atmospheric pollution in this part of Derbyshire, the prime source of each metal is the soil. We can thus rule out respiratory uptake as significant and this study is therefore a good test of whether these metals were biomagnified from the diet. Andrews and his co-workers compared their results against figures for an uncontaminated control site, an area of grassland in Hexham, Northumberland.

Firstly, we should note this data is only for October 1980 and that there were seasonal changes in each component for both metals. Here, we are bending the rules slightly in calculating CFs since we do not know whether the equilibria had been established between the organism and source. Secondly, whilst lead has no metabolic function, zinc does (it is needed at low concentrations for the operation of a number of enzymes) and is under some measure of physiological control. Because the structure of the trophic web is not known, we have to consider a range of possible CFs for the carnivores. Even so, there is no consistent pattern of biomagnification from one trophic level to another for either metal. Neither lead nor zinc is concentrated by the top carnivore (the shrew *Sorex*) at the contaminated site, though shrews at the control site display biomagnification of

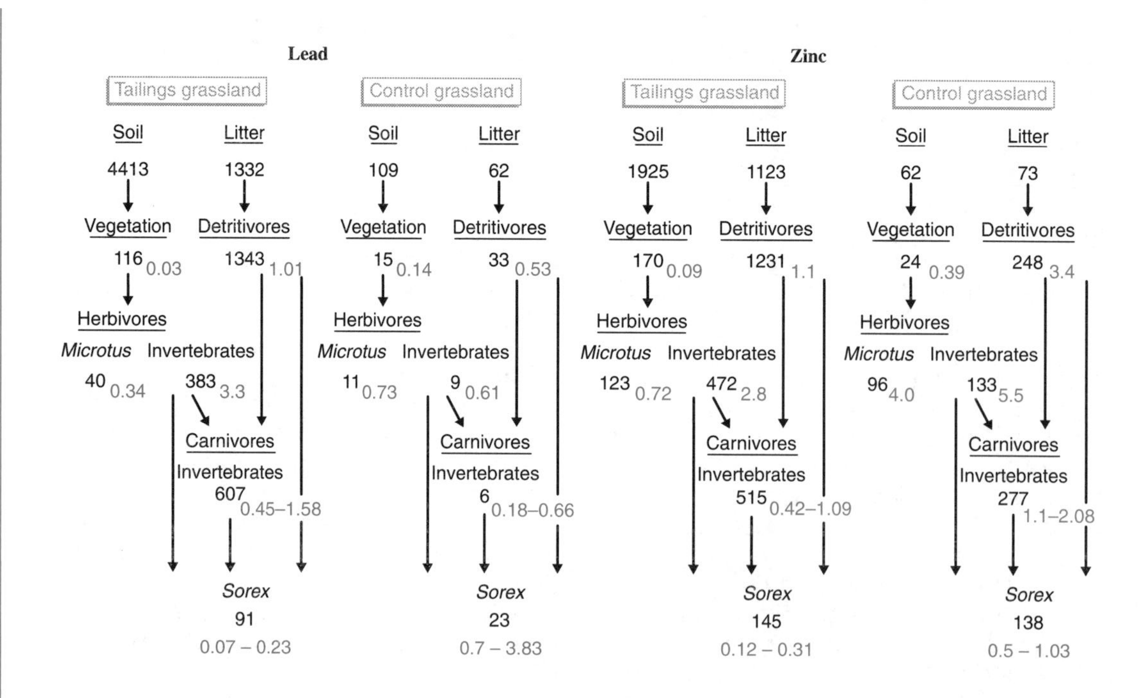

Figure 5.15 Concentrations of lead and zinc in two grassland ecosystems, showing the amount of lead and zinc (μg per gram dry weight) in soil at various trophic levels of two grassland ecosystems, one of which was established on highly contaminated residues ('tailings') from fluorspar mines in Derbyshire. The control site was at Hexham in Northumberland. Concentration factors are shown in blue. The data here is for October 1980 only, but note that levels in the soil, and more particularly the vegetation and litter, varied with the season of the year. Invertebrates were sorted into three broad feeding categories.

zinc. It may well be that the low environmental levels of zinc here cause it to be retained to meet metabolic needs. If this is the case, biomagnification of zinc is determined by physiological factors, rather than a simple function of the food chain.

Biomagnification is not the rule: our example shows that the passage of elemental pollutants through a simple trophic structure does not produce a consistent pattern. Indeed, there is now considerable doubt whether the most cited example of all, DDT, can rightly be said to be biomagnified. Whilst biomagnification does occur for some pollutants under some circumstances, we have to understand the system fully in order to show that it is a product of its trophic structure.

contrast, intensive poultry production returns just half of the energy put in.

Nomadic pastoralists are found over a much wider range of habitats, primarily because they follow or lead herds of semi-domesticated or domesticated grazers searching for suitable forage. The Lapps of northern Europe follow the movements of reindeer to exploit the brief summer of the tundra (section 8.1). The Bakhtiari of Iran and Iraq herd their domesticated sheep and goats over mountain ranges to avoid summer droughts and to find sufficient grazing.

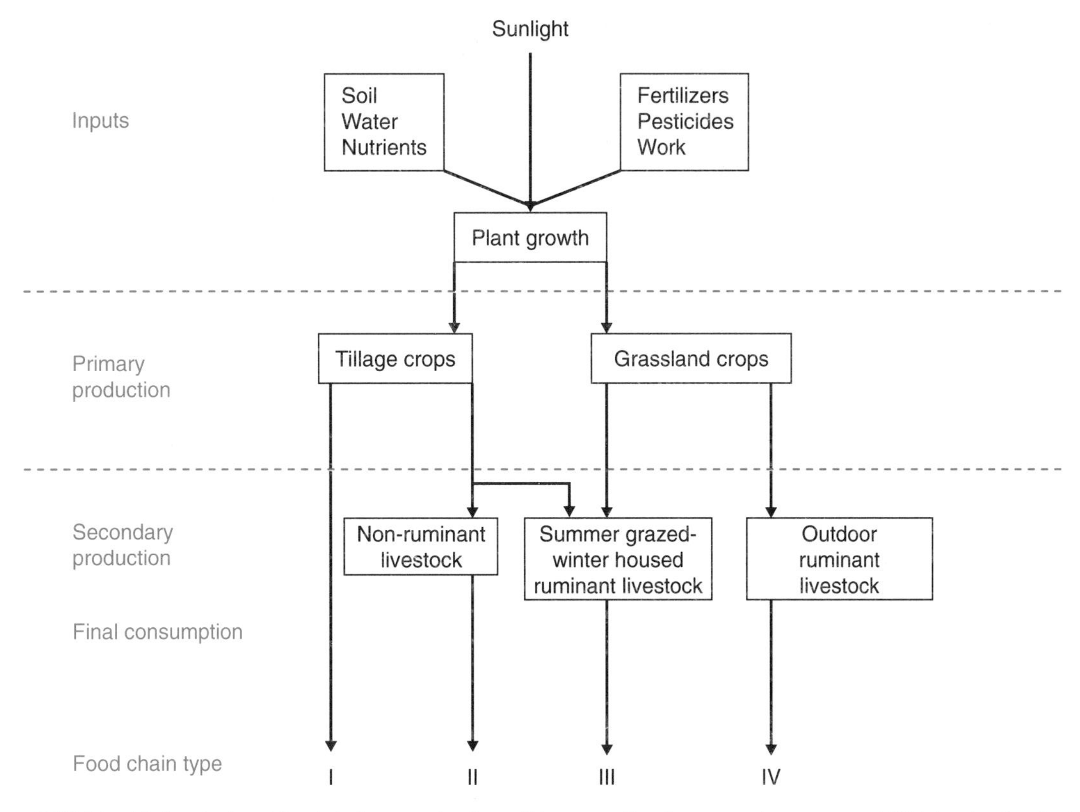

Figure 5.16 A 'web' of agricultural food chains, ranging from (I) humans directly consuming crops and (II) intensive livestock rearing to (III, IV) the grazing of ruminants.

This seasonal movement of livestock is termed **transhumance** and, like hunter-gathering, it has a long tradition; it was practised by some very ancient cultures. It is also used in some more recent agricultures, such as the ranchers of the Rockies. Again they exploit the high mountain pastures that briefly flourish after the winter snows have melted. Alpine pastures have been traditionally managed in this way, with herdsfolk moving their stock (and their homes) to the higher slopes during the summer months, returning to the valleys for the winter. This procedure requires a highly integrated human society where those who herd stock are supported by others in the valley who cultivate crops and make hay for use in the winter months.

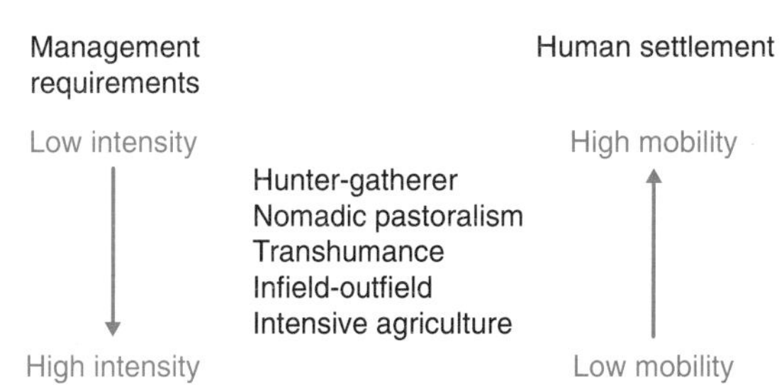

Figure 5.17 A checklist of agricultural systems. The five principal agricultural systems are ranked in terms of their degree of intensity and the amount of mobility in the societies in which they occur. These range from the highly mobile, low-intensity lifestyle of hunter-gatherers through to the highly settled, intensive practices of modern agriculture.

Cultivation of crops probably developed alongside the domestication of the first agricultural animals and almost certainly followed the evolution of a series of heavy-seeded grasses (Box 2.4). Indeed, the skills of sowing, harvesting, threshing, storing and trading of that seed prompted the settled existence and the societies that dominate most cultures today. As agriculture has developed and we have learnt to play the energetics of ecosystems, so our relations with the living world have become increasingly sophisticated, if somewhat strained.

Grazing and the cultivation of plants dominate world agriculture. Of the 36% of the Earth's land surface in agricultural production, 11% is used for cultivation and the remaining 25% is given over to grazing. The most established small-scale form of animal husbandry is the **infield–outfield** system. Before the development of modern intensive methods this was characteristic of highly settled village communities and has a long tradition in various parts of the world. Generally it operates on a roughly 10:1 ratio of outfield to infield. The outfield is relatively unproductive land that is unfenced, but used for grazing by domesticated livestock. The infield surrounds the settlement and is intensively cultivated, growing both food for the villagers and winter fodder, but primarily as meadows for high quality grazing.

Using this system, farmers can obtain a return from the outfield, exploiting its productivity and nutrients, though making little investment in its management. In contrast, cutting or mowing is central to the management of infield grasslands. While the livestock are grazing the extensive lands of the outfield, the hay meadows are allowed to grow ungrazed for several weeks. After their grasses have set seed, the standing crop is cut for hay.

As the vegetation dries, its seeds fall to the ground allowing new seedlings to establish themselves in the grassland sward. Meadows (and other grazed grasslands) can have a very rich flora – a square metre of chalk grassland can support more than 40 species of flowering plants. The clue to their diversity is the disturbance and release of nutrients that grazers facilitate. Grazers remove some of the biomass of the faster growing grasses, recycling its nutrient content in their faeces and urine. After the hay has been collected, livestock may be allowed to graze the infield and are then fed on the hay during the winter.

New methods of intensifying grass production are bringing about changes in these meadows. Old pastures are oversown with fine agricultural grasses, such as perennial rye grass (*Lolium perenne*), supported by the use of inorganic fertilizers. These outcompete the other species and result in a loss of the less competitive wild flowers that do not respond so well to the increased nutrients (section 4.3).

Even greater changes follow when the meadow is used for silage production. By mowing several times during the year, the farmer both collects the net production of the meadow and stimulates further leaf growth in its grasses. The clippings are then fermented to break down part of their cellulose, raising the nutritional value of the fodder and helping to reduce its bulk. Unfortunately, frequent cropping means that the grassland and its wild flowers get little chance to set seed and, inevitably, floral diversity declines as a result. In Britain alone, 95% of flower-rich hay meadows have been lost within the last 50 years (Plate 17), endangering a large number of wild flower species. Many temperate countries now have programmes to protect the flora of their meadows by promoting traditional grassland management.

5.5.1 Why grow animals?

The metabolic costs and variable efficiencies of each trophic level mean that energy is lost with each transfer along a food chain (Boxes 4.1 and 5.3). Why then do we not dispense with grazing land and its animals, and instead grow crops we could consume ourselves? Surely we would then recover more of the energy entering the system?

One simple answer is that, as omnivores, we do not have a physiology that can effectively assimilate the energy available in plant tissues. To make use of the primary productivity of a grassland human beings must rely on animals with the capacity to release its energy and convert into a form we can assimilate (Figure 5.16). Grazers represent an energy-efficient means of harvesting primary productivity: they range over many hectares of land, concentrating part of its energy in their tissues and in a form we can digest. This is particularly true of ruminants, who, with their symbiotic microorganisms, are able to digest cellulose, the major product of primary production which would otherwise be unavailable to us (and other carnivores: Box 5.5).

There are also sound ecological reasons why cropping is often not the most appropriate system. Much of today's pasture became grazed because it could not be used for anything else. Outfields and rangelands are frequently rough, stony places, with steep hills, deep valleys and poor soils. This makes them impractical and costly to cultivate and to maintain, so it is no accident of history that these areas have escaped cultivation.

Despite their limited structural complexity, grassland communities can be highly productive. Alpine meadows, for example, can produce as much as 1200 g/m^2 a year from each square metre and this rises to 1,500 g/m^2 for prairies. These figures are almost double that of crops grown for our consumption or for fodder (Table 5.1). Because our crops are in the soil for only part of the year, they cannot match the primary production of permanent pasture.

The soil beneath the sod is also better able to maintain its structure and decomposer community. With minimum disturbance, the grass sward provides year-round protection from severe weather and its roots bind topsoil, preventing erosion. The deposits left by grazers and the residues of the overlying vegetation help to raise its organic content, enabling the soil to hold water in its upper layers where it is of most use to plant life. Compared with other crops, grass has a more efficient water budget and there is a large range of grass species able to grow under all but the most extreme rainfall regimes. In contrast, the water requirements of crops can be considerable (Table 5.5) and they may require regular irrigation. In these situations, grasslands may be the only viable option.

Box 5.5

RUMINATION

Some herbivores have a symbiotic relationship which allows them to recover energy from cellulose which would otherwise be excreted. All animals have communities of bacteria in their gut, but ruminants (sheep, cattle, deer, antelope, gazelles and giraffe) have developed this association to exploit the powers of cellulose degraders. They are fermentation vessels on legs, operating a highly controlled continuous reaction as an anaerobic ecosystem within their gut.

The ruminant stomach is divided into four compartments. It contains between 100 million and 1000 million bacteria per millilitre and roughly half that number of protozoa, which live off the bacteria – a food web within the host.

The animal maintains a steady environment by its own body heat, by keeping the pH of the reaction vessel constant (by producing copious amounts of alkaline saliva) and by providing a supply of vegetation for the bacteria to work on. After it has been pre-processed by chewing, food is passed into the rumen for fermentation. From here it is regurgitated at regular intervals to be re-processed by ‘chewing the cud’. In this way, the ruminant breaks the vegetation into progressively smaller fragments, providing the bacteria with a greater surface area on which to work.

The microbes can be grouped into guilds that specialize in attacking particular components. Cellulolytic bacteria break down cellulose, whilst the amylolytic bacteria work on starch. Some are able to deal with both. The main products of the fermentation are volatile fatty acids, such as lactic, butyric, proprionic and acetic acids which the host is able to absorb across the rumen wall into the bloodstream. As with all anaerobic fermentation, the process produces large amounts of carbon dioxide and methane which the ruminant ‘vents’.

Proteins within the food are broken down into their constituent amino acids in the final chamber: the abomasum. This differs from the others since it is highly acidic, which means death to microbes that pass into it. Eventually, the bacteria themselves become part of the diet of the ruminant and represent an important part of its protein intake.

The energy costs of carrying and maintaining a large bacterial culture in an enlarged stomach are considerable (Figure 5.18). Forage, such as hay, can contain as much as 30% cellulose and, being able unlock an energy source denied to most other herbivores, offers a key advantage to the ruminant. The partners in the mutualistic association, the bacteria, benefit from the benign

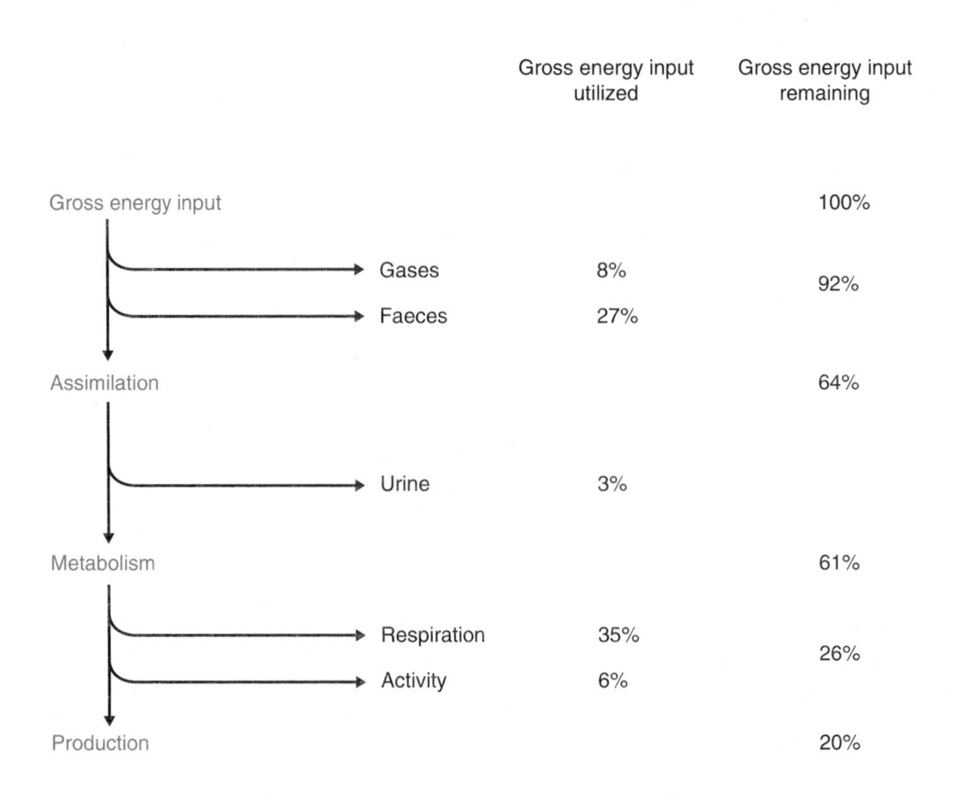

Figure 5.18 Energy budget of a ruminant, charting the fate of energy taken as food (gross energy intake). Losses occur at various stages in the process, with the largest being due to metabolic costs. The last column shows how much of the original energy is left at each stage. Eventually, 20% becomes fixed in the cow's tissues.

conditions in most of the animal's stomach.

There is a third, more recent beneficiary partner in this relationship: the many people for whom ruminants are the mainstay of their agriculture. Compared with pigs and poultry, which are able to convert 14% and 12% of their gross energy intake into production, ruminants are much more efficient (around 20%). Cattle benefit from our protection and husbandry and that has enabled them to become one of the most populous mammals on Earth. We, in turn, benefit from their capacity to convert low-grade plant material into protein and fat.

5.5.2 Getting the grazing right

In many areas, it is the grazing activity of the livestock that maintains it as pasture and prevents it from becoming scrub or woodland. To this extent, the grazers act as keystone species (section 3.5). However, as we saw earlier with the browsers of the African savannah, above a certain level their feeding can damage the system's capacity to recover.

Grazers are selective and will continue to graze their preferred food, even when it becomes scarce. This happened on the upland hill pastures of northern Britain. Hill-farming has always been a tough existence for farmer and stock alike. To keep sheep-farming viable (and prevent rural depopulation and loss of Britain's open moorland) financial incentives were used to support this upland grazing. Unfortunately, because of the way the incentives

Table 5.5 Water requirements of crops: amount needed to produce 1 g of dry weight

Crop	*Water (g)*
Rice	710
Potato	636
Oats	597
Wheat	513
Maize	368

were paid, this led to widespread overstocking of hill country.

Sheep prefer bent and fescue grasses (*Agrostis* and *Festuca* species) and these were the first species to decline with overgrazing. Mat grass (*Nardus stricta*), on the other hand, is avoided by the sheep. Tough and unpalatable, it forms thick patches, spreading readily over large areas where other grasses are absent. The deterioration in the grazing quality of the hill pastures was a cause for concern to not only the farming community but conservationists too. Hungry sheep started to overgraze heather. Heather plays a key role in the moorland ecosystem and loss of such an important primary producer had severe implications for the animals it supported further along the food chain. Once the problems were identified, measures were taken to adjust stocking densities to suit the carrying capacity of each area.

In some cases, undergrazing can be as much of a problem as overgrazing. As we saw earlier, moderate grazing controls and rejuvenates the plants within the sward, raising their productivity. It also helps to accelerate nutrient cycling within the ecosystem, promoting an active decomposer food web. Small patches of bare ground, dung heaps and other localized disturbances provide conditions for young seedlings to become established.

In the absence of grazing, changes start to occur. There is an increase in the standing crop, a gradual accumulation of biomass, and the community may then slowly start to revert to scrub and, possibly, woodland. Traditional agriculture therefore has techniques to compensate for low or no grazing. Most of these involve some form of biomass removal, principally cutting and burning. Farmers in the American Midwest now recognize that wildfire played a key role in the maintenance of the prairie ecosystem and they incorporate it into their management. In this way, they create localized disturbances to maintain the prairie's high floral diversity, preventing competitive species from becoming dominant. A wide range of herbivores, from buffalo to insects, are important in the turnover of organic matter between the standing vegetation and the soil.

5.5.3 Added energy

Agriculture uses energy to manipulate food chains and food webs. We try to prevent some species from using the energy entering the system and to ensure that most goes to those we are cultivating. We make an investment in time and effort, most easily measured as the energy applied to the system. This is termed its **energy subsidy**, and refers to everything from one person turning the soil with a hoe to the waves of combine harvesters that pass over our wheatfields (Figure 5.19).

We use energy to remove competition from non-crop plants by tilling or by the application of herbicides. Fossil fuels are used to manufacture and apply fertilizers, to side-step nutrient limitations on crop productivity. The energy we use to manufacture agricultural equipment, or the fuel used in farm machinery, all represent indirect costs, subsidizing our cultivation.

These inputs vary with farming practices. Figure 5.20 shows how types of energy input vary with different farming systems. Feedstuffs, for example, are a principle source of energy for intensive livestock

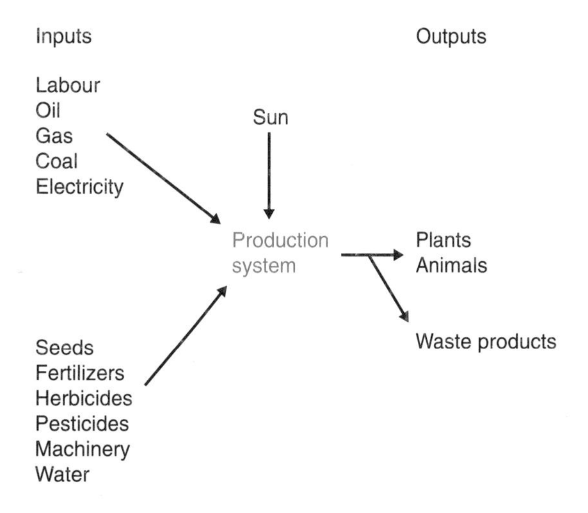

Figure 5.19 Energy subsidies can be direct energy inputs in the form of fuel, or indirect inputs from materials (e.g. fertilizers and pesticides).

production. In Table 5.6 we see the effects of energy input on the protein output of four agricultural systems, ranging from extensive farming through to intensive crop and animal production. (Note that the link between protein output and energy input is problematic since not all organisms will produce protein in the same proportion to their energy intake.)

The returns are greatest in intensive crop production where the food chain is shortest (chain I in Figure 5.16). The three other systems (chains II–IV) involve a third link – the livestock – and this leads to a reduction in the energy available for consumption.

Intensive animal production is costly, but it continues because it is a reliable and convenient means of producing protein. Not only is energy added with the food supply; animals being reared indoors often require a highly controlled internal environment, so that fuel is burnt to maintain lighting, heating and ventilation. Western dairy farming also demands a large energy subsidy in terms of handling the animals and the milk they produce.

Extensive grazing in the form of rangeland and pastoralism remains the most widespread means of producing animal protein. About 25% of the Earth's land surface is grazed under this system. Whilst protein returns are comparatively small, grazers can go where no plough can and are able to capture the productivity of land that might otherwise be unproductive. Sheep-farming incurs the lowest costs since the bulk of the sheep's grazing is confined to the extensive outfield or unmanaged pasture. These animals only need an occasional food supplement and the principal costs involve the management of periodic grazing within the infield.

The energy subsidy buys us out of some of the constraints which limit agricultural productivity, but not all of them. Eventually the costs start to outweigh the benefits and either returns on the investment or a deterioration in environmental quality follow.

Table 5.6 Energy inputs and protein yields of four major agricultural systems

Agricultural system	*Total energy input (10^6 kJ/ha)*	*Protein output (kg/ha)*
Hill farming (sheep)	0.6	1–1.5
Mixed farming	12–15	500
Intensive crop production	15–20	2000
Intensive animal production	40	300

Not only is the cost of the energy important, but also the way we use it. For example, phosphate fertilizers allow us to bypass the limits on productivity imposed by the low availability of phosphate in most soils (section 6.2). These fertilizers are manufactured from phosphate-rich rocks at considerable energy cost. However, a significant proportion of what we apply to soils does not fertilize the crop at all but instead is lost with run-off and can become a major cause of eutrophication (section 6.3). This is the unwanted enrichment of aquatic systems or soils and can lead to dramatic changes in the species composition of a community.

In a different way, simply adding more water to an ecosystem can cause long-term damage. Pumping groundwater, perhaps to irrigate crops more suited to wetter climates, may deplete underground reserves and, without due care, it may also disrupt the nutrient balance of the soil. This is especially a problem in arid areas (section 6.2), where high levels of evaporation cause salts to accumulate in the surface layers. Around a third of all irrigated land in India has been degraded by becoming too alkaline or saline.

Intensive cultivation can damage the soil in other ways. Frequent ploughing promotes the rapid breakdown (oxidation) of its organic matter. The soil texture and its capacity to hold nutrients and water depend upon its organic component. As this disappears, with it goes much of the fertility of the soil and its capacity to support a productive vegetation. Without substantial plant cover, soils are more mobile; and without a significant organic fraction, soil particles do not bind to each other. Vast amounts of soil are blown or washed away: some estimates suggest that the loss of topsoil amounts to about 1% of the world's cropland each year (section 9.5).

Around 80% of soil erosion is directly attributable to human activity and cultivation practices which are unsympathetic to local conditions. All farmers seek to achieve the highest levels of productivity consistent with being able to sustain this production over the long term (a principle we have met before in section 3.3). Many traditional practices have evolved to respond to the local conditions of climate, slope and soil properties. Techniques such as shifting cultivation may often be the best suited to habitats where nutrient supply and soil need time to recover. Unless we adopt cultivation techniques that are sympathetic to the local ecology, we can quickly create problems for ourselves in every part of the

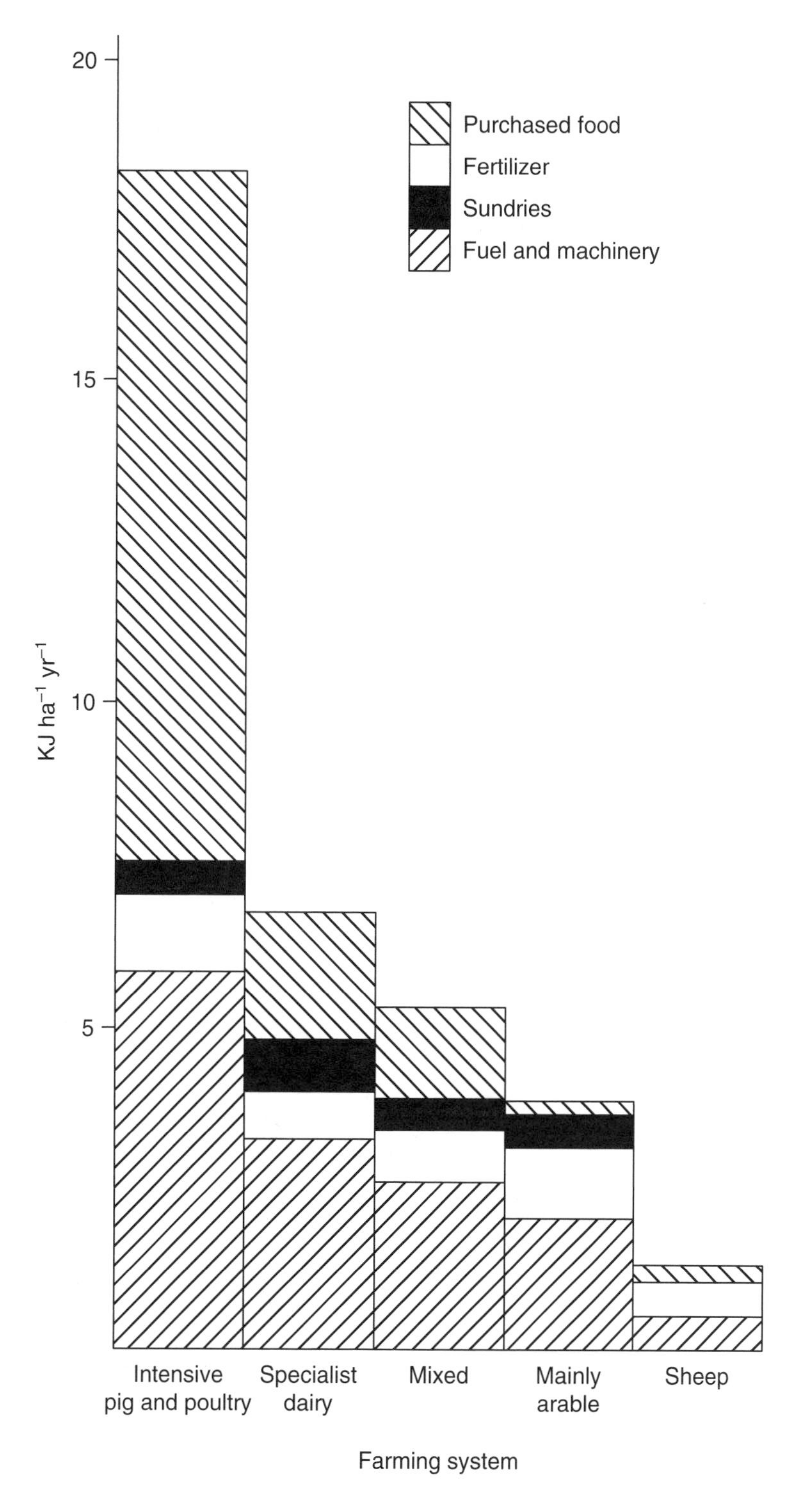

Figure 5.20 An energy subsidy breakdown for selected agricultural systems. The graph illustrates the main energy inputs for five major agricultural system, from intensive 'factory farming' of livestock, through to the extensive grazing of sheep.

world. Within a few seasons, a good quality soil can be reduced to little more than a poorly structured dust lost with the first rains or high wind.

Agricultural productivity is so central to our society that it has always been a focus for technology and innovation. Following the dramatic successes raising productivity with cheap artificial fertilizers in the 1940s and 1950s, high hopes were held for solving food shortages in the 1960s and 1970s. Improvements in farm management techniques, plant and animal breeding programmes to improve strains and advances to combat pests and disease amongst our cultivated species, all seemed to promise a way of feeding a rapidly growing world population.

The **'green revolution'**, as it became known, hinged on the development of new strains of crop plants. These were supported by a range of agrochemicals, pesticides and fertilizers and, in some places, new irrigation schemes. Cultivars were produced that allowed for easier mechanical harvesting and, most importantly, increased productivity. High-yielding varieties were selected for particular characteristics such as high leaf:stem and low carbon:nitrogen ratios; that is, they were bred to produce as many leaves and with as little stem as possible so as to pack protein into leaves, fruits or seeds. High-yielding varieties of rice could divert a massive 80% of net production into their seeds, compared with the more normal 20%.

However, the promise of many of these new strains rarely materialized. Crops which produced high yields in experimental trials failed to live up to expectations once out in the field. Sometimes the reason was blindingly obvious: strains bred by developed countries to help feed developing countries ended up being grown without western technology (or wealth) to support them. Local farmers found that the energy subsidy in terms of fuel, machinery, fertilizers and pesticides was too costly.

Important lessons were learnt from the mistakes of the past, and today's breeding programmes have adopted a more ecologically sound approach that aims to fit crop varieties and cultivation techniques to the environment in which they will be grown. Applying ecological principles to agricultural systems enables us to regard agricultural land as a habitat like any other and to compare it with other, more natural communities (Figure 5.21). In the process, we have acquired a greater respect for local traditions and practices, from established patterns of crop rotation to pest control methods. Increasingly, technologies are being used that are more suited to local economies and ecologies.

Many people see land degradation, and its significance for food and fuel production in developing nations, as the major problem facing the globe over the next 50 years. Over the last 50 years, as much as 9 million hectares have become severely degraded, mainly in overpopulated areas where most people make their living from the land. Perhaps the most significant development at the United Nations Convention on Environment and Development – the Rio Convention of 1992 – was Agenda 21, and its 'nonbinding blueprint for sustainable development'. This, at least, was recognition that future economic development had to be sustainable – both agricultural and industrial development. Although it will inevitably be interpreted very loosely, the Commission on Sustainable Development set up by UNCED will enable us to gauge progress. In the end, the test of sustainability will happen at the level of the ecosystem: whether local agricultural practices maintain productivity without signs of increased species loss or soil erosion (Plate 18).

SUMMARY

Primary producers convert sunlight into chemical energy through the process of photosynthesis, absorbing light with the pigment chlorophyll and

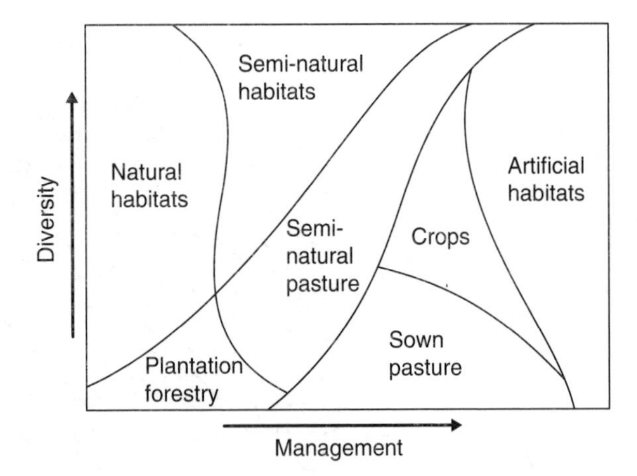

Figure 5.21 Ecosystems: a template of diversity vs. management, showing that natural, managed (semi-natural) and artificial ecosystems form an intricate jigsaw when viewed in terms of their diversity and scale of management. The template relates both of these to a range of ecosystems.

Plate 12 Meercats, like many communal animals, share the task of looking out for predators. One individual literally stands guard while the others forage outside their underground burrow.

Plate 13 Horns serve several purposes. Their size and shape are primarily determined by the battle between males for a female, in attack and defence. They are also an indication of the status of the male in the hierarchy of the herd, recognized by both the males and females. Pictured above are musk oxen (*Ovibos moschatus*) in a head to head confrontation.

Plate 14 A typical garden lawn.

Plate 15 Chickweed (*Stellaria media*), a weed of cultivation.

Plate 16 *Cistanche*, a close relative of the broomrape, has no chlorophyll of it own and is totally dependent on its host plant.

Plate 17 Well-managed meadows can be both productive and diverse. This traditional meadow in the south east of England has been managed for centuries by grazing and cutting and supports a wide range of plants including, in this particular case, green winged orchid (*Orchis morio*).

Plate 18 Monoculture: the familiar face of intensive agriculture.

Plate 19 The damaging effect of acid rain. Acid deposition causes direct damage to vegetation and also makes long-lived species, such as the conifers in this eastern European forest, more susceptible to drought, physiological disorders and disease.

Plate 20 Carnivorous plants like this *Sarracenia* have adapted to life in the nutrient-poor conditions of swamps and bogs by obtaining additional nitrogen from insects trapped within the pitcher-like body of the plant.

Plate 21 Seabirds like this male eider *Somateria mollisima* are among the first casualties of marine oil-spills.

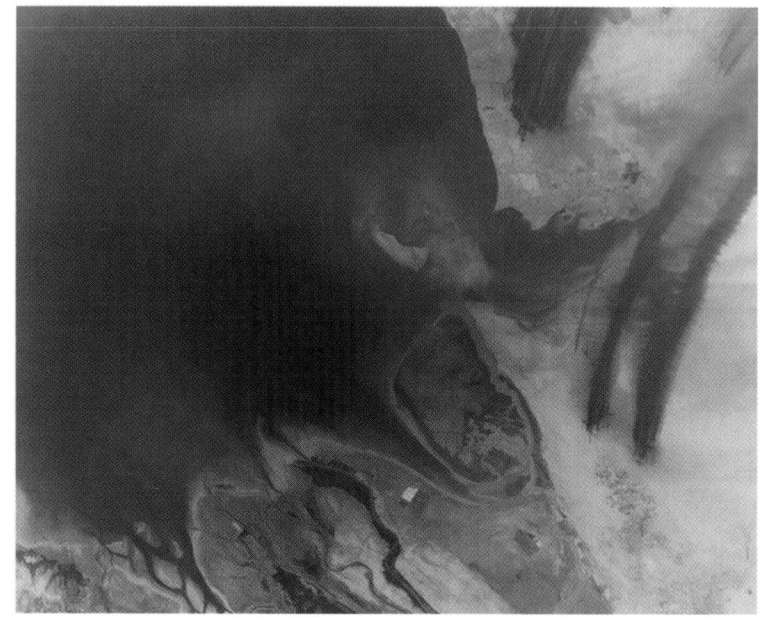

Plate 22 Burning Kuwaiti oilfields sabotaged during the 1991 Gulf War, as seen from the Space Shuttle.

Plate 23 Restoration in progress. Vegetation being used to stabilize chalk spoil from the Channel Tunnel.

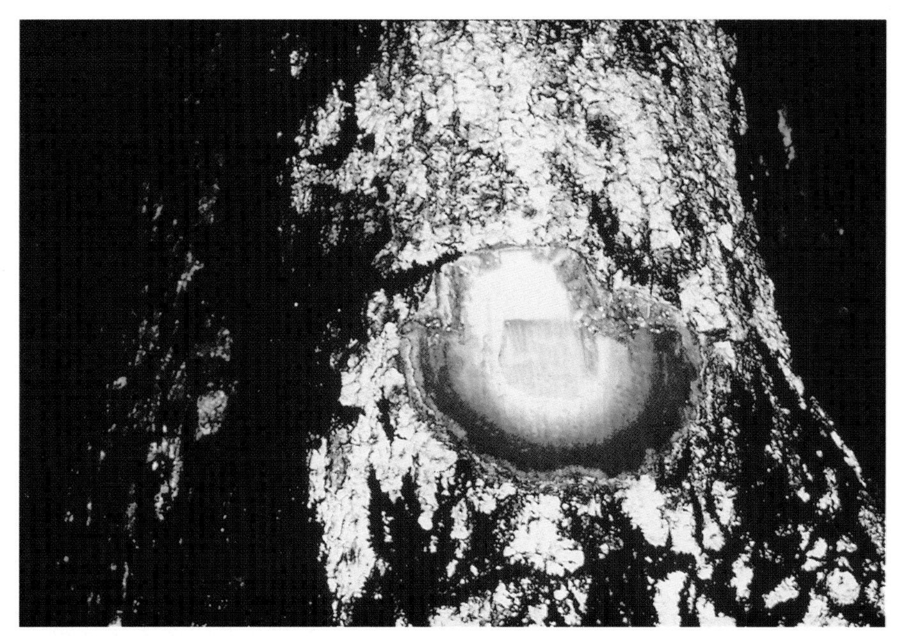

Plate 24 The New Caledonian nickel tree *Sebertia acuminata* grows on nickel-rich soils overlying serpentine rock and secretes a blue-green nickel-rich latex.

Plate 25 Heathland translocation in action. When it comes to habitat translocation, such as this patch of the West Pennine Moors in Lancashire, speed and accuracy are required to remove and replant the plant material with the minimum of damage.

using the energy to fuel chemical reactions which convert carbon dioxide and water into sugars with oxygen as a by-product. The overall efficiency of the process is around 2%. Some plants (termed C_4 and CAM) have modified the way in which they capture carbon dioxide to help conserve water in dry environments.

Much of the energy fixed in gross primary production (GPP) is expended in respiration used in the normal metabolic processes of cells and tissues. That which becomes fixed into the tissues represents net primary production (NPP). Herbivores consume this primary production, and begin a line of consumers that form a trophic level or food chain. A food chain marks one pathway through which energy moves through a community. All of the possible energetic pathways of a community constitute a food web. Within a web, organisms are classified according to their position in a food chain, the trophic level upon which they feed. Carnivores feed on herbivores and top carnivores feed on other carnivores. Current evidence suggests that most food chains are rarely more than four or five links long. This may be because they are limited by the inefficiency of the energy transfers from one trophic level to another, but evidence suggests that it is more closely tied to the structure of the food web. This raises questions of how closely linked food webs are and the extent to which we can regard the community as a functional unit.

Human beings make use of the systems associated with primary production. A range of agricultural strategies have been developed, and some of the most ancient, from hunter-gathering to transhumance, are still practised in different parts of the world. Although our agriculture seeks to improve the productivity of an ecosystem by applying energy subsidies (to remove competitors, supply nutrients and so on), these only produce sustainable systems where they are sympathetic to local ecology. In the last 50 years we have learnt valuable lessons from our attempts to manipulate the energetics of agricultural ecosystems.

FURTHER READING

Salisbury, F.B. and Ross, C.W. (1985) *Plant Physiology* (3rd edn), Wadsworth, Belmont, California.

Tivy, J. (1990) *Agricultural Ecology*, Longman, Harlow.

Whittaker, R.H. (1975) *Communities and Ecosystems* (2nd edn), Macmillan, New York.

EXERCISES

1. Complete the following paragraph by inserting the appropriate words, using the list below (some words may be used more than once, some not at all).

   Photosynthesis converts __________ into __________ energy. It does this through the action of the light-absorbing pigment known as __________ which is found in the __________ membranes within the chloroplast. The light reaction is the first part of photosynthesis and is used to fuel the __________ cycle, which is part of the dark reaction and which fixes __________ and manufactures __________. Some plants, known as __________ plants, have evolved alternative mechanisms to help them fix carbon more efficiently to avoid __________ loss. Others have developed this still further using __________ __________ metabolism so that they do not have to open their __________ in the heat of the day.

   C_3, C_4, Calvin–Benson, carbon, carbonic acid, chemical, chlorophyll, Crassulacean acid, flowers, kranze anatomy, leaves, nitrogen, respiration, oxygen, stomata, sunlight, sugars, thylakoid.

2. In terms of net primary productivity, the most productive ecosystem on Earth is:
   (a) the tropical rainforest
   (b) the open ocean
   (c) intensively cultivated areas
   (d) algal beds and reefs
   (e) swamps and marshes.

Table Ex.5.1

Organism	*Type of feeder*	*Diet*
Aphid	Herbivore	Live plant material
Blackbird	Omnivore	Worms, slugs, caterpillars, spider, centipedes and plants
Caterpillar	Herbivore	Leaves
Centipede	Carnivore	Millipedes, aphids, woodlice and springtails
Earthworm	Detritivore	Detritus
Ladybird	Carnivore	Aphids
Millipede	Detritivore	Detritus
Slug	Herbivore	Leaves, stems and roots
Spider	Carnivore	Springtails and caterpillars
Springtail	Herbivore/ detritivore	Living and dead plant material
Woodlouse	Detritivore	Detritus

3. Construct a simplified food web of a garden lawn ecosystem using the information provided in Table Ex.5.1 about the feeding preferences of common garden inhabitants.

4. Look at Table 5.5, which shows the water requirements for five of the world's staple crops. One of these has a substantially lower requirement than the rest. Which one is it and why might it be more efficient in its use of water?

Tutorial or seminar topic

5. What constitutes agricultural sustainability?

The hydrological and major nutrient cycles – the role of living systems in nutrient cycling – microorganisms and fungi in nutrient release and uptake – pollution as nutrient imbalances – eutrophication – marine oil pollution – restoration ecology – re-establishing self-sustaining ecosystems.

6 Balances

Geologists delight in describing soil as 'rock on its way to the sea'. This is not just evidence of a wry sense of humour but a very accurate picture of one of the long-term processes of the planet. Rocks will inevitably abrade through the action of wind, water and ice. Their particles, though briefly held in soils, eventually find their way to the bottom of seas, perhaps to be resurrected as new sedimentary rocks. Some of their elements may be released into solution and be diverted, through plants and microorganisms, into other parts of the biosphere (Figure 6.1).

In the last chapter we saw the central role that energy plays in driving living systems and how it fuels metabolism and the building of new tissues. For this to happen the raw materials have first to be acquired. Elements such as carbon, hydrogen, oxygen, nitrogen and phosphorus are needed to make the macromolecules upon which all life is built. For the Plant Kingdom, the only source of these nutrients is minerals and for most animals their supply ultimately comes from plants. Without life, elements move in geochemical cycles of erosion and deposition. With life, they move through biogeochemical cycles on their way to the sea.

You will remember from the Prologue that Biosphere 2 tried to mimic these global biogeochemical cycles, but failed to match their planetary fluxes. As we shall see, the global flux of nutrients is largely determined by the energy available to drive key processes, both biotic and abiotic. Energy captured in photosynthesis ultimately powers the metabolic processes that capture the nitrogen, phosphorus and other components needed

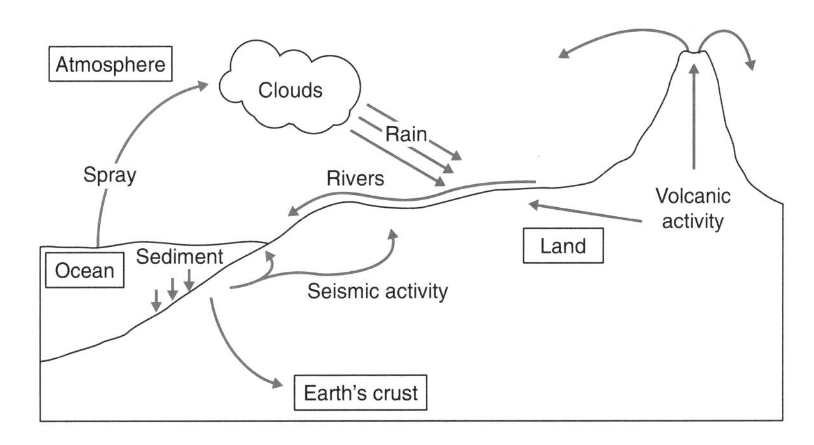

Figure 6.1 Major geochemical processes. Minerals move between land, sea and the atmosphere. Whilst some elements may be lost to the Earth's crust, others are returned to the ecosystem by volcanic and seismic activity.

to build organisms, primarily through the activity of microorganisms within the soil or sediments and through plant uptake. However, this cuts both ways – nutrient shortages may in turn limit primary productivity. Less energy will then be available to fuel nutrient movements through the rest of the community.

Our understanding of how metabolism and nutrient capture interact is central to our attempts at restoring self-sustaining communities. On the one hand, excess nutrients can lead to changes in the structure of a community and later we shall examine some examples. But we shall also see how some degraded habitats, short of nutrients, may need help to re-establish the complex machinery of their biogeochemical cycles.

6.1 NUTRIENTS AS LIMITING FACTORS

We can think of life on Earth as being something like a production line, building new tissues and new organisms, with the energy for the conveyor belt coming from the abundant radiation the planet receives from the sun. However, a number of factors limit or regulate this ecological machine.

Even if energy is plentiful and conditions are equable, the speed at which the production line can work is determined by the availability of key components: how much and how quickly each can be supplied. These elements are needed to construct the molecules required for living tissues. Without them, production grinds to a halt or proceeds very slowly. Conditions that are not equable – not warm and wet – will also slow ecological processes. Polar regions are bathed in a weak sunlight, adequate for photosynthesis but the low temperatures reduce production and decomposition considerably. In the taiga or northern forests (section 8.2), nutrients, such as calcium, cycle through the soil and trees in about 43 years, compared with an average of just 10.5 years in tropical rainforests. The taiga is therefore operating at a quarter of the speed of its tropical counterpart.

These rates have immense economic significance. For example, the nutrient capital of an undisturbed tropical forest is held in the standing crop, its vegetation and animal life, with only a small fraction in the thin tropical soil. For centuries, the indigenous peoples of these forests have worked with these nutrients cycles, establishing agricultural practices and traditions that limited their land use according to the nutrient budget of the forest – never clearing too much too quickly and regulating the size of their own group by social conventions. As a result, the forest recovers quickly when these people move on.

More recent colonizers have ignored the ecological reality at considerable cost. Clear-felling the forest or burning large areas and then applying western agricultural practices may produce a bumper harvest for one or two years, but thereafter yields drop. The small nutrient capital of the soil has been depleted and farmers then have to apply inorganic fertilizers to maintain economic returns. The nutrient capital of the forest was lost when the above-ground community was destroyed.

Tropical grasslands (section 8.2) also show how nutrient supply regulates entire ecosystems. Working in the Serengeti, Sam McNaughton found that the availability of phosphorus, sodium and magnesium indirectly controlled the density of wildebeest. Areas where the vegetation had higher levels of these elements were able to support greater numbers of large herbivores. Like all living organisms, the wildebeest act as nutrient pools accumulating nutrients, redistributing some as excretory products, passing some on to the future as young and surrendering the balance to the environment on death – either directly through decomposition or indirectly via carnivores and scavengers.

6.2 THE NUTRIENT CYCLES OF LIFE

Different elements move through the biosphere at different rates. These are known as their **flux rates** and their speed depends upon the physical and chemical properties of each element, and the use to which living organisms put them (Figure 6.2). We can classify nutrient cycles by the spatial scale over which they move. Phosphorus tends to move over short distances, usually within a very localized cycle, primarily because of the insolubility of its compounds. Nitrogen is far more mobile because its salts are highly soluble and these are rapidly taken up by plants or leached out by percolating groundwaters. Microbial processes also release nitrogen back into the atmosphere (Figure 6.9).

The global nutrient cycles include those elements which have a gaseous or aqueous phase. Conversely, sedimentary deposits tend to be very localized, often concentrated in particular locations

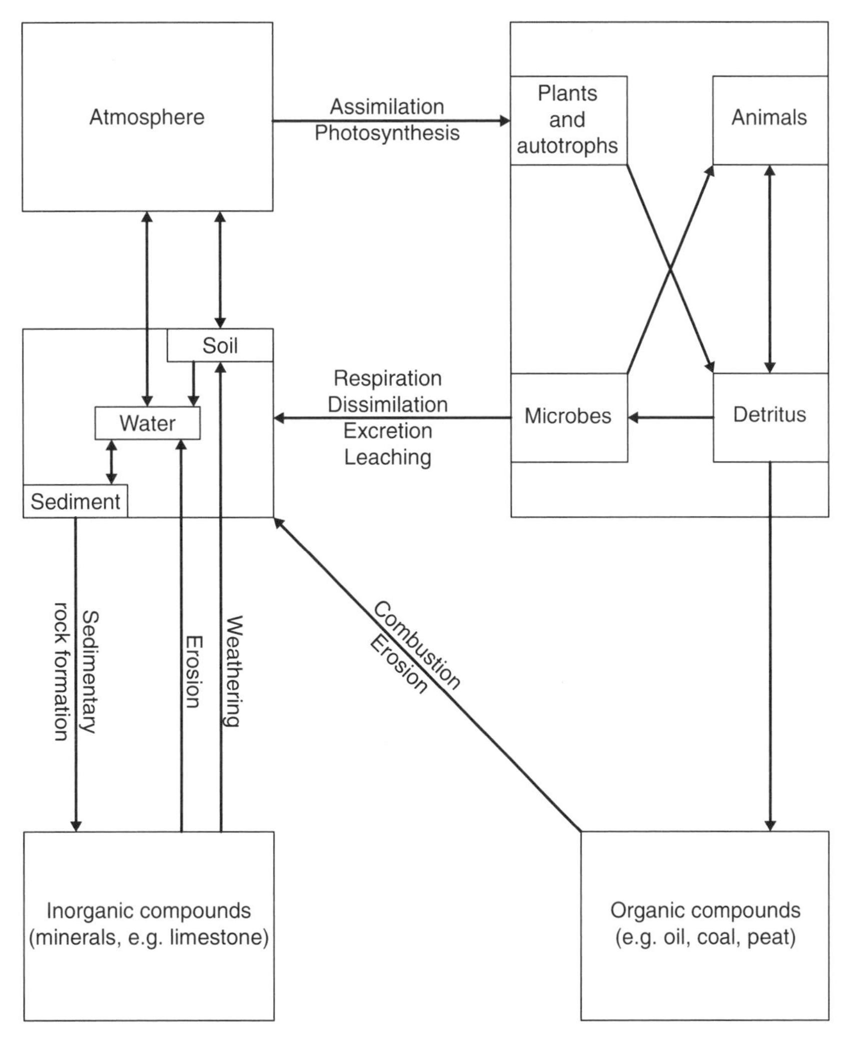

Figure 6.2 The ecosystem as a series of compartments. Minerals move between the atmosphere and organic or inorganic reserves via living and non-living systems. Movement can be either one-way or two-way depending on the element concerned and the nature of the compartment.

by living organisms. The nutrients in these deposits form a reservoir or pool, though it may be tens of millions of years before they return to the biosphere, through erosion or human activity. Some plant and animal remains enter the soil or aquatic sediments to form a labile (changeable) fraction and their nutrients recycle at much faster rates, especially if these deposits are regularly disturbed, say by an ocean current.

Water plays a key role in liberating and moving many of the important nutrients, but is itself driven through a cycle – the **hydrological cycle** (Figure 6.3) – by solar power. Water enters the atmosphere by evaporation and from the transpiration of plants, both of which depend upon solar energy as heat and radiation. The movement of water is one of the main components of the engine that disperses heat around the Earth (Box 8.1). High inputs of energy in the tropical regions raise temperatures and vapour pressures, forcing the moist air to higher altitudes or latitudes, where it cools and its water vapour condenses. The differential heating of land and sea create regular patterns of air movements (Figure 8.5), so that winds sweep moist air on to land, where it cools and its water precipitates as rain or snow.

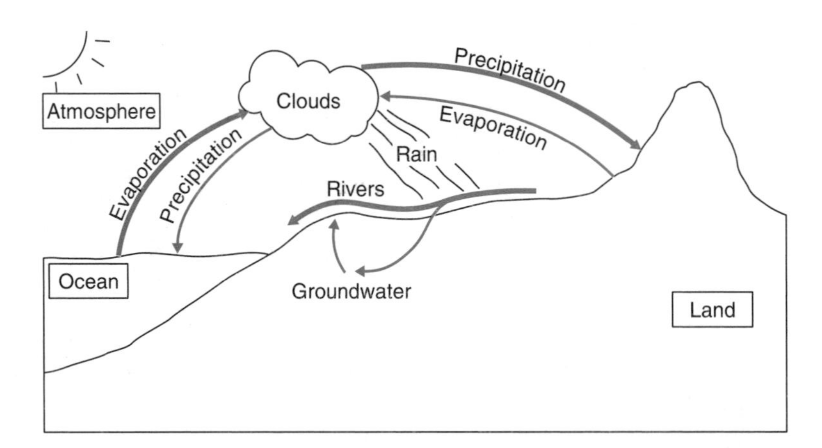

Figure 6.3 The hydrological cycle. Water evaporates from the sea and is deposited on land, where it enters the ground and eventually returns to the oceans via rivers. The thickness of the lines denotes the relative amounts of water moving from one compartment to another.

Bands of precipitation may also form where cold and warm air masses regularly meet. Patterns of precipitation also depend on distance from the sea and topography (the form of the land). Continental interiors, distant from any large mass of water, are typically deprived of rainfall. Similarly, land on the leeward side of mountains may receive very little precipitation because the incoming air has shed its moisture when it was forced to rise and cool over the mountain range. The area is said to be in a 'rain shadow'.

An estimated 97% of the planet's water is held in the oceans with just 0.001% in the atmosphere at any one time. Less than 1% occurs on land, in groundwater or soil-water or as rivers and lakes. The rest (around 2%) is frozen in polar and glacier ice. Much of the water we use comes from streams and rivers, or from groundwaters – that is, aquifers of water-laden rocks. Humankind has developed a variety of technologies for exploiting aquifers, using natural springs or bore-holes to draw up water from deep rocks. When our abstraction exceeds the natural recharge rate the aquifer starts to be depleted. Our increasing dependence on deep water – 'fossil water' that fell as rain many centuries ago – means that we are today reducing our major aquifers and can no longer afford to regard them as a renewable resource.

But the problem does not end there. Even if water is plentiful, its overuse can precipitate ecological disasters. This is especially true in arid and semi-arid areas where evaporation exceeds precipitation. Under these conditions, applying excessive water through irrigation systems can draw salts to the surface, where they crystallize as a saline crust. The capacity of water to dissolve and displace nutrients leads to salinization – the build-up of some elements (especially sodium and magnesium) to concentrations that are toxic to plants and microorganisms, rendering the soil useless for agriculture. This can happen very rapidly, especially when we use modern technologies to irrigate large areas without due regard for their natural ecology: in section 8.1 we describe the recent example of the Aral Sea. However, this is not a new problem. Archaeological evidence suggests that a number of ancient civilizations declined because their soils became saline through mismanagement (section 7.1).

6.2.1 The principal nutrients

Carbon

The patterns of nutrient cycling and the proportion fixed in their reservoirs change with each element and with time. The carbon content of the Earth's atmosphere has been far from stable over geological time. At times in the past, most notably the Silurian era, high atmospheric concentrations of carbon dioxide have promoted high primary productivity and this eventually translated into the thick deposits of carbonates and hydrocarbons of the Carboniferous era. As we shall see later (section 8.3), the rate at which carbon cycles, and its residence time in the atmosphere, regulates the planet's temperature.

Carbon makes life on Earth possible. Without its capacity to form molecular chains and rings, life would not exist. Dry a human being, or most other living things, and almost half of the weight will be carbon. Marine plankton and the major forests are the principal routes by which carbon is removed from the atmosphere and also added to reserves in the rocks and the sediments (Figure 8.15). A large proportion is locked away as calcium carbonate in the shells and skeletons of marine organisms, and is eventually compressed into chalk and limestone. Elsewhere, the remains of the long-dead exist as hydrocarbons, the gases and oils used to fuel today's industrial societies. Closer to the surface, the more recently dead plants and animals form a sizeable fraction as soil organic matter and peat. Overall, perhaps as much as 40 000 times the size of the atmospheric pool is fixed in the oceans and these crustal deposits.

Ultimately, all the carbon of living systems is derived from the atmosphere, which is surprising when we consider that it only exists here at very low concentrations – a mere 0.03% for its most abundant gas, carbon dioxide. Carbon fixed in sedimentary rocks has a slow turnover time, taking perhaps 100 million years before it re-enters the biosphere. Unfortunately, the rate at which we have been releasing carbon from these sinks over the last 200 years has led to changes to not only the carbon balance of the atmosphere, but also its energy budget (section 8.3).

The impact of the planet's photosynthesis on global atmospheric concentrations of carbon dioxide leads to a distinct twice yearly pulse of around five parts per million. This planetary equivalent of inhalation and exhalation is shown in Figure 2 of the Prologue and is due to the disparity in the land area between the northern and southern hemispheres. Terrestrial processes in the north dominate the pattern, so that carbon dioxide levels are low when its photosynthesis is high, but rise during its winter when its respiration exceeds primary production.

Microbial breakdown of the organic matter in the soil is one major source of carbon dioxide for the air. Ploughing and forest clearance expose the lower soil layers to the oxidizing atmosphere and provide soil bacteria with the oxygen needed for aerobic respiration. Overall, agricultural methods based on regular soil disturbance and short-lived crops serve to deplete the organic content of the soil and accelerate the generation of carbon dioxide.

Oxygen

With the advent of photosynthesis about 2.5 billion years ago, the increasing abundance of oxygen in the atmosphere selected physiologies which were adapted to its corrosive presence (section 1.5). Now, all higher organisms respire aerobically – that is, they use the capacity of oxygen to give up electrons and to bind with hydrogen to pull apart carbon molecules which, in a controlled way, can release their stored energy for metabolism. For this reason, the oxygen content of the atmosphere is closely linked to the biological activity of the planet, both in its production by photosynthesis and its depletion by respiration. As a result its movement is tightly linked to the carbon cycle. However, because oxygen forms such a large proportion of the atmosphere (21%) we do not detect annual shifts in its concentration, only changes through geological time.

Oxygen is also tightly linked to the other nutrient cycles of the planet. Where anaerobic (oxygen-poor) conditions dominate, some bacteria are able to use sulphate and nitrate to respire. As they do this, they release carbon dioxide, so liberating the oxygen in these compounds back into the atmosphere.

Sulphur

The fire and brimstone of the sulphur cycle makes it a truly elemental cycle – linking earth, air, fire and water (Figure 6.4). Significant amounts are added to the atmosphere from non-crustal sources, through volcanic eruptions, while sediments and the soil are again the main reservoir. Sulphur forms only a small fraction of living tissues, but plays a crucial role in regulating the structure of many amino acids and proteins. In the atmosphere its compounds play several important roles: dimethyl sulphide, produced by marine phytoplankton, functions as atmospheric nuclei around which water droplets condense during cloud formation.

Sulphur is also released as hydrogen sulphide through the action of sulphate-reducing bacteria in the anaerobic conditions of waterlogged soils. Each year, around 100 million tonnes of this gas are produced from soils, muds and sediments by species such as *Desulfotomaculo* and *Desulfovibrio*. Anaerobic muds are characteristically stained black by the reduced iron sulphide and give off the rotten-egg smell of hydrogen sulphide. Other microorganisms from the upper oxygenated layers will oxidize sulphur compounds back to sulphates (for example, *Thiobacillus*). These chemoautotrophs are using these inorganic compounds as their source of energy

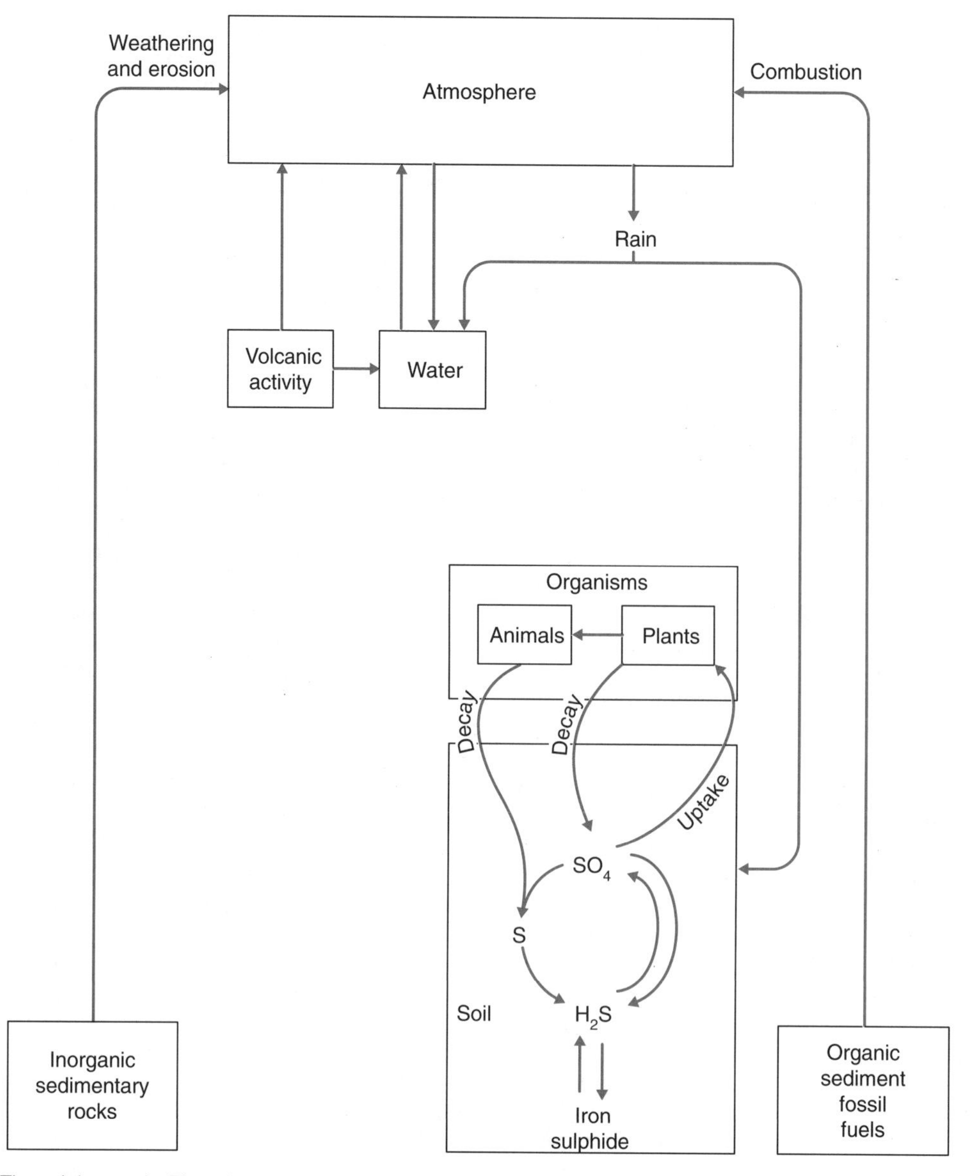

Figure 6.4 The sulphur cycle. The primary sources are releases from sediments and volcanic eruptions. The burning of fossil fuel has added considerably to localized sulphur pollution and the problem of acid deposition.

instead of organic (carbon) compounds and therefore they function without the intervention of photosynthesis.

The concentration of sulphur in the air has been increased on a local scale with the burning of fossil fuels, particularly coals rich in sulphate minerals. Although largely confined to industrial areas, sulphur dioxide may travel hundreds of miles from its source, contributing to the regional problems of acid rain (Box 6.1). Sulphate aerosols are known to be important in the energy balance of the atmosphere, producing regional cooling by their effects of cloud formation near their source (section 8.3).

Phosphorus

Phosphorus is an earthbound element (Figure 6.5) lacking any significant atmospheric component. Geological deposits (mainly in the form of apatite,

Box 6.1

ACID RAIN

For a long time, the problem of acid rain was thought to be largely a consequence of the sulphur-rich gases produced by burning some fossil fuels (especially poor quality coals) and the smelting of metallic ores. We now know that this picture is far too simple. Acid rain should be more properly described as acid deposition (it consists of both wet and dry deposition) and its effects are not attributable to sulphur alone.

The other significant element is nitrogen and especially the nitrogen oxides (NO_x) produced by the internal combustion engine. NO_x now account for 30% of the acid depositon in Europe. With the reduction in sulphate emissions from power stations over the last two decades, NO_x have become increasingly important. Acid deposition is a regional pollution problem and needs international agreements to limit atmospheric concentrations. Uncontrolled, the emissions from power stations and motor vehicles can cause considerable damage to forests, lakes and streams. Not only does acid deposition add large amounts of sulphur and nitrogen to these systems, upsetting their nutrient balance; the fall in pH also increases the availability of some key toxic metals.

Rainfall is naturally acidic because it dissolves carbon dioxide in its descent through the air. However, an increase in the acidity of rain in Britain was noted within 100 years of the start of the Industrial Revolution. Today about two-thirds of the acid deposited over Britain is dry deposition. Wet deposition occurs when SO_x and NO_x compounds are resident in the air long enough to combine with moisture to form dilute sulphuric and nitric acids. This typically produces a rainfall around 1 pH unit more acidic (a 10-fold increase in its concentration of reactive hydrogen ions). It can be much more than this.

Nitrogen and sulphur inputs have a fertilizing effect in some ecosystems. Nitrogen inputs from the air in Europe represent a 500% increase over background, leading to blooms in some plant communities, in areas where the nutrients are in short supply. Unfortunately, while the plants may begin to flourish, other changes soon follow. Many plants growing on sandy or chalky soils rely on rainfall as their principle source of nutrients (**ombrotrophs**) and are thus adapted to low nutrient levels. Abundant nitrogen means other species can invade and out-compete the native flora. Nutrient-poor grasslands and heathland may change – especially those near major sources of nitrogen, such as densely trafficked roads.

Acid rain damages ecosystems not only because of the nutrients it adds, but also due to its acidity, the reactive hydrogen ion input. Generally, the acidification of the soil mobilizes soil nutrients, promoting an initial surge of growth. However, this acidity displaces key nutrients and cations (calcium and magnesium) which may then become limiting to further growth. Nitrogen oxides entering through plant stomata can have more direct effects on plants, creating localized acidity and damaging leaves and new shoots (Plate 19).

The flush of nutrients will eventually find its way into water courses and cause algal blooms, comparable to the eutrophication described below. One of the longer term effects of acidification is to increase the availability of several metals, especially aluminium, which many plants and animals find toxic in significant concentrations. Scandinavian lakes have suffered particularly from acidification. The predominant wind directions means that this area receives much of the atmospheric pollution of north-western and central Europe. The region is dominated by coniferous forests and thin soils, with little capacity to buffer the incoming acidity. The result has been a large number of lakes becoming devoid of fish since the 1960s and much of the forests scorched and dying back. Sweden, Finland and Norway have had to resort to adding large amounts of lime to these ecosystems to try to protect or restore them.

a calcium salt) account for a global resource of 2000 billion tonnes, but its rate of release by weathering and erosion is very slow (just 100 million tonnes per year – less than one 500 000th of the total reserve).

The sea is the richest source of available phosphate. Indeed, phosphate-rich rocks tend to be of marine origin. However, the main route back to land for phosphate is through seabirds, feeding on phosphate-rich fish, then roosting and defecating on

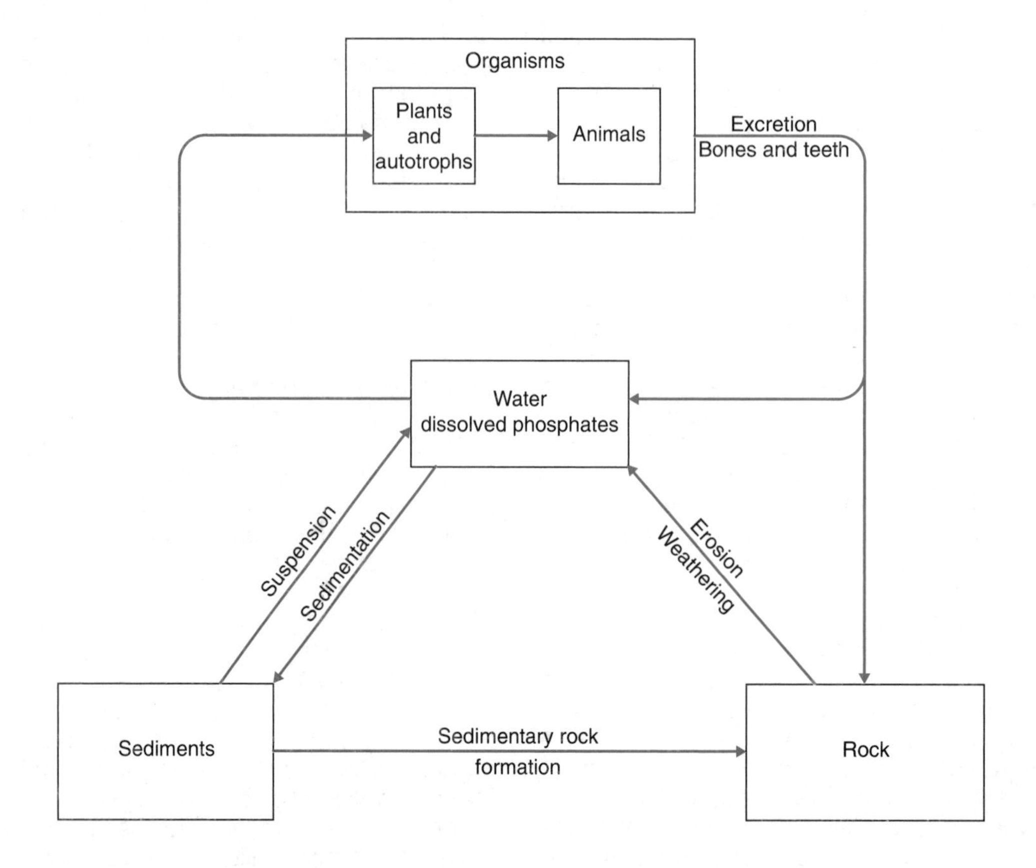

Figure 6.5 The phosphorus cycle. Phosphorus occurs mainly in the form of phosphates – insoluble compounds which are slowly weathered from rocks or resuspended from living organisms. This cycle differs from most of the others as it lacks a significant atmospheric phase.

land. Their deposits form guano, which in some locations has accumulated to a considerable depth, so much so that various deposits around the world are commercially extracted as fertilizers.

Plants and animals hang on to phosphate and concentrate it to many times its level in the surrounding environment. It enters the food chain primarily through plants and phytoplankton. An atom may remain in an animal for many years and within a tree for centuries. On death, it can be recycled into the living system or become immobilized in the soil. Here it might remain for many thousands of years, eventually finding its way into the sea through erosion. There, if it is liberated from the sediments, it will begin the cycle all over again.

Large amounts of iron sulphide in a soil will rapidly lead to the immobilization of phosphorus by forming insoluble compounds. Insolubility is the fate of most free phosphorus and the reason why this element is the main limiting nutrient in many ecosystems. Phosphorus is essential for proteins, nucleic acids, cell membranes, teeth and bones, but under normal conditions it is nearly always in short supply. It plays a central role in many cellular processes, not least because it is the key component in the currency of biological energy transfer – adenosine triphosphate (ATP).

As before, microorganisms and fungi also play an important role in maintaining the circulation of this most sought-after nutrient. A number of plant groups have got in on the act by forming mutualistic (symbiotic) associations (section 4.2) with fungi to secure additional phosphorus. In these associations, known as **mycorrhizae** (*myco-*, fungus; *rhiza*, root), the fungus either grows around the plant root (ectomycorrhiza) or around and within it (arbuscular mycorrhizal fungi, or AMF). In the latter the fungus penetrates the root, forming structures within it (Figure 6.6). The plant supplies sugars and other nutrients to the fungus in exchange for the ready supply of phosphorus scavenged by the fungal network extending beyond the root. The fungus also helps by supplying moisture, conferring drought resistance to the host plant. The importance of these associations has been highlighted in the development of restoration ecology (section 6.5), since we now know they are essential for the establishment of some plant communities. Much remains to be discovered about mycorrhizae, particularly the hidden costs and benefits behind some specialized associations where a single plant species has its own particular strain of fungi.

Nitrogen

Another well-known mutualistic association in the soil is formed to capture the other major plant nutrient, nitrogen. The flux of nitrogen again exerts considerable control on ecosystem productivity and processes (Figure 6.7). After carbon and oxygen, nitrogen is the third most abundant element in biological molecules; it is a major component of proteins and nucleic acids. But it is so difficult to coax reactive nitrogen out of the atmosphere that some plants have again resorted to symbiosis, this time with some very ancient genera of bacteria and actinomycetes, collectively known as the nitrogen fixers (Box 6.2).

The bulk of the atmosphere (around 78%) is nitrogen and it is from this reserve that living systems have to derive sufficient to support their production of proteins. Yet atmospheric nitrogen is largely in the molecular form (N_2), which, with its three shared electrons, is difficult to split apart. Thunderstorms and the massive discharges of energy they produce can do this, though only relatively small amounts of nitrate enter the soil in this way. Even so, most organisms cannot afford the energetic cost of splitting the nitrogen molecule and only a few microorganisms have the enzymic equipment necessary. Some, such as the cyanobacteria and *Thiobacillus*, are free-living. Others form associations with plants where these costs can be shared.

Nitrate is the main form of nitrogen taken up by plants, but this is highly soluble and readily leached from the soil. In moist climates, soils are easily depleted and competition for available nitrate shapes many communities of plants and animals. This has lead to some extreme strategies for supplementing an individual's nitrogen supply. Carnivorous plants (Plate 20) like the Venus fly trap (*Dionaea muscipula*) and pitcher plants (*Sarracenia* and *Nepenthes* species) are found in waterlogged bogs and swamps, conditions under which nitrogen is lost very readily. Insects caught by the carnivorous plants help to make good this loss. Directly or indirectly, all animals ultimately depend upon plants for their nitrogen – so this is something like justice, with these plants, at least, getting their own back.

Plants find themselves in competition for nitrogen not only with each other but also with the microorganisms in the soil. Different genera of bacteria use different nitrogen compounds to derive energy from its bonds. **Heterotrophic** bacteria,

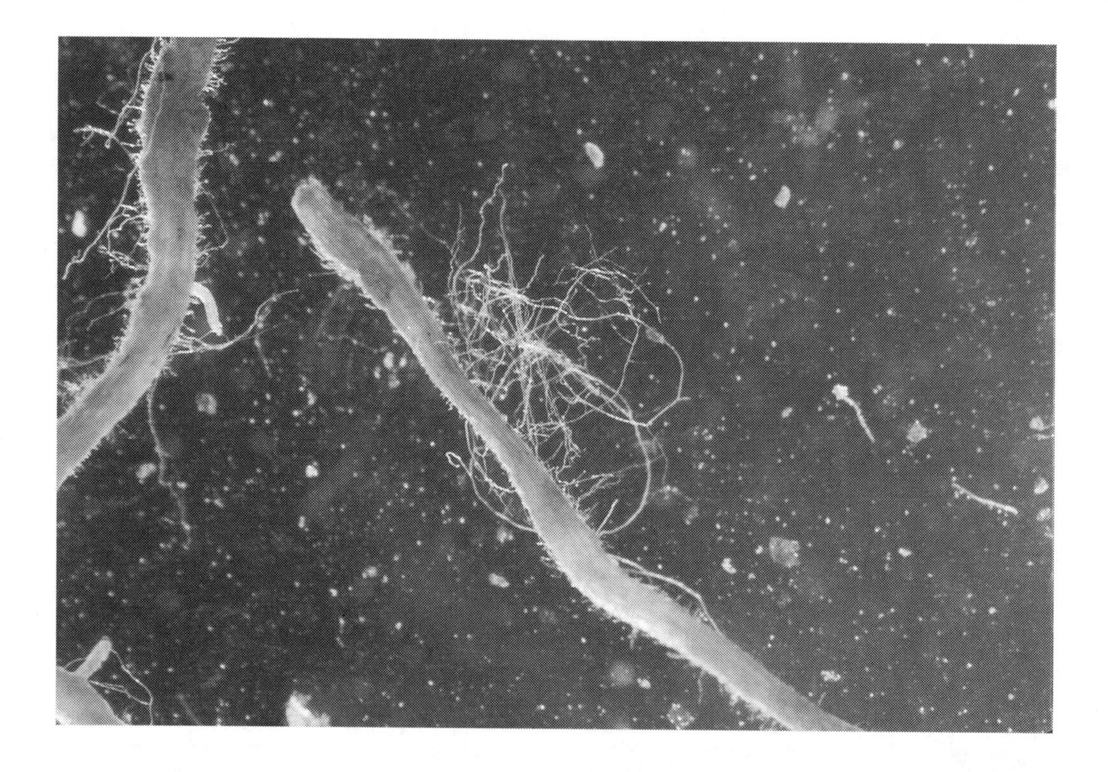

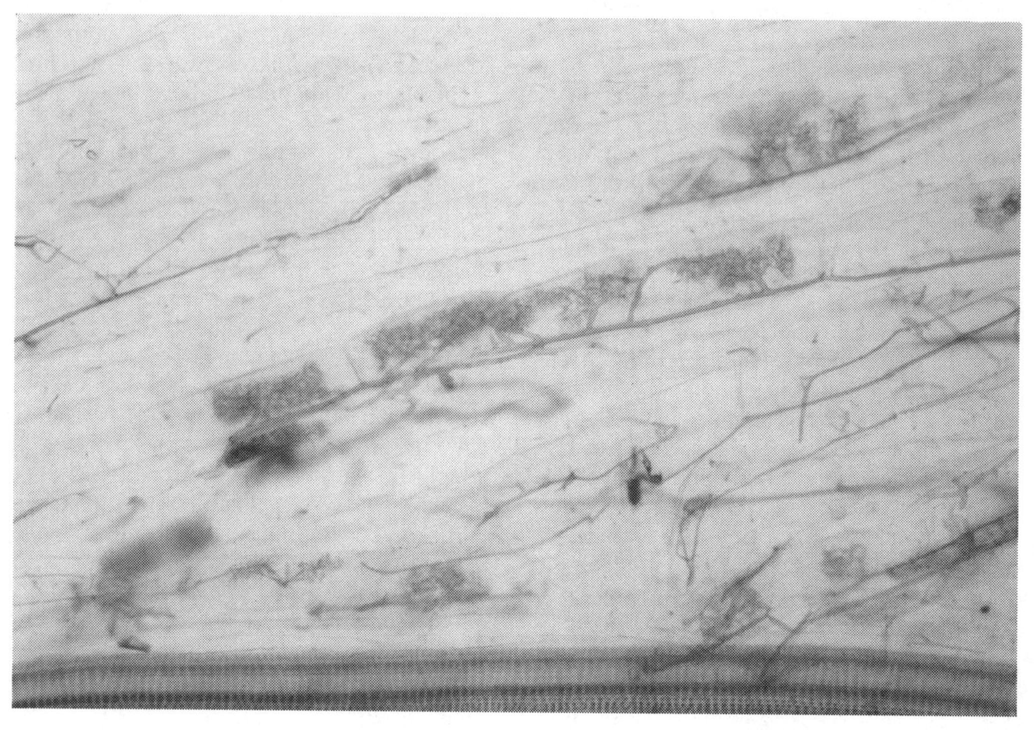

Figure 6.6 Mycorrhizae: plant and fungus in partnership. Arbuscular mycorrhizal fungi (AMF) growing (a) around and (b) inside the root.

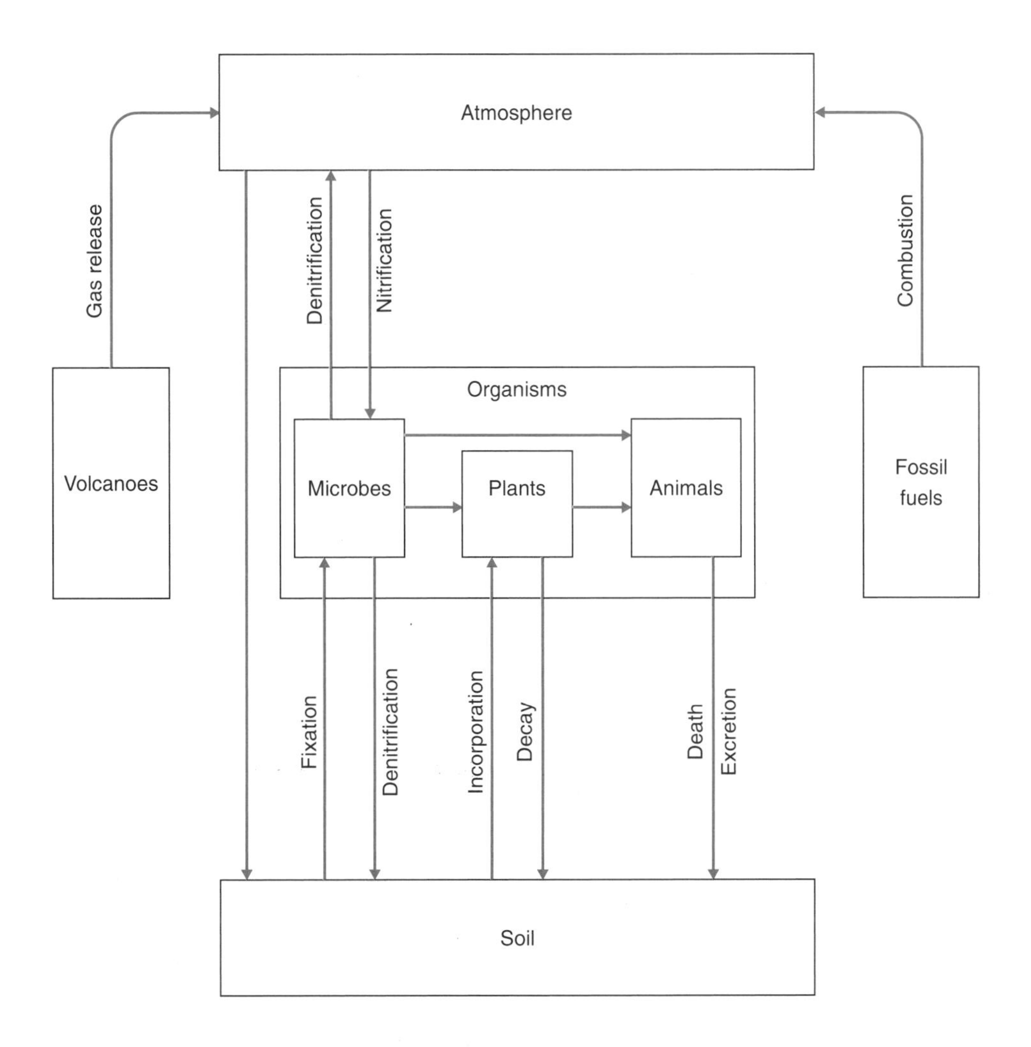

Figure 6.7 The nitrogen cycle. Nitrogen passes through living organisms on its way to the atmosphere from the soil. Additional sources are oxides of nitrogen produced by volcanoes and the burning of fossil fuels.

Box 6.2

NITROGEN FIXATION

Bacteria and cyanobacteria (blue-green algae) are responsible for the bulk of biological nitrogen fixation from the atmosphere. They occur in both terrestrial and aquatic environments and can either be free-living, using organic detritus as their energy source, or have a symbiotic relationship with certain plants. All possess the enzyme nitrogenase, which is responsible for the processes by which molecular nitrogen (N_2) is split and combined with hydrogen to form ammonium.

The most studied nitrogen fixers are the bacteria belonging to the genus

Rhizobium. These survive in the soil as saprophytes – that is, until they can infect the roots of leguminous plants, those belonging to the pea and bean family. Here they are housed inside root nodules (Figure 6.8), as symbionts with their host. The plant provides shelter (especially an anaerobic environment) and also sugars that serve as a carbon source for the *Rhizobium*. The abundant nitrogen fixed by the bacteria makes this a worthwhile investment by the plant – their private income of nitrogen gives many legumes a competitive advantage over their neighbours in nitrogen-poor soils. Even so, if these costs can be avoided, both the *Rhizobium* and its host will switch, taking up any abundant nitrate that might become available in the soil.

Nitrogen fixation using symbionts is not only found in legumes. Root nodules are found in a number of other plants, especially alder (*Alnus glutinosa*), which has a symbiotic relationship with a nitrogen-fixing actinomycete, *Frankia*. There are also various of free-living bacteria that fix nitrogen for their own use, as well as cyanobacteria which have this capacity. This can give the cyanobacteria a competitive edge in waters where nitrogen levels are scarce but other key nutrients are abundant. For this reason, waters which become enriched with phosphorus often bloom with a flourishing but toxic blue-green population.

That nitrogen fixation can occur in a variety of plants encourages us to try, through genetic engineering, to introduce the association into the major crop plants. Considerable effort is being directed at cereals in particular which, ordinarily, need considerable amounts of artificial fertilizers. In the process we shall learn more about the mechanism of nitrogen fixation and perhaps how this symbiotic relationships might have evolved.

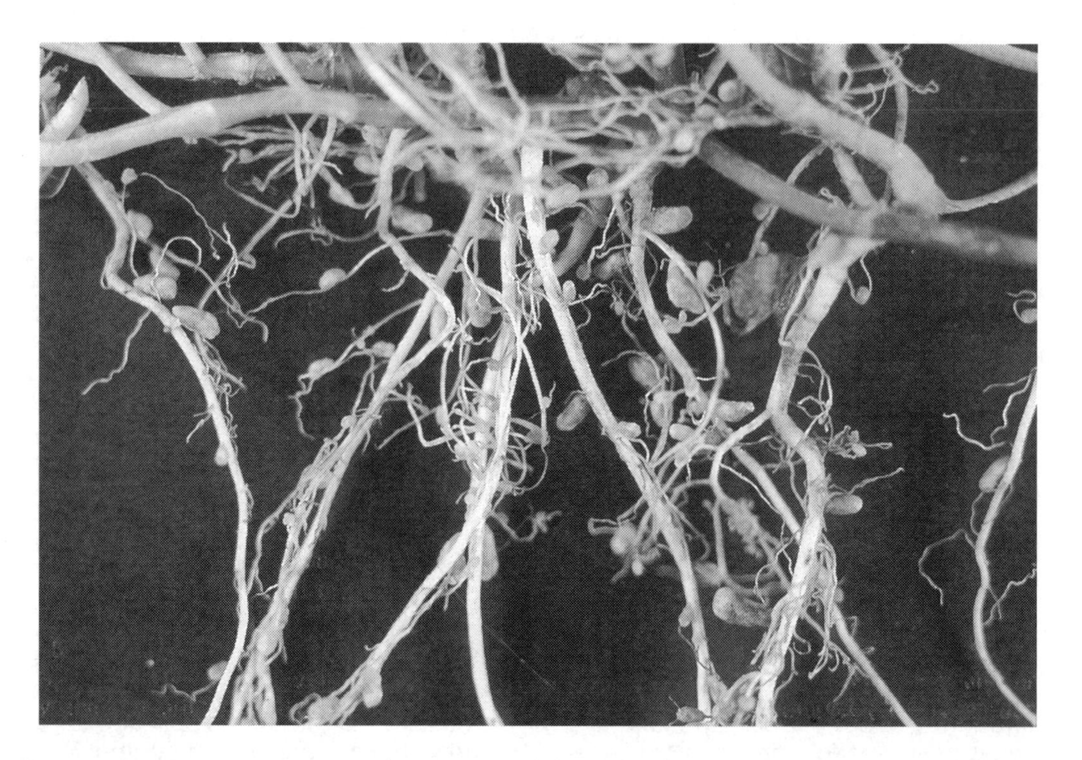

Figure 6.8 Root nodules. (a) Root nodules of clover (*Trifolium* species) which have formed from root epidermal cells as a result of infection by *Rhizobium* bacteria. (b) Using electron microscopy, the y-shaped bacteriods are clearly visible within the nodule.

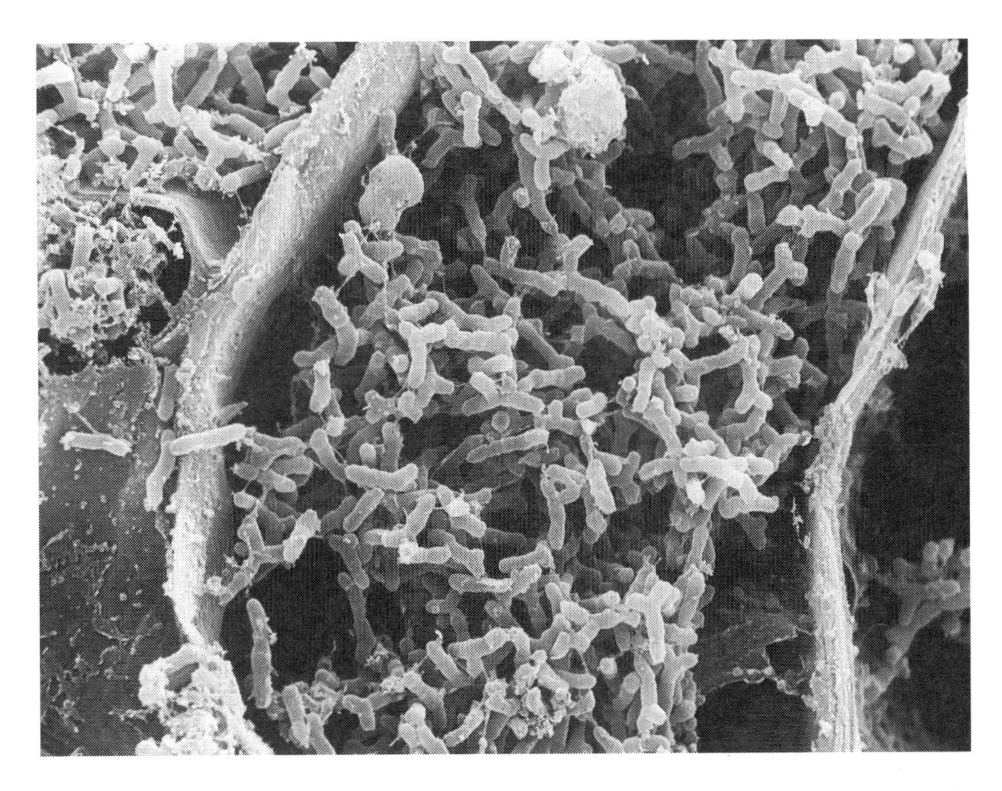

Figure 6.8 *continued*

those using organic compounds as their energy source, mineralize organic nitrogen to ammonium (Figure 6.9). This can lead to losses if the ammonium evaporates, or it may be used by other bacteria, the nitrifiers. These are **chemoautotrophs,** deriving their energy from the oxidation of inorganic compounds, in this case soil nitrogen. *Nitrosomonas* converts ammonium to nitrite (NO_2) which may then be further oxidized to nitrate (NO_3) by *Nitrobacter*. Together these two processes are called **nitrification**. Any nitrate not taken up by plants or microorganisms may undergo **denitrification**. This is a series of reductions carried out by a variety of bacteria (including *Bacillus* and *Pseudomonas*) leading to different gaseous forms of nitrogen. Denitrification is a very important process in oxygen-poor environments and one cause of the shortage of nitrogen in many waterlogged soils.

Large amounts of nitrogen also enter the atmosphere from the burning of fossil fuels, making a significant contribution to acid rain. Around 60 million tonnes of nitrogen pass into the air each year as nitrogen oxides. On the ground, inorganic fertilizers add a highly concentrated source to the soil in a form that can be readily leached away. These have produced highly localized concentrations of nutrients, upsetting communities adapted to nutrient shortages. Such surpluses have led to considerable changes in natural communities, some of which we examine now.

6.3 EUTROPHICATION – TOO MUCH OF A GOOD THING

It is apparent to any farmer or gardener that the fertilizer applied to the soil one year is not there several years hence. For those of us without green fingers, it might never have been there in the first place. Run-off and nutrient leaching mean that the money we spent trying to produce something on our patch of land has gone to fertilize something else somewhere else.

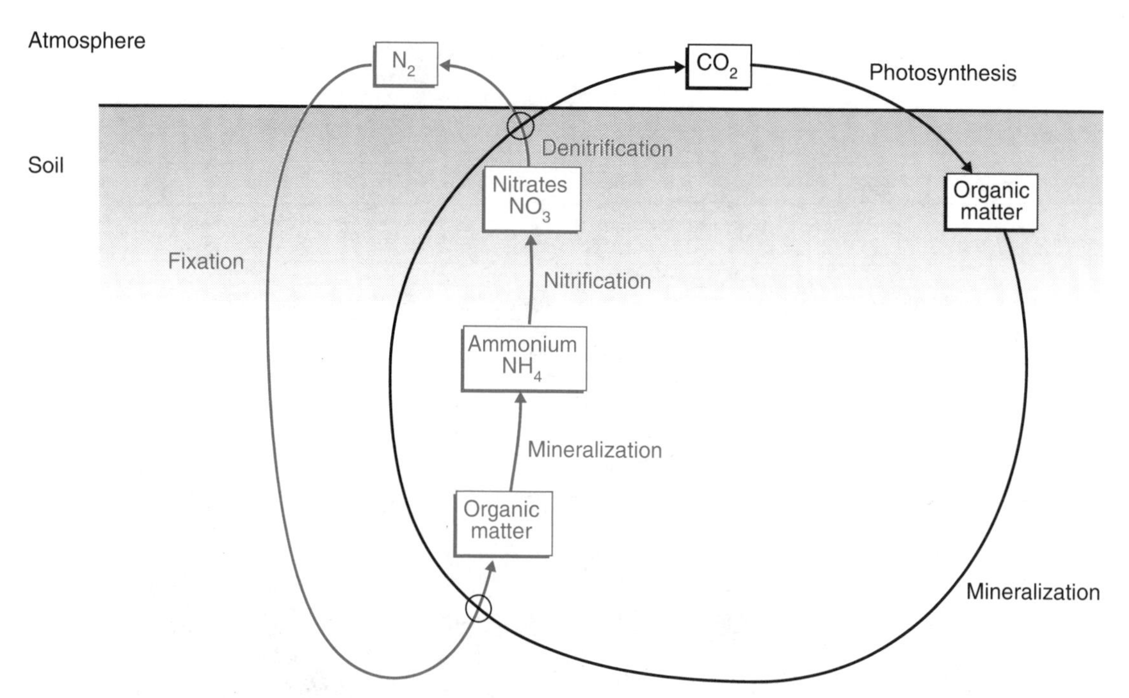

Figure 6.9 Microbial interactions between the nitrogen and carbon cycles. The carbon cycle fuels the nitrogen cycle in its fixation and denitrification phases.

Very often, it is primary production in the waters draining away that snatches up these nutrients. Enriched by fertilizers or other sources of nutrients, rivers, streams and lakes change their community structure and patterns of cycling in a process termed **eutrophication**. This is nutrient overload – an example of how a nutrient in the wrong place at the wrong time and the wrong concentration can become a serious pollutant. Eutrophication can be a natural process too, when freshwaters become enriched with age.

Human activity is by far the greatest cause of such imbalances in freshwaters and can also cause eutrophication in soils, when fertilizers stay where they are put and a build-up occurs. The main culprits are phosphorus and nitrogen, with the former being the most serious because it is, ordinarily, so unavailable in natural systems.

Eutrophication leads to a dramatic deterioration in water quality that is all too obvious at certain times of the year. The water becomes green and soupy as the phytoplankton density increases (sometimes forming bright green flocs – floating mats of blue-green algae). It often smells foul. This is a sign that the oxygen content of the water has been depleted and that other life in the water will be struggling to respire aerobically.

We can easily measure the decline in the oxygen content of the water and also its potential oxygen demand – that is, the amount needed to oxidize all the organic matter in the water. This is the BOD (**biochemical oxygen demand**), a fair indication of how unbalanced the system is, and how this affects the fish and aquatic invertebrates needing this oxygen. When these die their tissues add to this oxygen demand. So, the system can quickly spiral down to a community reduced to a few species swimming in turbid, oxygen-poor waters.

Whilst BOD represents a useful snapshot of current conditions, it says little about past pollution events. Evidence for these can be found recorded in the community of invertebrates in the water. Some species will only survive in oxygen-rich waters and these can act as **indicator species**, whose presence or absence tells of the water quality. For example, stonefly and mayfly larvae are only found in well-oxygenated waters. A community dominated by species such as oligochaete worms – groups that can tolerate very low oxygen levels – indicates that the water has a high oxygen demand and will be carrying an unbalanced carbon load. Different **biotic indices** are used in different parts of the world, but many refer to the species composition of the freshwater community to judge past levels of pollution, particularly organic pollution from sewage discharges.

6.3.1 Restoring eutrophic waters

When people think of pollution and effluent they picture industrial units billowing fumes and happily forget the number of times they flush yet more nutrients down the toilet. Likewise, we tend not to connect the food on supermarket shelves with the fertilizer run-off or oceans of slurry from livestock housing on intensive farms. Even free-range livestock can add to eutrophication when areas are overstocked and their wastes enter the drainage system.

Clearly, the best way of dealing with any form of pollution is to control it at source. Domestic and industrial sewage account for over 80% of the phosphorus input into the environment, primarily from human sewage and domestic detergents. Fortunately in the last 25 years, technology and legislation have led to the marketing of low-phosphate detergents – increasingly supported by consumers consciously choosing less damaging products. That said, we still have to contend with the mistakes of the past, along with the more intractable problems of agricultural pollution.

A classic example of eutrophication is Lake Washington near the American city of Seattle. As the city expanded in the 1940s and 1950s its sewage discharges into the lake led to increasingly poor water quality – a 75% reduction in clarity and a massive explosion of the filamentous cyanobacteria *Oscillatoria rubescens*. Public concern led to a campaign to reduce effluent inputs into what had become known as 'Lake Stinko'. The phosphorus levels in the water were shown to be the key to the problem, though it took seven years to reverse these trends.

Experiments carried out in Ontario, where whole lakes were deliberately fertilized, confirmed that phosphorus plays a key role in prompting algal blooms. More recently, experience of other eutrophication events suggests that changes to the plant and animal community mean that treating the problem requires more than simply halting the nutrient inputs.

This has become apparent from studies of the problem in the Norfolk Broads of East Anglia. These wetlands, the product of peat digging and reed growing in the Middle Ages, have developed over several centuries into a specialized community. These fenlands have a number of rare and distinctive plants and animals, including the water soldier plant (*Stratiotes aloides*) and the fen raft spider (*Dolomedes plantarius*). However, the area borders some of the richest and most intensively farmed soils in Britain. Over the last 60 years these soils have received immense inputs of fertilizer. In the 1970s, the Broads began to show signs of eutrophication but early attempts to address the problem by controlling discharges failed. One reason was the large capital of phosphorus that had accumulated in the sediments which, under anaerobic conditions, could be mobilized and released back into the water. One strategy, therefore, has been to pump out the enriched sediments, but even this failed to return clear water to the Broads.

Now the Broads Authority uses a management strategy designed to combat eutrophication at several levels – firstly at source, by designating Nitrogen Sensitive Areas in which incentives are used to encourage farmers back to more traditional, less intensive agricultural practices. Domestic sewage is also treated to reduce phosphate before discharge. On nature reserves, ecologists use different methods according to the severity of the eutrophication. In mild cases, reeds are harvested and removed, to 'crop off' excess nutrients. In more grossly polluted areas the 'mudsucker' (Figure 6.10) is used, sometimes in conjunction with phosphate precipitation techniques by which iron and calcium sulphates are introduced to produce highly insoluble phosphate salts in the sediments.

Other strategies, used on lakes in Holland, have manipulated the animal community, with some success. The idea of **biomanipulation** is to promote a phytoplankton and rooted plant community typical of a clearwater system by manipulating its species composition. In this case, it meant introducing certain fish and removing others, especially bottom feeding species, such as bream (*Abrama bramis*), that stir up the sediments and whose tissues represent a significant pool of phosphorus. They also added a freshwater mollusc (*Dreissena*) that can feed on the cyanobacteria, which are poisonous to most other planktivorous animals. It seems that once the ecosystem has shifted to the turbid state, it needs such wholesale manipulation of the community to restore it.

Figure 6.10 The Broads Authority mudsucker in action. This machine is used to remove nutrient-rich sediments from eutrophic bodies of water as part of the wetland restoration programme.

6.3.2 Eutrophication in terrestrial ecosystems

Nutrient enrichment is not confined to small bodies of freshwater. Many enclosed seas and coastal regions also suffer from seasonal blooms from localized enrichment. So too do many soils, especially those receiving fertilizers from agricultural sources.

Nutrients essential for life can create problems when abundant as well as when in short supply. Sometimes ecologists have to restore a eutrophic habitat by removing excess nutrients. Species characteristic of heathland or chalk grasslands, for example, are adapted to low nutrient levels. Most established natural communties with a high diversity are habitats where competition for nutrients is intense (section 7.1). These communities change when nutrients become abundant, as some species are able to increase their numbers or growth faster than their neighbours. Typically, nutrient-rich habitats are dominated by a small number of fast-growing species at the expense of others. This has been a particular problem in fenland habitats, where wetlands receiving nutrient-rich run-off from surrounding farmland are dominated by reeds and rushes, while many fenland plants decline. The rarities are adapted to a particular nutrient balance and would only return if this was restored.

As we saw in the last chapter, high stocking rates and the abandonment of traditional grazing practices have brought major changes to many temperate communities. Excessive grazing and trampling make life very difficult for some plants, while the enrichment from fertilizers applied to pastures mean that slow-growing resident species often lose out to highly competitive invasive weeds. Species diversity has declined in many lowland pastures as a result. In less than 50 years over 95% of British lowland flower-rich hay meadowland has been lost through these and other changes.

Residual fertility is a challenge to ecologists involved in habitat restoration. Cropping-off excess nutrients can be a very hit-and-miss technique. Another technique, perhaps extreme, is to bury the rich topsoil or remove it completely. The problem is not restricted to farmland. Amenity areas such as parks, gardens and public open spaces may also have high levels of nutrients following years of fertilizer applications along with the inputs from frequent visits by domestic pets. Recent research has shown that the highly manicured golf greens are so rich in fertilizers that they could be bagged up and legally sold as phosphatic fertilizer!

Oversupply of nutrients can also result from a build-up of biomass. Habitats traditionally managed by cutting, grazing or burning can suffer if this management is suspended and the standing crop is allowed to recycle back into the system. The nutrient input can then be sufficient to produce effects similar to the direct application of fertilizers that we have just described. In grossly neglected sites total removal of the standing crop – and, on occasions, litter and top soil too – may be the only way of restoring the habitat to its pre-neglected condition.

6.4 OIL POLLUTION

In water, excess carbon quickly depletes water of its oxygen. On land, dead and decaying plant and animal material (detritus) accumulates as litter in soils, to be slowly oxidized by the processes of decomposition. Where oxygen is in short supply, decomposition will produce energy-rich compounds which, with enough time, heat and pressure, may form hydrocarbons. The industrial revolution was powered by the energy released from this fossil carbon. So dependent are today's industrial economies on oil that they are prepared to go to war to maintain their supply. We transport oil many thousands of miles and occasionally spill it (Table 6.1).

There are few pollution incidents as dramatic or distressing as the sight of a broken tanker and the oiled seabirds in its wake (Plate 21). More mundane and far more common is the continual pollution associated with the loading and unloading process, or the surreptitious washing-out of tanks at sea. Still more deplorable is the deliberate targeting of marine and terrestrial oil installations in times of war (Plate 22). The 1991 Gulf War ended with the worst spill both at sea (800 000 tonnes) and on land (between 5 and 21 million tonnes).

The damage caused by oil spills has been studied extensively, so with each incident we have learnt more about their causes and consequences. Depressingly, each spill provides us with yet another opportunity to test our techniques. As a result, our methods of dealing with oil spills have changed considerably over the last 30 years.

Faced with a major spill, the obvious response is to contain the oil and to clean it up with whatever means are available. Back in 1967 when the *Torrey Canyon* broke its back on the Cornish coast, a variety of techniques were used, including bombing the vessel with napalm to set fire to the oil. The remainder of the slick was treated with more kitchen-sink technology – in the form of detergents and dispersants, to break it up. Over 10 000 tonnes of these chemicals were used, but they turned out to be more toxic to the wildlife than the oil they were meant to combat. Some parts of the coastal community took around 10 years to recover from their effects.

A slick on the water surface limits oxygen diffusion into the water, leading to anoxic conditions beneath. Oxygen levels also fall as the microbial community begins to decompose the oil. As a macropollutant, oil physically obstructs living organisms, smothering and weighing them down. More direct poisoning follows if an animal ingests it. Oil is a complex cocktail of hydrocarbons and various trace compounds – by and large, the most soluble hydrocarbons tend to be the least toxic. Hydrocarbons damage cellular metabolism and also affect cell membrane function. They become

Table 6.1 Principal marine oil spills during the last three decades

Vessel	*Date*	*Location*	*Extent polluted (km)*	*Size of spill (tonnes)*	*Bird deaths*
Torrey Canyon	March 1967	Cornish coast, England	225	117 000	30 000
Amoco Cadiz	Winter 1978	Brittany coast, France	360	233 000	4 600
Exxon Valdez	March 1989	Prince William Sound, Alaska	700	35 000	100 000–300 000
Braer	January 1993	Shetland Isles, Scotland	235	86 000	6 313
Sea Empress	February 1996	Milford Haven, Wales	100	75 000	22 132

integrated into the membrane and inhibit its ionic communication with the outside.

Detergents too are toxic and also interfere with cell membrane function. Because membranes are partly composed of lipids (fat molecules) any compound capable of breaking up petroleum hydrocarbons will also cut through cell membranes. Ecologists have learnt from their early mistakes and today detergents and dispersants are relegated to the armoury of last resort.

The next great challenge came with the *Amoco Cadiz* spill. This was by far the largest accidental spill, releasing a quarter of a million tonnes into the English Channel. Reluctant to use detergents, physical mopping up of the oil was used along with a reliance on natural degradative processes. The worst of the pollution was weighed down with sawdust and chalk and allowed to settle on the sea bed. Meanwhile on the shore, oil was physically removed. Detergents were restricted to economically important areas, such as harbours. Relying on the resident microorganisms paid dividends with 80% of the main hydrocarbon groups being degraded within the first seven months of the spill.

A similar integrated clean-up operation was used after the *Exxon Valdez* ran aground in Alaska. Pollution control experts tailored their response to the nature of each beach, some of which were left to recover by themselves, though with the help of fertilizers to encourage microbial growth. Others areas were washed down with high-pressure water jets and some were mopped with a new absorbent cloth, in an operation called ‘rock-polishing’.

Sometimes nature finishes the job almost completely. This happened when the *Braer* struck the rocks off the Shetland Isles in winds reaching Storm Force 11–12, and gusting to Hurricane Force 13–14. The oil was quickly whipped into an oil–water emulsion (known as a mousse) which assisted in its dispersal. Although the severe conditions meant that little damage occurred to the coastal communities, the wind did carry the oil further inland than in previous spills.

The received wisdom now seems to be to take things easy, to contain the spill as far as possible, but to rely on natural degradative processes with any oil that leaks out. A range of microoganisms can be relied upon to attack these hydrocarbons. Interestingly, their response (the growth in their populations) is faster if they have been exposed to spills in the past. Certainly this, along with the high water temperatures, helped the Gulf to recover from the deliberate spill of 1991. To these bacteria, oil is just another carbon source and they may only need a supply of limiting nutrients, nitrogen, phosphorus and abundant oxygen, to utilize it. Human intervention is perhaps best confined to mopping up the oil that comes ashore in the most severely contaminated areas.

6.5 RESTORATION – THE ‘ACID TEST OF ECOLOGY’

Restoration ecology helps us to redress some of the damage we so frequently inflict on our environment. This is ecology as technology – a science applied to solving real-world problems. As a science, restoration ecology answers questions about ecosystem function and uses these insights in the techniques of restoration. It is something of an art too, since restoration requires creative skills to design entire landscapes and their ecosystems.

Much of modern science is based on a reductionist approach, understanding the system by taking it apart and in some cases rebuilding it again. We do this because we are trying to describe a complex world and it is often easier to consider one thing at a time and gradually form a picture of how things work together. Whilst there have been many elegant experiments which have examined the nuts and bolts of the environment, we cannot take whole ecosystems apart – natural ecosystems are too precious to be used in destructive experimentation.

However, there are numerous habitats that have suffered degradation and which offer ecologists a convenient opportunity to study the ecological processes in a self-sustaining community, as we try to put each one back together again. The scientific knowledge gained from such projects is considerable, from nutrient cycling to species interactions and population dynamics. They present ecologists with a chance to test their theories in the real world.

All this has led Tony Bradshaw (Box 6.3) to call restoration the ‘acid test of ecology’: a means of testing theories and hypotheses and adding to the sum total of our ecological knowledge.

6.5.1 The three Rs of restoration

Taken in its most literal sense, restoration means returning something to the way it was before a change took place. Indeed, the aim of many projects

Box 6.3

RESTORATION ECOLOGY: PHILOSOPHY AND PURPOSE

Professor Anthony D. Bradshaw, FRS
Emeritus Professor of Botany at the University of Liverpool, UK

In 1992, the world's Heads of State met in Rio to consider the future of the world's environment. Their now famous declaration promised to work towards sustainability in the use of the Earth's resources. To biologists and ecologists there is nothing startling in this, because it is all too obvious that if we do not manage the world's ecosystems properly – by preventing direct destruction, pollution and overuse – the human population will have a very murky future indeed.

Part of this management involves caring properly for what we have and what we are using at the moment. But we need to first face the fact that, because we have managed things so badly in the past, there are many areas in which we have done a great deal of damage – and even destruction. These areas, and their lost ecosystems, cannot just be given better management: they have to be rebuilt, sometimes from totally raw starting points in which all soil has been lost, as well as plant and animal communities.

This is what ecological restoration is all about – the reconstruction of lost or damaged ecosystems. To provide the answer, ecologists have to get into the construction business, facing the same need to meet design, cost criteria and deadlines just like those of engineers. But such ecologists are working with a different set of materials, ones that are alive and have the power to grow and to interact, not only with each other but also with their physical environment. This leads to complications, but also to great advantages. What engineer ever managed to plant a baby factory and leave it to grow, at no cost, into a full-sized one?

The secret of ecological restoration, therefore, lies in it being ecological. The science, and the art, of ecological restoration is that, wherever possible, it should harness the natural processes that are the basis of ecosystems. These can be ordinary processes such as growth and organic matter production, or more complex processes such as nitrogen fixation and soil nutrient accumulation. In doing this, it meets the very practical objectives of repairing the ecosystems of our damaged planet.

is to reinstate lost habitats and the plant and animal communities they support. In practice, restoration ecology often encompasses several aims:

- stabilization of land surface;
- pollution control;
- visual enhancement;
- creation of amenity value;
- increase of biological diversity;
- improvement of ecosystem function;
- establishment of productivity.

Much of our knowledge and techniques has come from **reclamation** projects on sites degraded by extractive or industrial processes. Mining or quarrying often results in large amounts of waste material (spoil) that can be unsightly and a significant source of pollution. Our aim is make use of these wastes to create something more aesthetically pleasing and ecologically sustainable (Plate 23).

A number of options are available. Restoration might involve returning a site to its pre-industrial state, perhaps in some cases even to a natural or semi-natural habitat. Restoration to these states can be a costly and complex procedure and it may be that we choose, instead, a partial restoration or **rehabilitation**. This might, with time, mature to a full restoration. Other possibilities include making the land fit for agriculture or housing. These do not require a natural community and are more properly described as a **replacement** habitat, a compromise somewhere between the existing degraded state and a restored site.

Figure 6.11 shows the range of alternatives available to ecologists through reclamation, rehabilitation and replacement. In some situations, it may be best to let nature take its own course and for the site to recolonize naturally. However, this is not an option where a substrate is unstable or a significant source of pollution.

6.5.2 The time factor

Restoration techniques can change entire landscapes within a relatively short time, yet these changes can themselves be long-lasting. Our decisions at the planning stage are, therefore, critical: we need to see into the future and visualize the consequences of our decisions over ecological time. It is often impractical or impossible to return the habitat to its pre-impact state and we need to consider the options available to us (Figure 6.12).

Our choice of endpoint is governed by a number of factors, including the history of the site. These include not only the existing conditions revealed by a baseline survey but also the resources at our disposal. If the site is stable it may be possible, and indeed preferable, to let nature take its course and simply allow it to develop through the successional process. In other cases, neglect may not be a viable option and some restoration work is needed.

The most complicated option seeks to anticipate future successional changes, predicting the way an ecosystem might develop. We then have to consider the time-scale involved in the restoration process, including the time required for establishment, and seek to place our restoration plans on that trajectory.

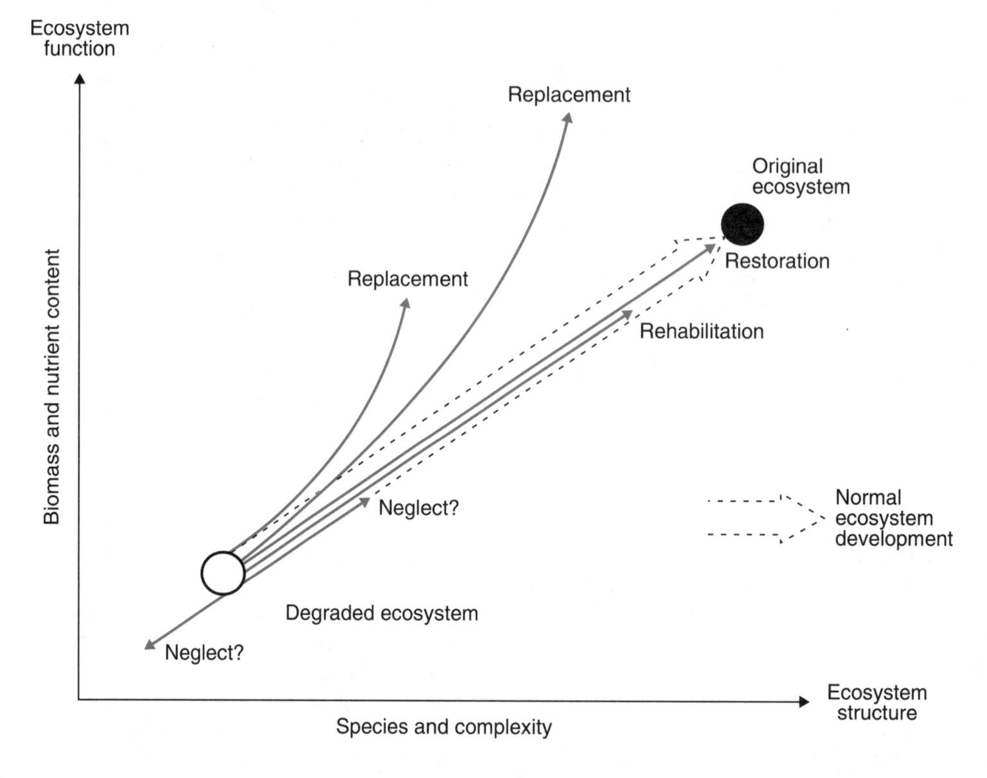

Figure 6.11 The options available in ecological restoration. As degraded sites are reclaimed they show an improvement in both the structure and function of the ecosystems. Complete restoration might not always be the aim of a project. Partial restoration in the form of rehabilitation or alternative replacement habitats may be appropriate endpoints for a project.

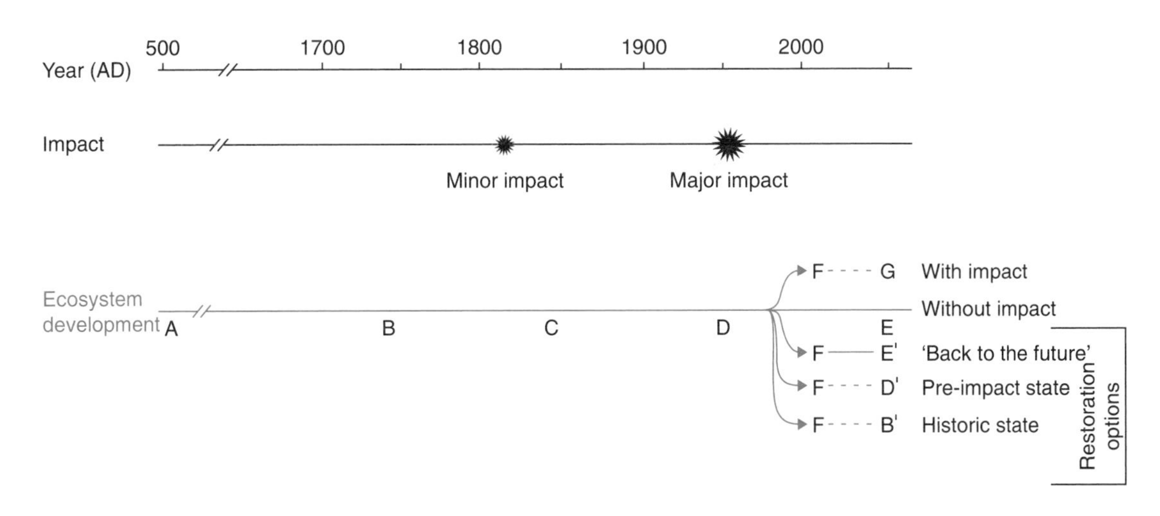

Figure 6.12 The restoration continuum, showing ecosystem transitions both with and without impacts and restoration work. Without the major impact (in this case around 1950) the ecosystem would have developed from stage D to E. The impact deflects it to F and if left unchecked this would develop into G (which might be undesirable such as a badly eroded site). Restoration offers a range of options to aim for. One is to return to the immediate pre-impact stage (D'); another is to return to an earlier successional (historical) stage (B'). A third option follows the transition which would have occurred if the impact had never happened (E').

Here we restore a site to some previous state with what might be called the 'back to the future' option.

6.5.3 Restoration methods

In ecological terms, restoration is a form of accelerated succession with the ecologist trying to establish the new habitat as completely and as quickly as possible. Most spoils are waste materials which tend to be devoid of organic matter and nutrients and, therefore, demand something close to a primary succession (section 7.2). In other cases, sufficient organic matter and even some topsoil may remain, with a reserve of nutrients that allows secondary succession to occur. Such a site, if near a source of colonizing species, may be left to restore itself naturally, provided there is no risk of the spoil moving or becoming a nuisance.

Most of the time, ecologists will help the process along by creating a suitable substrate and providing nutrients and colonizing species. In its simplest form, this may mean adding soil, fertilizers and seed. Elsewhere, more drastic measures are used. Limestone quarries in the Peak District of northern England have been restored using 'conservation blasting'. Firstly quarry faces were blown to create naturalistic rockfalls; then sewage slurry containing the seeds of lime-loving plants (calcicoles) was sprayed on the sides of the artificial cliff. The slurry added nutrients and provided a more favourable substrate as well as dispersing the seed.

Spoils vary considerably in their chemical and physical properties (Table 6.2). Consequently, they differ in their nutrient requirements and site preparation. Often sites need to be landscaped and the spoil broken and mixed to form a substrate with a suitable range of particle sizes. In some cases, ecologists may be working with a relatively inert substrate such as the sand from china clay operations, crushed concrete and brick rubble. Others, such as acidic colliery spoil or smelter wastes, may be excessively acidic or alkaline and rich in a range of toxic metals. Most have few, if any, plant nutrients. Initially, artificial fertilizers can provide these, although many spoils are free-draining and so without a proper soil structure these nutrients will be simply washed away. This is one good reason for raising the organic content of the waste, to help create a soil structure which will hold the material together. It also helps to bind the nutrients we add as well as to feed the microbial community, which is essential if a soil is to be capable of decomposition and nutrient cycling.

We can also use plants to improve nutrient capture – especially legumes, by exploiting their association with nitrogen-fixing bacteria (Box 6.2).

Table 6.2 Problems associated with spoils and their immediate and long-term treatment

Category	*Problem*	*Immediate treatment*	*Long-term treatment*
Physical structure	Too compact	Rip or scarify	Vegetate
	Too open	Compact or cover with fine material	Vegetate
Stability	Unstable	Stabilize/mulch	Regrade or vegetation
Moisture	Too wet	Drain	Drain
	Too dry	Organic mulch	Vegetate
Nutrition			
Macronutrients			
	Nitrogen	Fertilizer	Legumes
	Others	Fertilizer and lime	Fertilizer and lime
Micronutrients		Fertilizer	
Toxicity			
pH	Too high	Pyritic waste or organic matter	Weathering
	Too low	Lime or leaching	Lime or weathering
Heavy metals		Organic mulch or metal-tolerant cultivars	Inert covering or metal-tolerant cultivars
Salinity		Weathering or irrigation	Tolerant species or cultivars
Plants and animals			
Wild plants	Absent or slow colonization	Collect seed and sow or spread soil containing propagules or plants	Ensure appropriate conditions
Cultivated plants	Absent	Sow normally or hydroseed	Appropriate aftercare
Animals	Slow colonization	Introduce	Ensure appropriate habitat management

Similarly, species having mycorrhizal associations can be planted because of their capacity to scavenge phosphates. Both associations may require the soil to be inoculated with the microbial or fungal partner, perhaps by introducing some topsoils where these are known to be abundant. It is also possible to buy the appropriate *Rhizobium* strain for the legume being sown.

Dealing with the toxic components of a spoil requires a different set of techniques. Colliery spoils are often rich in iron pyrites (iron sulphide). When in contact with air, sulphides will slowly oxidize to produce sulphuric acid and lower the pH of the soil to levels that are toxic to many plants. It is possible with some wastes to remove large pyritic particles from the site, though more often it is treated on site by the addition of lime. This 'mops up' the acidity as it develops. In some cases it has been possible to solve the problem by combining two wastes: an acidic and an alkaline (such as basic slag). An alternative is to mix the waste with a clean material, or to bury it beneath a layer of imported soil. The danger here is that this soil itself becomes soured by acids or metals migrating upwards and that means the cover needs to be sufficiently deep to prevent this and to prevent deep-rooting plants coming into contact with the waste.

Using resistant species – plants that can tolerate high acidity or high levels of metals – can offset some of these problems. A considerable range of species is now available commercially, and they may also be collected from around ancient mineral workings or natural rock outcrops rich in copper, cobalt, lead, nickel and so on. These races or ecotypes (Box 2.3) have evolved mechanisms which limit toxic metal uptake and they are genetically distinct from nearby populations growing on normal soils. The locally adapted ecotype can be tolerant to several metals. For example, 'Merlin', a cultivar of the red fescue grass (*Festuca rubra*), is resistant to zinc, lead and copper and has been used in mining restoration projects throughout the world. Not only grasses but also trees and other flowering plants can be used in some locations. One of the most dramatic examples is the 'nickel tree', *Sebertia acuminata* (Plate 24), found on the nickel-laden serpentine soils of New Caledonia, which oozes a blue-green nickel-rich latex from its bark. These, and less obvious ecotypes, have been used by prospectors as indicators of mineral-rich rocks, an activity called geobotanical prospecting.

Perhaps the most extreme form of restoration is **translocation**. Here an entire habitat is moved from one site to another. This is neither easy nor cheap and is only attempted when a valued habitat is under severe threat. Some habitats are easier to translocate than others. Grassland, for example, can be lifted as turves with extra topsoil and re-laid elsewhere. The same is true for heathland (Plate 25). Woodlands too have been moved, though this is more problematic, given the scale of the operation and the species involved.

As well as the expense, restorationists have to consider whether the community is likely to survive in its new location (the local geology, topography, drainage and so on) and carefully have to match up donor and receptor sites. Clearly there is little point in transplanting a nutrient-poor habitat where it is likely to receive nutrient-rich run-off from neighbouring farmland. Equally, we need to know what impact the transplant will have on the existing communities. There are also ethical questions about moving habitats around and the extent to which a translocated habitat 'belongs' to its new site. In some cases partial translocation is possible, with material (soils, plants and animals) being taken from one site to assist the development of another. Whatever the scale of the restoration undertaken, an established community will need proper aftercare, including site monitoring and a review of its management as the community develops.

Restoration ecology is playing an increasing role in civil engineering where it offers alternatives to concrete barriers and embankments. Figure 6.13 compares the options for 'hard' and 'soft' engineering, in terms of cost and performance. More often than not, inert structures cost more to construct and have a limited lifespan. **Ecoengineering**, on the other hand, makes use of sustainable, self-renewing materials, such as trees, shrubs and other vegetation. Plant roots are mechanically very strong and provide increased anchorage and stability as they grow through the soil. Similarly, a mat of vegetation intercepts rain and shelters the ground surface, so protecting the structure from erosion. These habitats are not only cheaper to construct; they can also

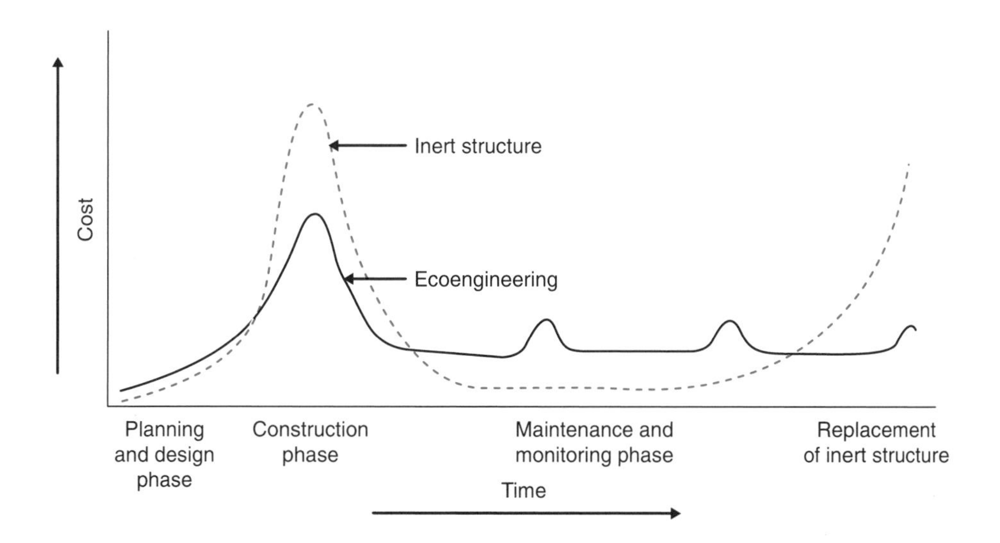

Figure 6.13 Cost analysis of 'hard' vs. 'soft' engineering. A comparison of relative costs and time-scales involved in planning, construction and mature operation of ecoengineered and inert structures.

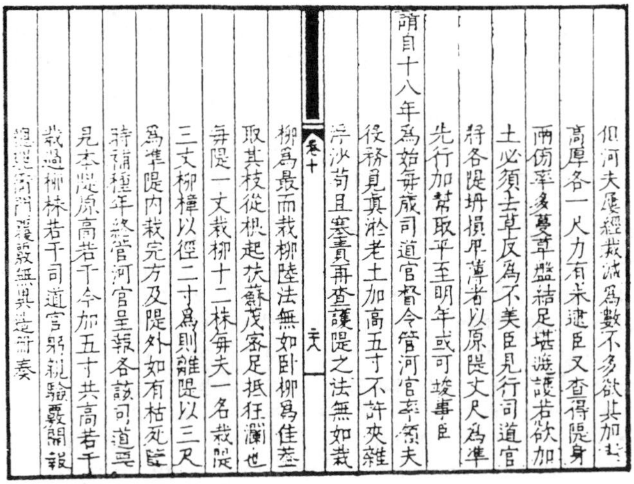

Figure 6.14 Ancient Chinese manuscript – possibly one of the first documented examples of ecoengineering. The translation is as follows.

It is observed that both sides of the earth dam embankment surfaces are covered with creeping sage grasses, which provide sufficient surface protection. If it is intended to raise and thicken the embankment, the grasses have to be removed and the surface protection condition could become worse ... It is revealed that the best method for protecting the embankment is planting willows. Among the six planting methods of willow, the lateral planting is best. Since this method permits the willow's branches to grow much closer from the root system, thus it allows the willow to have much blooming branches to resist the impact of imponding water. Every 1 zhang [320 cm] of embankment should be planted with 12 willows ... The willow should have a minimum girth of 2 cun [3.2 cm] and stick out from the embankment for 3 chi [96 cm]. The planting should start from the inner to the outer portion of the embankment. Any dead willows found should be replaced immediately ...

be maintained indefinitely, provided they are well managed.

People seem to prefer trees to a concrete wall. They have a range of other advantages: the acoustic and mechanical properties of vegetation, along with its capacity to trap dust, absorb rainfall and bind soil, all make it preferable to a simple, constructed barrier.

Ecoengineering is not new. Back in the Middle Ages the Chinese used willow in their flood protection schemes (Figure 6.14). Then, as ever, necessity was the mother of invention with people using materials which were readily to hand – in this case the willow trees growing along the riverbank. Restoration ecology has rediscovered other, far older land management techniques, many of which disappeared with the advent of the Industrial Revolution.

6.5.4 The limits of restoration

The overall aim of restoration ecology is to produce ecologically viable ecosystems. However, decisions about endpoints have to be taken in a political and social context, including the need for local amenities and the wishes of local people. Some of these non-ecological objectives can present us with difficult choices.

Past successes have led to sweeping claims for restoration ecology, even though our ability to restore lost and damaged habitats is limited, not least because of our lack of knowledge and the restricted time-scales to which we work. The best we can hope for is an approximation of natural ecosystem. For the ecologist, restoration is the use of natural ecological processes to create and recreate ecosystems and, in the process, perhaps learn something of how they operate.

SUMMARY

The physical and chemical processes by which crustal rocks are degraded and rebuilt are supplemented by the activity of living organisms in the biosphere, creating the major biogeochemical cycles. In these, essential nutrients are diverted through living tissues, initially by autotrophic organisms – bacteria and plants. Organisms use the chemical properties of the different elements in their metabolism and tissues. These properties also determine the range and speed over which these elements cycle through the biosphere.

The principal cycles are of water, carbon, sulphur, phosphorus and nitrogen. The energy for the biological phases of each cycle is derived from the carbon-rich compounds fixed by primary producers or, to a much smaller extent, in the chemical reactions of some specialized microorganisms. Bacteria, cyanobacteria and fungi all play a crucial role in scavenging and fixing key nutrients and their subsequent chemical transformations. In the warmer latitudes, it is shortages of some of these nutrients, especially phosphorus and nitrogen, which serve to limit ecological activity. Most ecological communities are adapted to these shortages and a variety of plant species have formed mutualistic associations with soil bacteria or fungi to improve their supply of nitrogen and phosphorus.

Considerable changes can occur in communities when key nutrients become abundant. Eutrophication – nutrient enrichment – can occur in both aquatic and terrestrial habitats and leads to major changes in their species composition and may require treatment to reduce or eliminate the pollution. In contrast, oil spills are often best treated by containment, leaving nature to restore itself. At the other extreme, on severely degraded sites (such as industrial wastes) recovery happens slowly or not at all, requiring ecological restoration and a range of techniques to accelerate succession. Other options include partial restoration in the form of rehabilitation or the recreation of an entirely new replacement habitat. Although restoration is used to improve surroundings and combat environmental degradation and species loss, it has limitations and should not be regarded as a true recreation of the original habitat.

FURTHER READING

Bradshaw, A.D. and Chadwick, M.J. (eds) (1980) *The Restoration of Land: the ecology and reclamation of degraded land*, Blackwell, Oxford.

Freedman, B. (1995) *Environmental Ecology: the ecological effects of pollution, disturbance and other stresses* (2nd edn), Academic Press, San Diego.

Jordan, W.R., Gilpin, M.E. and Aber, J.D. (1987) *Restoration Ecology: a synthetic approach to ecological research*, Cambridge University Press, Cambridge.

EXERCISES

1. Select the correct answer to the following statement.
   The most limiting nutrient in the majority of terrestrial ecosystems is ...
   (a) nitrogen, because its salts are highly soluble and leaches out of the soil.
   (b) phosphorus, because its salts are highly insoluble and unavailable.
   (c) nitrogen, because it constitutes almost 80% of the atmosphere.
   (d) phosphorus, because its salts are highly soluble and leach out of the soil.
   (e) sulphur, because it forms only a small fraction of living tissues.

2. Complete the following paragraph by inserting the appropriate words, using the list below (some words may be used more than once, some not at all).

   The oversupply of nutrients is known as __________. It can affect both __________ and __________ ecosystems, causing destabilization of plant and animal communities. Domestic and industrial __________ are the principal causes of eutrophication in aquatic environments. Agricultural sources are also a significant source of eutrophication, with nutrients coming from __________ and __________, from livestock and cultivated land respectively. The main nutrients responsible for eutrophication are __________ and __________, with __________ being the most problematic because of its short supply in nature.

aquatic, effluent, effluent slurry, eutrophication, fertilizers, industrial, large, nitrogen, oligotrophy, phosphorus, potassium, small, sulphur, terrestrial.

3. Examine the map below (Figure Ex.6.1), then consider the following scenario. The imaginary fishing village Safe Haven is situated close to areas famed for their wildlife and landscape value. A supertanker has run aground on rocks 6 km offshore and most of its 100 000 tonnes of oil are being driven towards the coast by strong winds. As coordinator of the emergency control centre, what are your priorities?

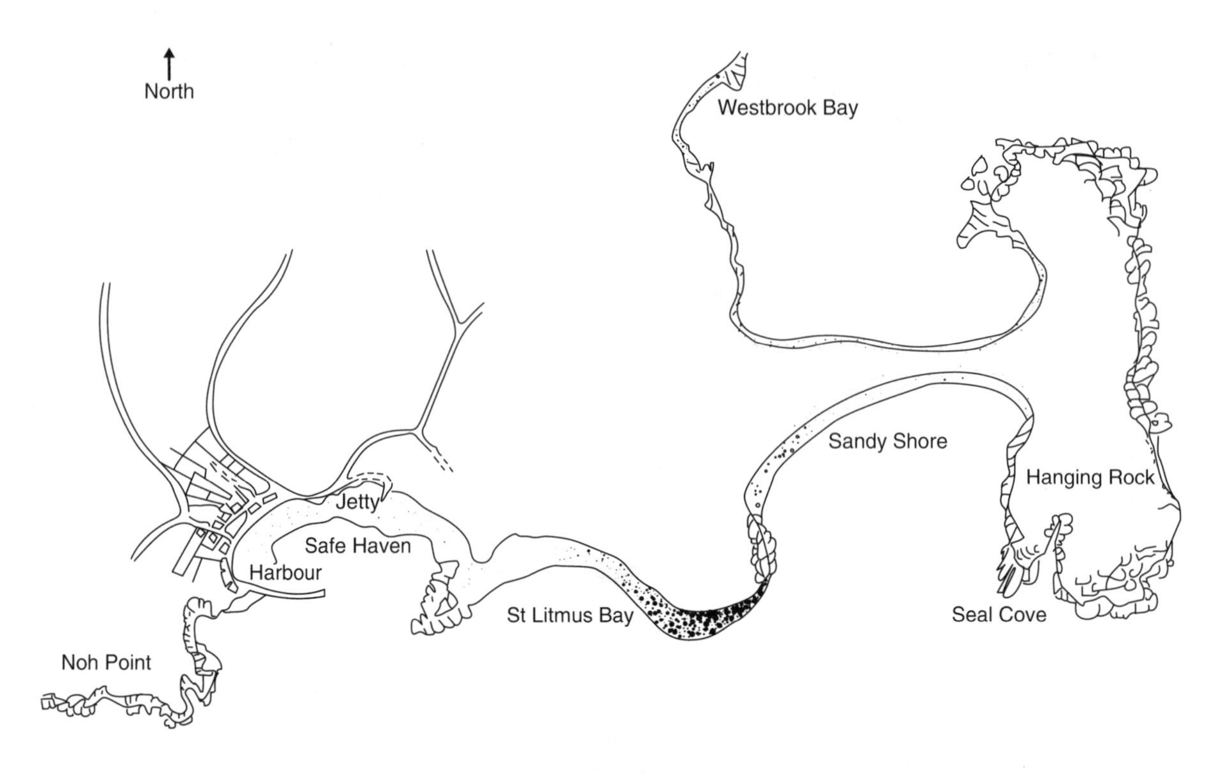

4. What are the three Rs of restoration ecology and how do they relate to the aims and objectives of restoration projects?
5. The following experiment was carried out on nutrient-poor spoils from china clay mines in Cornwall. The graph (Figure Ex.6.2) shows differences in the growth rate of sycamore (*Acer pseudoplatanus*) growing in the vicinity of alder (*Alnus glutinosa*, *A. incana*) trees. Examine the graph carefully. What is happening here? Has it any use in other restoration projects?

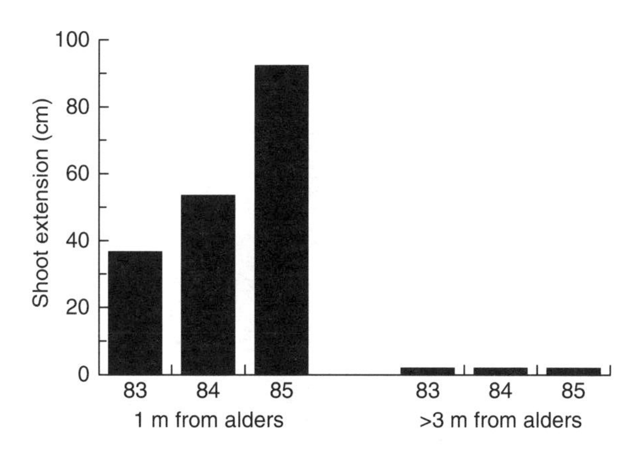

Tutorial or seminar topic

6. *In the wrong hands, restoration ecology can be a rogues' charter.*

   Discuss the above, giving careful consideration to what restoration projects can and cannot do, and pointing out the possible risks of misuse.

The nature of the community – a comparison of five distant mediterranean-type communities – how communities are organized – the role of species interactions and abiotic factors in community structure – how communities develop with time – succession and conservation.

7 Communities

You are left ... with something rather like the skeleton of a body wasted with disease; the rich soft soil has all run away leaving the land skin and bone.

The Critias, Plato

Emerging from an aircraft after flying any distance, we always compare the temperature of our destination with that we left behind. For those of us from the colder and wetter temperate latitudes, this is nearly always a pleasant change.

On leaving the airport, it is soon apparent that the weather is not the only difference. Beyond the cultural, social and architectural contrasts, those with an eye for the plantlife around them will quickly spot species very different to those back home. Some growing here in the wild may be found only in greenhouses there. Others we may have never seen before.

Of course, this is no surprise to us. We expect this wonderful climate to support a different plant life. If we give it rather more thought, we would also expect the plants to show adaptations to the local regimes of warmth and moisture. We may anticipate the presence of particular types of plants associated with a particular climate – the almond blossom in the Mediterranean or the heather of the Scottish moors. Once we start looking, other patterns emerge. We often find that species distributions reflect the presence of other plants and of animals, forming distinct groups or assemblages.

This chapter examines these assemblages and asks whether a community is governed primarily by the abiotic conditions under which it develops or the interactions between its species. We have to decide if a community is simply a loose collection of species adapted to the same environment or something more tightly integrated. This is important for our understanding of ecological systems, how they function and how they are held together. A self-sustaining community, able to re-establish itself following disturbance, is taken by many to be evidence for a 'balance of nature'.

Ecologists too need to know how communities are regulated to help to protect those under threat. If the same combinations of species are found under particular conditions, we might suspect that only those assemblages could persist together. Equally, different species organized into similar functional relationships would again suggest that only some community structures were viable in those habitats.

An experiment allowing us to examine these ideas has been operating on the Earth over the last million years. Across the globe, where the tropics give way to temperate regions (Figure 7.1), five discrete areas have established the climatic conditions typical of the Mediterranean. Each started with its own distinct collection of plants and animals, yet these now form plant communities that today share a characteristic appearance. This prompts us to ask whether they also share the same community structure. Below we shall examine the extent to which the same patterns have been recreated in each region and what this tells us about the functional relationships that govern an assemblage of species.

For many people, the mediterranean climate is the closest to the ideal. The Mediterranean basin was the birthplace of western civilization and the cradle of the European package holiday. Under its benign influence some of the most influential cultures have grown up, from the Nile to Jericho, Athens and Rome. Going further back, remains of modern humans are found in the near and middle east, dating from before the present climate developed. Human hands shaped the ecology of these lands and, conversely, the fate of several major

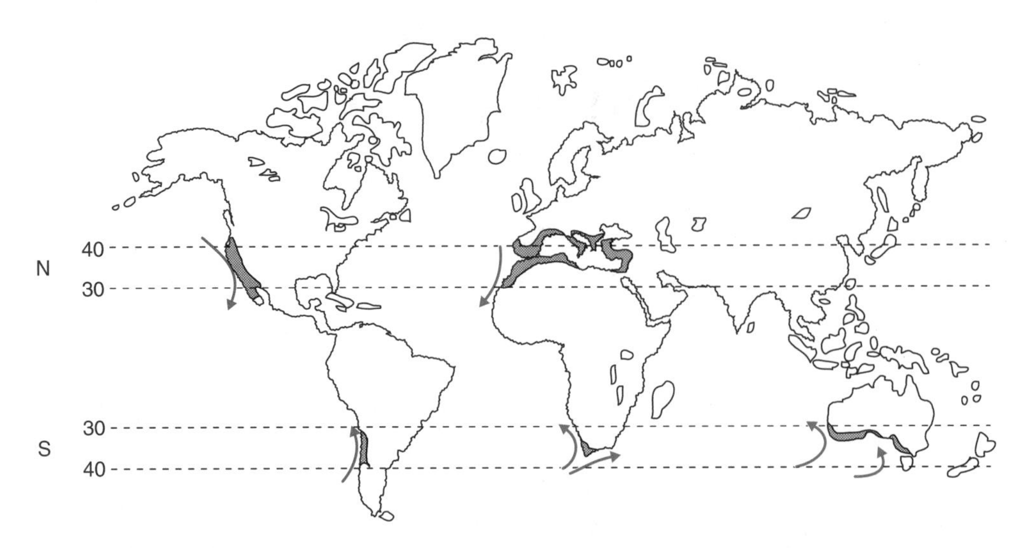

Figure 7.1 The location of the five mediterranean biomes around the globe, between 30° and 40° north and south of the Equator. The arrows indicate the cold oceanic currents which help to generate the drought conditions during the summer, a key feature of their climate.

cultures was written in its soils – from the annual rise and fall of the Nile to the doomed irrigation schemes around Babylon. Plato observed the wasting of the hills of Greece 2500 years ago, but even before that time the Greeks had taken to the sea, leaving Arcadia to trade and to conquer when its agriculture failed to feed its people.

Today, all mediterranean-type communities throughout the world, including the Mediterranean itself, are under threat, whether it be from tourism, agriculture or industrialization. Despite developing alongside human cultures for hundreds and thousands of years, the recent disturbance of these communities may signal more drastic change.

7.1 MEDITERRANEAN COMMUNITIES

Mediterranean climates represent transition zones between the moist temperate areas and the semi-arid regions of the subtropics. They are the most restricted of all climatic zones, confined to narrow bands around 30–40° either side of the Equator (Figure 7.1), on the western edges of continents. Here cold currents induce coastal fogs and mists so that moist air never progresses very far inland. Although the coastal area may be moist and green, beyond this there is often a desert. The prime climatic feature is an extended drought during a hot summer. Plants have to conserve water, so many cease to grow and some even shed their leaves in the summer. Winters are cool and wet, though rainfall is highly variable from one year to another.

Not only is the climate very similar in the five regions; three of them also share similar topographies. California, Chile and the Mediterranean basin have rugged landscapes with steep-sided valleys close to the coast. This creates gradients of moisture and exposure, pockets with different soils and contrasting plant communities, forming a finely textured mosaic of habitat patches. In South Africa and Western Australia the landscape is more ancient and rounded by a long history of erosion. Here there are fewer contrasts and their soils are also well leached and much less fertile. The vegetation of the five areas shares a common physiognomy (structure); that is, their plant species have similar appearances and physiologies, a consequence of their adapting to similar abiotic conditions. Not only do the plants have to withstand summer drought, they also have to survive the periodic fires which race through the tinder-dry vegetation. Typically, the community is of an open woodland dominated by short, evergreen trees (Table 7.1.) which ecologists call sclerophyll evergreen shrub; local names include maquis, chaparral and mattoral. Trees range from 2 to 5 m tall and have tough, small and leathery (sclerophyllous) leaves which they

Table 7.1 The main forms of mediterranean-type vegetation communities

Group	*Description*	*Location*	*Name*
General type	Typically, low woodland (trees 2–5 m), evergreens with sclerophyllous leaves, beneath which is an understorey of annual and herbaceous perennials.	Mediterranean	Maquis (France) Macchia (Italy)
		California	Chaparral
		Chile	Mattoral
		South Africa	Renosterveld
		Australia	Mallee
More arid or distributed types	Low and open communities (trees 0.5–2.0 m or low tussock bushes), often with drought-deciduous species, thorn bushes and aromatic species.	Mediterranean	Garrigue (France) Phrygana (Greece) Batha (Israel)
		California	Coastal sagebrush
		Chile	Jaral
Low nutrients soils supporting a heathland-type community	Low and open communities (between 0.2–1.5 m high), frequently showing a high degree of diversity and species endemism. Dominated by species of *Protea* in Africa and *Banksia* in Australia.	South Africa	Fynbos
		Western Australia	Mallee heathland

retain throughout the year. These are principally **xerophytic** plants; that is, they are adapted to minimize water loss through transpiration (Plates 26 and 27).

Where conditions are drier still, a lower plant community is found, often consisting of tussocks or low shrubs, or even tight cushions of thorns or **chamaephytes** (Plate 28). Their thorns prevent grazing and also help to create a microclimate of still air within the cushion. Some areas, notably the Californian coastal sagebrush and the Chilean jaral, have a variety of succulents that store water against the summer drought.

These similarities in plant form are not due to the same species being found in each region. Each has its particular collection of plants and animals which have colonized since the climate first developed over the last 1–1.6 million years and then re-established itself after the ice ages. Overall, each plant community appears to have accumulated individual colonizers from the surrounding areas, rather than an intact community moving in to occupy each region. South Africa and Australia were colonized from tropical communities to the north and west. The others have drawn species from adjacent temperate areas as well. Following this, each region enjoyed a rapid speciation as these new communities developed, resulting in a high level of endemism – especially amongst the plants.

The regions also differ in a number of key abiotic factors: soils and rock types, nutrient and moisture regimes, and the frequency of fire. In areas where there is a large altitudinal and latitudinal range we can find many community types within a short distance of each other, from the valley floor to the top of a hill, and this too promoted speciation. The regions also have different histories of human disturbance. Large-scale human influence can be detected in the Mediterranean basin about 7500 years ago, while its expanse of garrigue follows changes in land use in the thirteenth century. Francesco Di Castri shows how these factors produce plant communities which are characteristic of each region (Table 7.1 and Figure 7.2).

About half of the land in the Mediterranean basin can be described as undisturbed, confined primarily to upland areas. The cool and moister areas support cork oak (*Quercus suber*) and sweet chestnut forests (*Castanea sativa*). On the lower hills, much of the land had been used for grazing, resulting in soil erosion and the development of more arid conditions. As this has been abandoned, **maquis** scrub has developed, dominated by evergreen species, including kermes (*Q. coccifera*) and holm oak (*Q. ilex*) (Plate 26a). At lower levels, other trees and shrubs become important including broom (*Cytisus*), *Genista*, olive (*Olea europaea*) and strawberry tree (*Arbutus andrachne*, *A.*

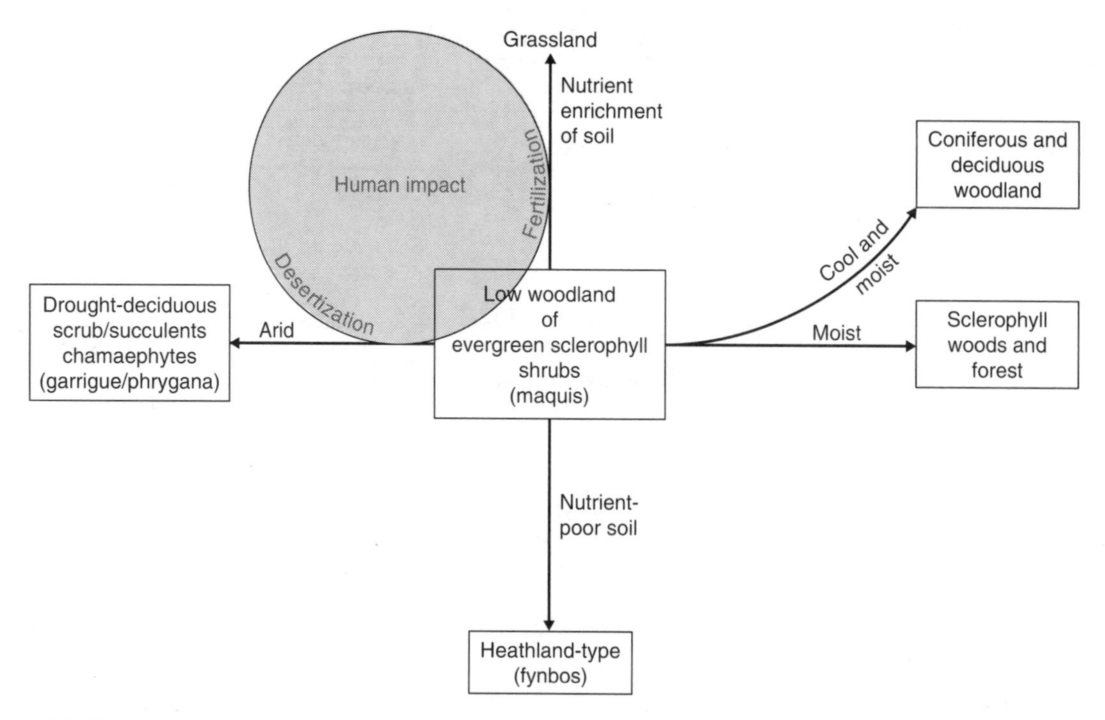

Figure 7.2 The main factors governing the plant composition of mediterranean-type plant communities. Human impact, primarily from the loss of cover (by fire and the grazing of animals), increases aridity and the fertilization of the soil – conditions that favour grasslands or, under drier conditions, low-growing chamaephytes.

unedo), usually with an understorey of xerophytic grasses.

With significant disturbance, grazing and trampling by sheep and goats, or the setting of fires, grasslands may develop or more often a **garrigue** – a low open scrub of evergreen oaks, juniper (*Juniperus oxycedrus*, *J. phoenicea*) and aromatic herbs, including thyme (*Thymus mastichina*), lavender (*Lavandula pedunculata*) and rosemary (*Rosmarinus officinalis*) (Plate 26b). The large patches of soil and rock separating the plants are evidence of its long history of erosion.

An equivalent community, but one with different species, has developed under similar pressures in Chile (**matorral**) and been subject to human disturbance. This is not so in California, where one or two species form large even-aged stands. Here, the open evergreen oak community has its counterpart as **chaparral** with scrub oaks (*Q. dumosa*) and aromatic shrubs (principally *Salvia mellifera* and *Artemisia californica*).

In south-western Australia, trees (primarily *Eucalyptus*) grow faster and taller than in any other mediterranean-type woodland and produce a multi-layered plant community. Where major soil nutrient levels are low and exposure is high, a short heathland community develops (heathland **mallee**, Plate 29). For the same reason, inland of the Cape region of South Africa a drier scrubland is replaced by a xerophytic heathland called **fynbos**. This is one of the most diverse plant communities in the world, with a vast range of endemic species. Both the fynbos and the heathland mallee are marked by having a predictable annual rainfall. This association of low soil nutrients, high environmental stability and high plant diversity is a pattern we shall encounter again in Chapter 9.

7.1.1 The role of fire

Along with summer drought, fire is a key feature in many of these habitats. This appears to be most important in Chile and to a lesser extent in the Mediterranean basin. Relatively little grows beneath the main vegetational cover and, given their evergreen habit, little litter is produced. Fires pass quickly through the vegetation, largely confined to its low canopy and, in the absence of fuel, are soon extinguished. Most plants, such as cork oak, are unharmed by the hot fast burn or sprout rapidly from crowns at soil level.

Many also produce seeds that need scorching to induce germination – a useful trigger for when space and nutrients are available. The matorral has species well adapted to frequent fires, with seeds able to germinate within days of the fire passing, so that a self-replacing community has developed. In contrast, species assemblages of the Mediterranean basin change as the frequency of fire increases: oaks are replaced by more resistant pine species, such as the Aleppo pine (*Pinus halepensis*) in France and Spain. At even greater frequencies, a true garrigue of aromatic scrub develops (Figure 7.3).

Fires have always occurred naturally in such regions, but ecologists recognize that human disturbance, including the deliberate setting of fires, is key to the physiognomy of these plant communities and is vital to their continuance. Much of the uplands bordering the Mediterranean were wooded until hominids increased the frequency of fire. Humans discovered the value of fire in clearing the scrub to improve both their hunting and their gathering (Table 7.2). The earliest indications of hominids using this strategy can be found from Northern Greece over one million years ago. By 10 000 years ago it was used routinely in the eastern Mediterranean. The gaps encouraged rapid regrowth by a range of edible plants and led to the early domestication of grasses for cereal production (Box 2.4).

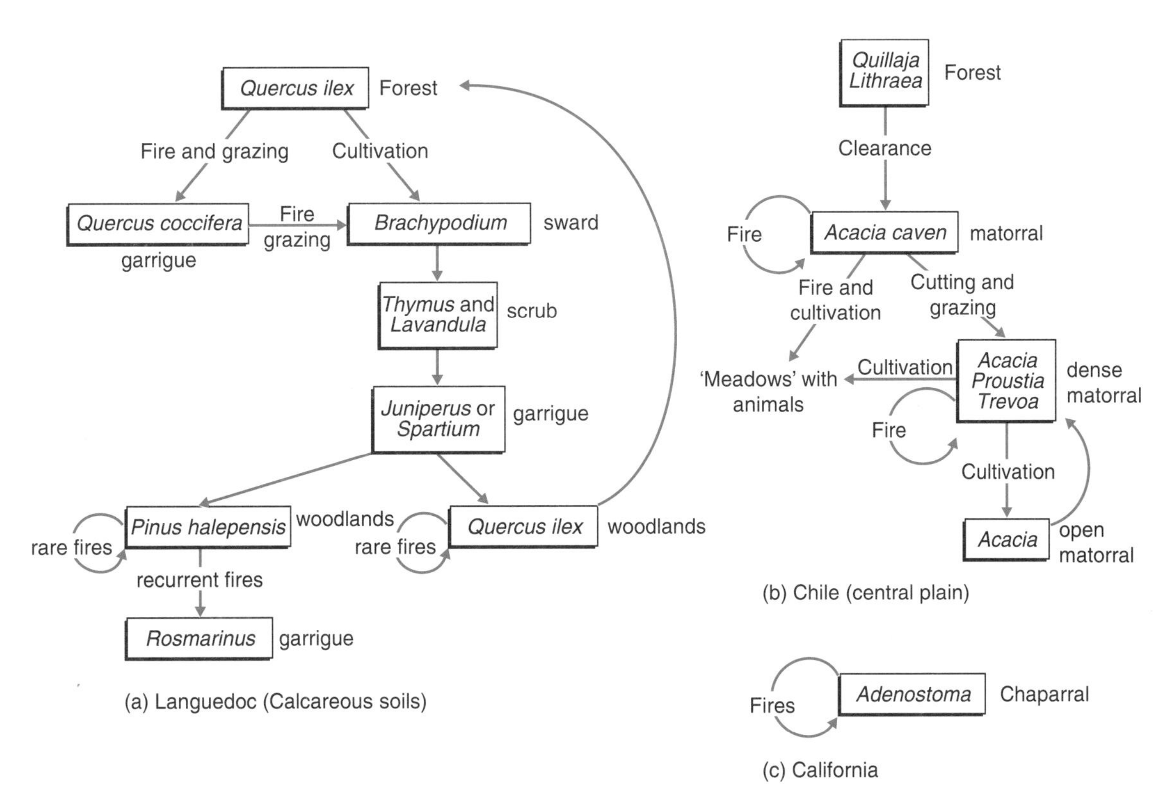

Figure 7.3 Changes in the composition of three mediterranean-type communities under different disturbances. In (a) the Languedoc of southern France and (b) the central plain of Chile, forest gives way to a series of more open communities depending on the nature of the disturbance and its frequency. The matorral of Chile is now largely confined to steep slopes where there is little grazing pressure. The abandonment of grazing in the Languedoc has allowed the native holm oak (*Q. ilex*) forest to return, especially in upland areas. (c) The chaparral of California is actually dominated by two species – either chamise (*Adenostoma fasciculatum*) as shown here or scrub oak (*Quercus dumosa*) – and each will tend to sustain itself under the influence of fire alone.

Table 7.2 The impact of three *Homo* species on the ecology of the Mediterranean basin during the period in which its climate became established. The development of agriculture, as hunter-gathering gave way to a more settled way of life, had a profound impact: *Homo sapiens* began to use fire and simple tools to expand food production. Use of certain selected grains promoted conditions that favoured the evolution of the earliest cereals. At the landscape level, human disturbance led to widespread soil erosion. It also led to established settlements and trade

Time-scale (thousand years BP)	*Homonid development*	*Technological advances*	*Ecological impact on the Mediterranean basin*
500	*Homo erectus*	***Lower Palaeolithic***	Relatively minor – mainly creation of gaps, later widespread using fire
		Hunter-gatherers using hand-axes	Gaps important for the spread of some grasses, especially *Avena sterilis* and *Hordeum spontaneum*, the precursors of modern cereals. Gaps also attract game. Fire may be used to hunt.
100	*Homo sapiens neanderthalensis*	***Middle Palaeolithic*** Hunter-gatherers. Flaked tools. First torch for carrying fire found in S. France	
40	*Homo sapiens sapiens*	***Upper Palaeolithic*** Bone/antler tools. Leaf blades. Fire can be kindled.	Fire used extensively creating gaps allow food flourish.
20		Animal husbandry: sheep and goats.	Forest clearance on a large scale for grazing and cultivation.
12	Population starts to increase rapidly	***Neolithic Revolution*** Agriculture, pottery weaving.	Primitive cereals found in Greece and Near East.
9		Agricultural cultures in Egypt, Greece and Persia.	
7		Grapes and olives cultivated.	
6		Shepherding in S. France.	
5		Sequence of cultures flourish in E. Mediterranean and Mesopotamia, later W. Mediterranean.	Increased aridity as plant as cover is lost. Widespread soil erosion begins
1.4	Population growth slows		Forest clearance stops.
0.1	Population growth confined to undeveloped nations	Industry and, more recently, tourism	Desertification in nations dependent on agriculture. Woodlands used for fuel. Overcropping and over-cultivation.

This had significant effects on the vegetation – removing cover that otherwise protected the fragile soil from the searing summer heat and drying winds. The increased grazing and trampling from livestock led to further soil loss. In effect, humans exacerbated the summer aridity, thereby favouring plants that could survive long droughts. A number of ecologists have noted that the Mediterranean basin represents a collection of communities that have developed in step with changes in human endeavour.

One of the most drastic changes was the switch to agriculture and direct cultivation (Box 7.1).

Upland areas were terraced to conserve soil and to create groves of olives, figs, almonds and pomegranates. Vineyards were planted. The grape, its cultivation and its fermentation have since been exported to each of the other mediterranean climatic regions. As viticulture has become more lucrative, the area under the vine has expanded, leaving the formerly grazed uplands abandoned, to regenerate a cover of maquis or garrigue.

7.1.2 Animal communities

Besides the regime of disturbance that is, to a greater or lesser extent, a feature of each region, other characteristics of each community result from the interactions between its species. Obviously, an animal community is closely tied to the collection of plants upon which it relies.

Of all the animal groups it is, perhaps, the insects that show the greatest similarities between the

Box 7.1

MANKIND AND THE MEDITERRANEAN

A large part of western culture has its origins in the Mediterranean basin – Arabic, Jewish, Christian and others can trace their heritage back to the lands of the 'middle Earth'. Many authors have pointed to the influence of Mediterranean ecology and geography on their cultural evolution. Francesco Di Castri suggests that 'co-evolution features are present in a number of ecological and cultural characteristics of these regions'.

To decide how true this is, it would be interesting to make such comparisons between cultures and ecologies for all biomes. Certainly the plant and animal communities of the Mediterranean basin have been selected, directly and indirectly by a series of human technologies, for the 10 000 years since the last ice age (Table 7.2). The current climatic conditions first appeared after the first glaciation of the Pleistocene, around 1.64 million years ago, as *Homo erectus* arrived in the area. A series of glaciations followed, as did (much later) a new species of *Homo* – the Neanderthals, a people adapted to cold conditions. Around 40 000 years ago these were joined in the Mediterranean by a new subspecies: modern humans. With the end of the last ice age, the coexistence of Neanderthals and modern humans comes to an end, and the older residents lose out.

Thereafter, from about 12 000 years ago, with the development of various tool technologies, the Mediterranean begins to change. This corresponds to a shift from a hunter-gatherer way of life to a largely sedentary and agricultural existence. We know that human beings had some measure of control over what was growing in different regions long before they actively cultivated the land. Wild barley may have been collected from the Nile Valley perhaps 18 000 years ago, and horses may have been used as pack animals from around the same time. Even before then, humanity left its mark on the landscape. Perhaps simply a sequence of burning and clearing to encourage useful grasses was used. Evidence that sheep, goats and gazelle were being husbanded exists from the Near East at the beginning of the Neolithic, around 10 000 years ago.

Cultivation proper began in the 'Fertile Crescent' of the Near East at this time and developed independently in China and Mexico at later dates. Primitive cereals were exploited for the first time in Greece and the Levant, around 8 000 years ago, and this encouraged a more settled way of life (Box 2.4). The division of labour promoted a culture and tradition which operated according to the seasons of the agricultural year. It may also have led to establishment of markets and towns, where produce could be traded.

The general trend in the Mediterranean basin was for technologies to originate in the east and slowly spread west. Technologies and cultures developed through trade and travel, with the sea being used as the highway between different centres.

The human population grew considerably and the expansion of agriculture took its toll. Evidence from deposits in Greece suggests that major soil erosion began about 1000 years after the onset of significant land use, a process that continued until about AD 600. These losses would have been well under way when Plato described them nearly 2500 years ago.

The history of the Mediterranean is of cultures meeting, often leading to confrontation and attempts to grab resources. These conflicts continue to this day. Because of the different traditions and languages packed into this small area, the Mediterranean is both blessed and cursed by its heritage. Yet the fruits of this treasury have been bequeathed to the rest of the world, from the grape and the olive to mathematics and philosophy, from the sublime to the ridiculous.

regions, possibly reflecting their ancient origins and also the problems of having to feed on tough, waxy leaves. One dramatic example of their community integration was examined in section 4.2: the mutualistic association between the protea *Mimetes cucullatus* and the ants which disperse its seed in the South African fynbos.

Rabbits are thought to have originated in the Mediterranean and were probably instrumental, along with goats and sheep, in preventing trees from dominating the drier areas. A number of similar grazers are found in the chaparral of California (brush rabbit and mule deer) and in the matorral (llama, alpaca). Rodents or their equivalents are important in all regions, in soils where it is easy to burrow. Additionally, the soil is a rich source of plant storage organs and a large range of invertebrates, and together these provide an extensive larder to their homes.

The communities of invertebrate decomposers within the soil are also very similar; they extend to a great depth and also move up and down the soil profile with changes in moisture levels. Their distribution reflect the depths to which some plants send roots in search of water.

7.1.3 Similarities

So, how much are the patterns of community structure repeated in these five locations? Some of the obvious comparisons are drawn together in Table 7.3.

Firstly, we should recognize that some regions have more in common than others. Two groupings stand out as distinct: Chile/California and South Africa/Australia (Figure 7.4). Their similarities stem from a shared geological history and also their latitudinal and altitudinal ranges (Figure 7.1). The Mediterranean basin stands out as different because of its large geographical extent, its great range of habitats and long history of human disturbance.

The plant communities of each region are typically determined by their soil types. In the Mediterranean basin, major differences emerge between the siliceous rocks and the more freely drained and highly eroded soils overlying lime-

Table 7.3 Ecological characteristics of mediterranean-type ecosystems

(a) Similarities across all the regions

* A high level of plant endemism which has arisen since the Pleistocene. Each region seems to have enjoyed rapid speciation of plants and some animals during this time.
* Each region has been disturbed by human interference which has opened up the plant community, allowing sclerophyllous shrubland to dominate large areas.
* Sclerophylly produces short, thick and tough leaves that are photosynthetically active throughout the year, though at a low level. This seems an adaptive response to low water and nutrient levels, especially phosphorous. However, sclerophylly also offers some protection from herbivores.
* The physiognomy of sclerophyllous plants seems to favour certain insect types.

* The nature of the insect community in turn appears to have favoured a particular collection of insectivorous birds and lizards.
* Counts of native bird species are similar for each region as are population densities amongst different feeding categories. This indicates that resources and resource utilization are comparable between the regions.
* The soil invertebrate community has a high endemism. This is attributed to rapid speciation during the Pleistocene, though there are greater similarities here than in the plant community. Giant earthworms are known from France and Australia. Mites are more dominant than Collembola (springtails) in mediterranean-type soils. Termites are found in all regions.
* The soil invertebrate community is formed into discrete layers in all regions and extends to a great depth with a large vertical movement.

Differences across all the regions

* Fire has a different impact on the species composition of each community. The Mediterranean basin changes radically as fire frequency increases, California and Chile less so.
* Most groups of plants and animals are phylogenetically distinct in each region, primarily because each was derived from different stock before the Pleistocene.
* Plant species richness differs markedly from one region to another and some communities change relatively little following fire or disturbance.
* Some ecologists suggest that sclerophylly may be a response to different factors in different regions – the diversification of sclerophyllous plants in South Africa and Australia may have been a response to poor soils rather than climate.
* The feeding ecology of South African birds is much less specialized than those of the other regions.

(b) Ecological characteristics

Chile and California

Similarities

* The plant community changes in a similar way with latitude and altitude in each region. Both have a similar seasonal growth pattern, concentrated in the spring.
* The plant communities have identical carbon-gaining strategies (patterns of photosynthetic activity). Similar leaf forms are found in comparable habitats in each region.
* Each region has been severely affected by weeds and other pests invading from the Mediterranean basin.
* Birds, lizards and mammals fill similar niches in each region, but with different species derived from different stock. Birds foraging in the vegetation for insects appear to fall into equally well-defined niches (which are also comparable to the Mediterranean basin).

Differences

* Californian chaparral has higher productivity and greater litter production. It is more prone to fires.
* Large areas of chaparral are dominated by a single shrub species, all of a similar age, and which does not develop into forest. The low shrubs seem to need fire to promote germination.
* Chile is more intensively disturbed by grazing. The matorral also has a well-developed herbaceous layer below it, which is missing in California.
* The matorral has a more fertile soil, a legacy of its recent deforestation. Its coniferous and deciduous forests are better developed. It also has a greater diversity of plants.

Australia and South Africa

* These woodlands are taller with a grassland understorey. South African woodlands have been invaded by several species from Australia.
  Both plant communities have a summer-dominated growth season that probably echoes the tropical origins of their floras.
* Both form distinct heathland communities on less fertile soils, with high levels of plant endemism.
* Much of their more fertile and flatter areas have now been lost to intensive grazing.
* Each region has high termite activity. They also share a large proportion of invertebrate genera. Australia has the highest soil invertebrate diversity of all regions.

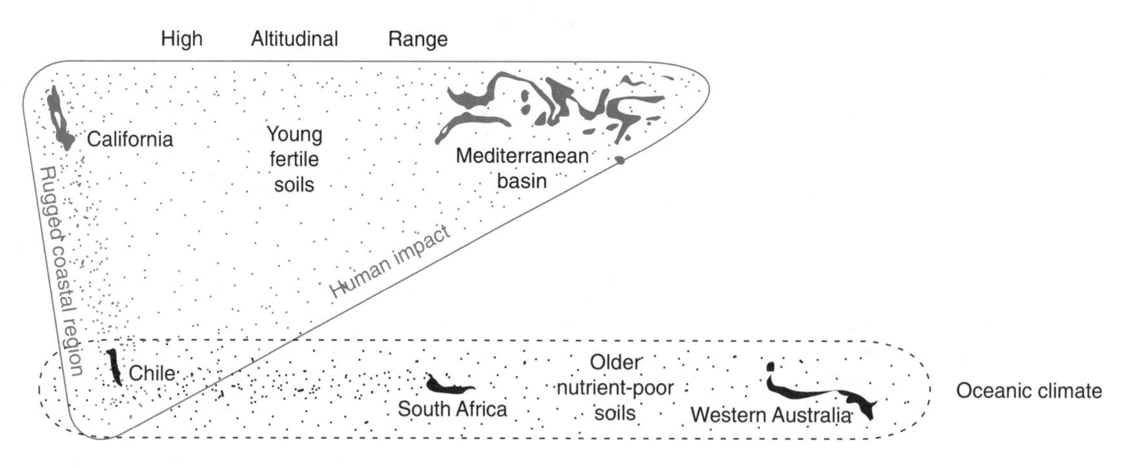

Figure 7.4 A division of the five mediterranean regions according to shared characteristics. The greatest similarities are between the matorral of Chile and the chaparral of California, and between the communities of South Africa and Australia. An oceanic climate sets apart the three southern hemispheric regions, but Chile has young and relatively fertile soils such as those found in California and the Mediterranean Basin. The latter is somewhat different from the rest because of its large east–west and north–south range and its long history of human interference.

stones. In South Africa and Australia the main distinction is between the more fertile soils, now largely used for grazing, and the poor soils that support fynbos or heathland mallee.

The main plant species of both Chile and California have evolved equivalent carbon-gaining strategies. Growth is confined to the spring, but the plants continue to photosynthesize at a low rate throughout the year. This is an advantage in soils devoid of nutrients, especially phosphorus. However, the response of the community in each area differs. California chaparral has little human interference and will restore its dominant species readily (Figure 7.3); the matorral is heavily grazed and now exists in its least disturbed form only on the steeper slopes. Animals both trample and fertilize the soil, encouraging different plants to grow, so these differences are entirely predictable.

Depending on the structure of the vegetation, animals show signs of producing equivalent communities in each region. A key factor for the birds feeding on the plants, or their seeds and insects, is the vertical development of the shrubs and the density of their cover (Figure 7.5). Similar numbers of bird and lizard species are found in each region and comparable population densities occur in equivalent positions in the vegetation. Feeding strategies are also comparable in each region.

There are differences. The fynbos, with its dominant plant group, the Proteas, provides more opportunities for nectar-feeding birds. Likewise, lizard diversity and abundance is higher in Chile because of the well-developed herbaceous layer absent in California. Overall, the complexity of the vegetation in all mediterranean-type regions seems to be a good predictor of bird diversity and the degree of niche separation of lizards within communities.

There is also some measure of similarity in their insect and other invertebrate faunas, perhaps because these are relatively ancient groups. This is particularly evident in their soil communities, especially in the southern hemisphere. Australia has the most distinct soil fauna, probably stemming from early isolation, but all regions are notable for their lack of beetles and high diversity of woodlice. Similarly, mites are more prevalent than springtails (Collembola) though the reasons for this are not clear.

A rapid speciation seems to have been a characteristic of each region as the modern climate became established. This was probably aided in some regions by the isolated habitats created by a fragmented landscapes. With a large altitudinal range, populations would have been isolated in discrete patches of habitat, so promoting local adaptations. From a starting point of a variety of species, each region was able to experiment with different species combinations. In most cases, human disturbance and interference may have prevented that experiment reaching a conclusion.

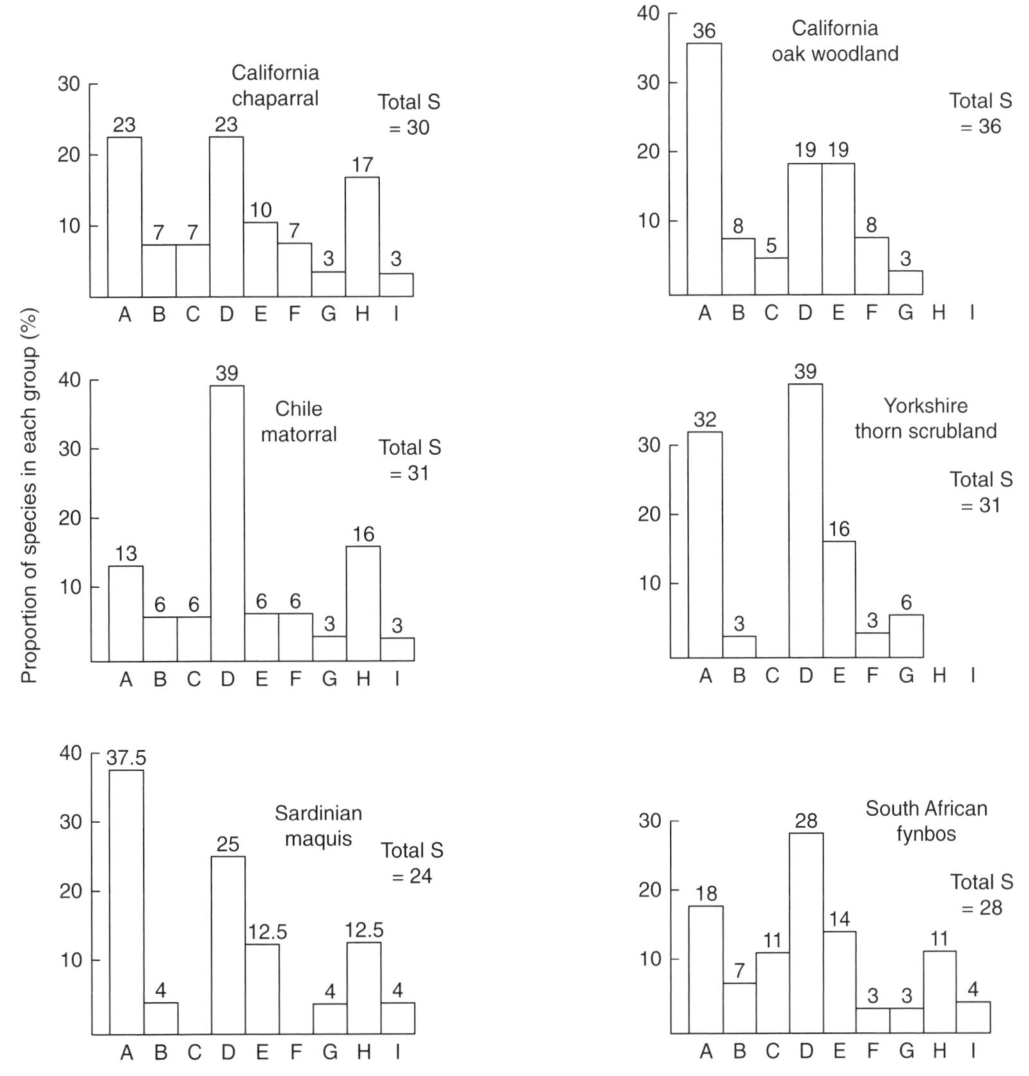

Figure 7.5 A comparison of bird communties in four mediterranean regions and in two very different woodland communities: Californian chaparral; Californian oak woodland; Chile matorral; Sardinian maquis; Yorkshire thorn scrubland; South African fynbos. In each case, the birds are grouped according to their feeding habit (A, foliage insectivores; B, sallying flycatchers; C, nectarivores; D, ground foragers; E, seed/fruit eaters; F, trunk/bark foragers; G, aerial feeders; H, raptors and scavengers; I, crespuscular insectivores). Notice that the proportion of bird species in each category shows the same sort of profile in all communities. However, the mediterranean-type communities do have groups (H and I) that are not found in the other habitats, and there is a greater similarity in the profiles between these regions. This is some indication that the bird community has developed to a similar result in each region, with similar proportions of species in roughly equivalent niches. The most different mediterranean community is Sardinia, which has the smallest number of species and an absence of birds in two categories, most probably because it is the one island among this group.

7.1.4 Integration

Does all this point to a unified community structure, following a well-defined plan in each of the five regions? A fair test of whether these communities converge on the same configuration would require us to account for all the variables of geology, fire frequency and so on. Many of their differences can be traced to variations in each of these factors between the regions (Table 7.3). Perhaps the simple truth is that no two habitats or regions are sufficiently alike, even though they have equivalent climates: it seems they are separated as much by their histories as their ecologies.

We should also recognize that these assemblages may be too young, or are too disturbed by human influences, for us to decide. Perhaps with a longer history and more consistent pressures, these communities would have greater similarities. Among the vertebrates we do see evidence of convergence in their niches, and the relatively fixed proportions of insectivorous birds (Figure 7.5), for example, does point to species–species interactions determining community structure.

A highly integrated community is one whose species are highly dependent on each other. One measure of the degree of integration is whether the loss of one or more species causes the rest of the community to collapse. Unfortunately this is not much of a test in communities with a long history of disturbance, where the connections between species might be looser. In fact, we find it difficult to predict how many or which species could be removed from any community before it changes significantly. Certainly many of the interactions are not critical for the survival of the larger community, even though several species may be very closely associated. While some species of orchid may need specific insects to pollinate them (section 2.1), orchids seldom play a central role in their community.

One of the old ideas of ecology – that the community functioned as some kind of 'superorganism' – is now largely defunct (Box 7.4). It is true that mediterranean communities have developed some comparable structures and relationships across widely dispersed regions, but in some measure these have arisen from different species sharing similar strategies to life in mediterranean climates. We need to decide now what constitutes a community and the degree to which its form and function results from the interplay between its species. The ideal experiment would allow communities to assemble themselves, starting with the same abiotic conditions and drawing from the same pool of species. Then, given long enough for species turnover to reach a minimum, would the result be the same collection of species, organized in the same way, or would chance push in them different directions? Of course, this experiment, carried out over realistic time and space, has never been performed. The five mediterranean regions have not produced identical results, in large part due to their different starting points.

7.2 WHAT ARE COMMUNITIES?

Because such an experiment cannot be performed easily, we are left speculating whether communities have clearly defined **rules of assembly** (section 5.4) determining their species composition and configurations, with ecologists arguing whether a particular community is inevitable for a particular climate. And argued they have, since the earliest days of the science. There are two main opposing views: one sees a community as being inevitable and predictable, determined primarily by species interactions and climate, whilst the other sees it as a loose collection of species adapted to that habitat, and which come together by chance (Box 5.4).

Central to this debate is the role of species within the community. To what extent does one species determine the presence of others? Clearly some species are crucial for the structure of the community and ecosystem function. In terrestrial communities, the **dominant species** are typically the larger plants and these govern what animals, and indeed other plants, form the rest of the community (section 8.2). **Keystone species** are those whose presence determines the nature of the community, without which the community would change to a different configuration. One example is the elephant of the African savanna, whose browsing activity maintains the open grassland and suppresses encroachment by scrub (section 3.5). There will also be a number of species that can be lost without an appreciable effect on the community as a whole. Sometimes no single species is important, but rather functional groups of several species, termed **guilds**: specialist insect pollinators or more generalist carnivores, for example, often play key roles. Added to this, some animals can be important in more than one community. Migratory birds, for example, visit two or more communities as guest workers in each ecological economy. This reminds us that communities should be regarded not as closed and fixed arrangements, but as open and dynamic associations.

A community is the collection of plants and animals that we observe persisting together and interacting with each other. They are often built around dominant species which, in terrestrial communities, are invariably plants. That many plant species tend to occur in association with other species inevitably suggests there are repeating patterns, tightly integrated species assemblages, in which species closely depend on each other. For plant ecologists, who find it convenient to work with identified species associations, these communities have a functional reality (Box 7.5). The immobility of plants means that, in temperate regions at least, we can predict these associations with some measure of confidence. This is because they are largely based on a plant's strategies – water relations, nutrient demands and life history – reflecting its adaptations to the abiotic characteristics of its environment. However, variations from one habitat patch to another, with some plants present and others absent, show that the associations between species are far from fixed.

Animals are more closely tied to the lives of their companion species and we often have to talk in terms of niche rather than the detail of the species inhabiting it. We expect to find certain configurations: without herbivores, carnivores are not possible; where there are insects, there will be insectivores; where those insects are found depends on the structure of the foliage; and so on.

While communities seem to be organized in particular ways, the example of the mediterranean regions show that we should not expect the same list of species. Their species do share comparable adaptations to life, but those found in each region also depends on history and chance. Communities are dynamic and we must expect that what we observe today will be different tomorrow. Nevertheless, the community can reasonably be used as a functional unit, which, rather like the species concept (section 2.2), is primarily a device in helping us to understand species interactions and ecosystem processes.

7.2.1 Change in communities

Human disturbance is a key to the structure of many mediterranean communities and we play an important role in maintaining these assemblages. But even in our absence, certain species would be lost and others would invade. The competitive battles and other interactions that would develop within a community left to its own devices would quickly lead to a procession of species changes. Eventually, perhaps, a community might develop in which invasion is difficult and species loss is small. Relative though it may be, this is as close as the community gets to a stable configuration and it is called a **climax** (Box 7.4). The process by which the community arrives at this point – the sequence of species invasions and replacements – is known as **succession**.

Studying these sequences shows how species interactions come to define a community and explain how they can change so readily. A **primary succession** develops on a site where there has been no previous occupation and where colonizing species must themselves establish ecological processes on a substrate previously devoid of life or its products. A dramatic example is the colonization of volcanic lava, which is often able to sustain a simple but viable plant community within a few months of cooling (Plate 30). Likewise retreating glaciers and encroaching sand dunes present material which contain no organic matter and few nutrients and they too can be colonized very quickly. **Pioneer species** assist in the accumulation of nutrients and organic material, eventually to become a resource for later colonists.

A **secondary succession** is a process of recolonization. Here the site has been previously occupied and may even retain some of its original species. Most importantly there is soil containing organic matter and a reserve of nutrients that can be used by invading plant species. This makes the process of recovery faster than in primary succession. However, which species colonize depends on not only what remains of the original ecosystem, but also the distance invading species must travel to colonize the site.

In both primary and secondary succession the sequence of species arrivals and losses tell us how the community is put together. Some species cannot colonize unless others are present and some will only colonize when others are absent. These sorts of interactions give us an insight into the degree of dependency and integration within a community.

7.2.2 Assembling communities

For a primary succession to start at all, some species must be able to survive conditions that are not conducive to most others. So what makes a good colonist? A prime feature is a capacity to disperse, to find a new site in the first place. Both plant and animal pioneers are often *r*-selected opportunistic species (Chapter 6) – typified by weedy annuals and prolific insects – capable of producing vast amounts of seed or offspring very quickly.

Colonizers need to survive a wide variety of conditions and not be tied to any particular resource. Good colonizers are also rarely dependent on other species. This is one reason why early-successional communities tend to consist of loose associations of opportunists, which are able to survive a range of habitats.

Clearly, plants have to become established if animal colonists are to survive. Although obvious, here is one simple rule of assembly: plants ultimately provide the resources needed by both herbivores and carnivores. Similarly, the development of the microbial and fungal communities, essential to the development of the soil community, will be assisted by a pioneer plant community. This assistance, known as **facilitation**, is one of the prime roles played by colonizing plants. It is not confined to providing energy or nutrients, but extends to modification of the environment – by providing shelter or hastening the release of nutrients from rocks or, more generally, by creating conditions amenable to new arrivals.

Nor is facilitation confined to the early stages of a succession. Later in a sequence, well-established perhaps keystone species create conditions or provide resources upon which much of the rest of the community depends. The long-lived oaks of the maquis or chaparral play this role in several mediterranean communities throughout the world.

However, early-successional facilitators often sow the seeds of their own destruction, so that pioneer species are rarely part of later-successional communities. Species which colonize later are typically more competitive and will tend to squeeze out *r*-selected species. In sand dune communities (Figure 7.6), marram grass (*Ammophila arenaria*) stabilizes the loose sand and adds organic material but this enables bent and fescue grasses (*Agrostis* and *Festuca* species) to invade and gradually dominate the community.

And so succession proceeds. As a general trend, competitive species become more dominant as the community develops. These are species whose use of resources or activity make conditions difficult for their neighbours, preventing the latter's establishment or growth whilst securing resources for themselves. **Inhibition** slows down colonization by other species or excludes them altogether.

Eventually, as competitive species dominate resources and space, colonization is reduced to a minimum and then species turnover is small. We could thus describe a succession as a process by which colonizers are replaced by a more stable community of competitors. Again, we could see this as a general rule of assembly so that late-successional communities, not subject to significant disturbance, become dominated by slow-growing competitive species.

However, these species changes are matched by changes in ecological processes. As nutrients and organic matter begin to accumulate, so a reserve of resources and a decomposer community can develop. As before, this will facilitate the arrival of other species and, in the process, will enable nutrients to

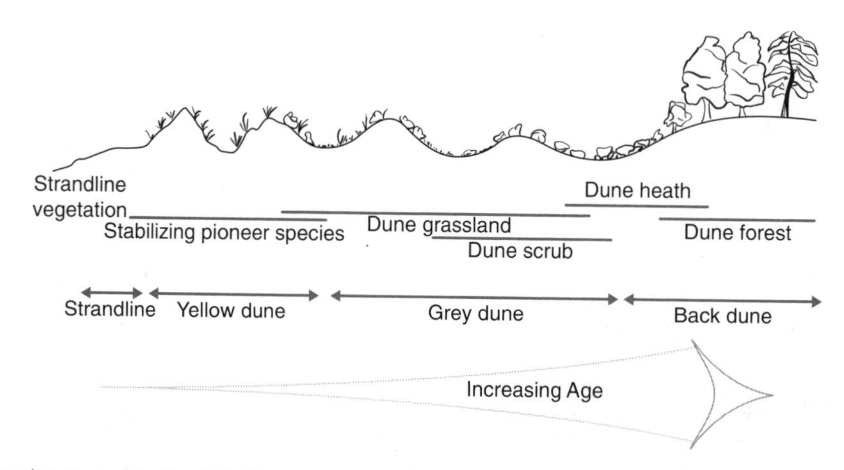

Figure 7.6 Succession on an idealized Mediterranean sand dune. New dunes are colonized at the seaward end whilst heathland and forest occupy the oldest part of the site, consisting of late successional species that have replaced the earlier colonizers. Note that changes occur as transitions rather than as abrupt zonations, with overlap between the various species assemblages within the cross-section succession.

cycle more tightly within the system. Towards the end of a succession, nutrients released by the decomposers are quickly used by the competitive species and all nutrients are cycled more efficiently. We thus expect ecological efficiencies to improve from early- to late-successional communities.

So great is the pressure on these resources that late-successional communities often favour plants which are able to survive when nutrients, light or water are scarce. This is known as **tolerance** and such species are typically slow-growing and infrequent reproducers (section 4.3). Theirs is a waiting game, maintaining themselves in times of shortage, ready to exploit times of plenty. As others fall away, so tolerators come to dominate. This again may be a rule of assembly in habitats where nutrient supply is highly restricted.

Where development is primarily driven by these internal community interactions, ecologists refer to it as **autogenic** succession. If some abiotic factor determines the succession, it is termed **allogenic**. For example, the periodic waterlogging of a soil in a sand dune slack (hollow) limits the sort of plants and animals which may colonize and leads to the slack developing entirely differently from the drier dune ridges. Another example, as we have already seen, is that the frequency of fire or length of summer drought in mediterranean communities favours certain species over others in each region. Their assemblages change with the frequency of fire (Figure 7.3). That these communities show similar physiognomies in different parts of the globe reflects this shared adaptation to disturbance.

Together, these species interactions and rules of assembly limit the number of possible outcomes to a succession. By comparing a large number of examples on equivalent sites in different locations we can identify repeating sequences. If chance associations are responsible for the community, then predictability will be low. Otherwise a volcanic eruption is like setting the ecological clock back, forcing a community to reconstruct itself, eventually returning to a similar endpoint or assemblage.

7.2.3 Disturbance and succession

Any disruption that causes the loss of most or all of the resident species will re-set the community clock. The rate of change can be dramatic. Volcanic ash and lava can wipe out a community in a matter of hours, while the advance of an ice sheet is much more pedestrian. In both cases, some species may escape, but others are lost or do not survive being displaced. Clearly the likelihood of extinction will depend on the scale of the disturbance and the availability of refuges. In some cases, a gradual environmental change can be tracked by the colonizing species, so that pioneers are continually advancing into nearby territory followed by late-successional species as sites become more established (Box 7.2).

During the building of a community, species come and go. Some become an essential part of a succession only to be lost later on. Even in the later stages, species are being lost and gained and so there is always a turnover of species. As a climax is established, species turnover is reduced to a minimum. Overall the number of species of any group of plants or animals may remain unchanged – a balance between losses and gains called the **species equilibrium**. The number of species depends on the area and the number of niches it contains (Box 7.3).

A succession will proceed to some sort of resolution only in the absence of disturbance although all communities are subject to some disruptions. Communities are often adapted to a particular frequency of disturbance. They are said to **incorporate** this disturbance. Mediterranean communities are dominated by plants able to withstand short fires, and a certain frequency of fire is needed to maintain their community structure. These communities never come to a stable unchanging climax configuration dominated by relatively few competitive species. Instead, fires continue to initiate a cycle of changes in the plant community. This is sometimes called an **arrested succession** or a **plagioclimax**.

The diversity of the whole community may depend on the frequency of such disturbances. Certainly the diversity of birds in the five mediterranean regions rises with increasing frequency of disturbance. In his **intermediate disturbance hypothesis,** Joseph Connell suggests that a community shows its greatest number of species when the disturbance is relatively frequent – not too frequent to cause major extinctions but frequent enough to prevent the competitive dominants squeezing out other species. Disturbance, in whatever form it takes – trees being blown over, periodic drought and so on – creates gaps which invasive species may occupy, quickly utilizing the resources which become available. The competitive species do not then have it all their own way. Some species, called

Box 7.2

STUDYING SUCCESSION

Ecologists who investigate succession are dogged by one problem: time. Its gradual, long-term change makes it difficult for one ecologist to follow its sequence (known as a **sere**) within a single site. One solution is to match the successional pattern over several comparable sites, each at a different stage in development. Some sites may consist of areas of several ages, containing communities at all stages of the sere. The most common example is the hydrosere, the transition which occurs from open water to dry land, where each stage of community development can be seen (Figure 7.7).

Shallow lakes have a short ecological life. Gradually the colonizing activity of reeds, rushes and sedges and the sediment that collects around their roots mean these lakes are inevitably shrinking. Away from the water's edge, moving on to land we encounter sediments deposited long ago and now further down the successional sequence. Close to the water, typical trees would be willows (*Salix* spp.) able to tolerate wet conditions. These species are excellent facilitators since their large root runs and transpiration pump water out of the soil. Further back on drier deposits, willows give way to late-successional species in the form of canopy trees such as oak (*Quercus* spp.) and a well-developed ground flora.

Sand dunes, too, show seral change as a linear sequence going inland (Figure 7.6). In the Mediterranean, the

Figure 7.7 The hydrosere: an area of open water gradually being encroached by vegetation.

stabilizing effect of marram (*Ammophila arenaria*) and the nitrogen-fixing activity of sea medick (*Medicago marina*) eventually give way to the grassland of the grey dune with its damp dune slacks and dry ridges. On the more sheltered landward side of the dune ridges, assemblages of xerophytes such as *Cistus* and lavender (*Lavendula stoechas*) form. This becomes more continuous further inland and on nutrient-poor sites the result is a dune heath dominated by heather and heaths (*Calluna* and *Erica*) gorse (*Ulex* spp.) and juniper (*Juniperus phoenicea*). On less disturbed sites with richer soils, trees colonize the oldest dunes and woodland of oak (*Quercus* spp.) and pines (*Pinus* spp.) develops. There is a similar pattern of development in other regions with a mediterranean climate. In California, a heathy sagebrush or pine forest may form, whilst in Australia the soil conditions determine whether a mallee heath or *Eucalyptus* forest will develop. Sand dunes can be problematic to study as they are prone to disturbance. In a matter of hours, a dune system may be engulfed in seawater or drifting sand, burying the plant community and pushing back or changing the course of a succession.

This can make the interpretation of a succession very difficult indeed. We can investigate the history of some sites by taking vertical cores of soil and dating each layer using radiocarbon dating and fossil pollen analysis. This can be correlated with tree ring analysis to look for climatic extremes which may have affected growing conditions in the past. If one is lucky, evidence of ploughing or burning may even be visible in the soil profile as disturbed banding or a blackened deposit of carbonized material.

Zev Naveh describes ash from a fire in the Tabun cave on Mount Carmel in Israel that dates a human presence in the region to the Early Upper Pleistocene. We can detect human activity in the region from the change in the composition of the pollen here and in deposits of many other regions. In the Mediterranean basin, pollen characteristic of recolonization following fire (so-called fire-followers) includes, for example, goosefoots (*Chenopodium* spp.), mayweeds (*Matricaria* spp.), bistorts (*Polygonum* spp.) and plantains (*Plantago* spp.) as well as the shrub species that begin the succession after disturbance (*Pistacia*, *Quercus*, *Pinus*, *Juniperus* and others). In the same way, we have constructed the rise of agriculture from the pollen grains left behind from the first cereal crops of barley, wheat and oats (*Hordeum*, *Triticum* and *Avena*). Sedimentary records like these tend to work best in anoxic environments, such as bogs, where the breakdown of organic material is slow and the quality of preservation is high.

A few sites have been studied for a sufficiently long period to record community change. These often use fixed-point photographs taken at the same place over the years. Here one can see movement in the extent of various species as the communities change with time. Until ecologists live for ever, this is one of the few ways we have of investigating succession in real-time.

fugitives, are able to maintain a population just by colonizing one gap after another.

One way of judging predictability of a succession is to measure the scale of disturbance it can withstand before it becomes diverted into a different configuration. A community which is largely a product of autogenic succession may be very elastic, always returning to a similar species configuration. In this case we might regard the community as being highly integrated with strong interactions amongst its members. In contrast, a change in a key abiotic factor may allow an allogenic succession to switch to a very different configuration. Allogenic factors are seen to dominate all present

Box 7.3

INVASION AND EXTINCTION

Succession proceeds through the arrival and loss of species. In time the number of species present tends to become fixed, or at least the number of invasions and extinctions slows. We might therefore define a climax community as one in which species turnover is at a minimum and dominant or keystone species do not change.

In fact, for most communities the number of species present at this stage is closely related to area – the greater the area, the larger the number of resident species. This means that later on in a succession, when most of the competitive battles have been fought, we may be able to predict the number of species present provided we know both the area of the site and the type of species involved (as this varies for different groups of animals and plants).

Robert MacArthur and Edward Wilson presented a simple model that shows how species equilibrium is achieved for oceanic islands. In fact, their **island biogeography theory** can be applied to any habitat isolated from similar areas by hostile surroundings. Such habitat islands include mountain tops, isolated cliffs, lakes, ponds and so on. We often see 'islands' of trees or a pond amid intensively farmed land (Figure 7.8) or open spaces amongst city sprawls, isolated from similar habitats and the potential colonists they contain.

Early on in the process of colonization, species are added quickly to the island communities, starting with those that can colonize rapidly. With time the rate of immigration (line I in Figure 7.9a) falls off as the species that are most able to establish themselves do so.

Species which are gained can also be lost. Some may colonize but never establish a breeding colony and the non-viable population dies out. Others become extinct by chance or because they lose competitive battles for the island's resources. Initially, competition is low, with few species present to compete for the limited food, water and shelter, but later, as niches become filled, the pace of competition steps up. As the island's succession proceeds, the extinction rate (line E in Figure 7.9a) therefore increases. Eventually immigration is balanced by extinction and the number of species on the island does not change.

MacArthur and Wilson pointed out that rates of invasion and extinction within island communities depend on an island's size and its distance from the mainland. Larger islands are bigger targets and stand a better chance of being colonized. They also provide more resources and niches, allowing more species to colonize. Extinction rates will be lower than for their smaller counterparts because they can support larger populations. Similarly, islands nearer the mainland will be more readily colonized. As a result, species number does not depend simply on area but on distance from the potential colonizers on the mainland (Figure 7.9b). Since powers of dispersal vary widely between various plants and animals, the equilibrium number differs between groups.

For this reason, island biogeography is applied to groups of species rather than entire communities. These ideas have also been picked up by conservation ecologists who find that an understanding of how species are lost and gained by a site is often central to its good reserve management.

Figure 7.8 A small fragment of woodland surrounded by a sea of intensively farmed land – a virtual island or habitat island.

mediterranean communities, and fire probably played a key role even before human disturbance became significant.

Few ecologists today would argue that succession can only have one outcome – a particular climax community for a particular climate. Most now recognize that chance, history and different starting positions can only lead to different results (Box 7.4). We can see recognizable and repeating patterns which allow us to identify broad community types, like the sclerophyllous evergreen scrub we have been describing from around the world (Box 7.5). But while they show similar configurations and collections of adaptations under similar conditions, there is little evidence to suggest that the resulting community functions as an organic whole.

7.3 SUCCESSION AND CONSERVATION

Clearly, some communities have been displaced a long way from their natural state. Mediterranean-type communities are seen as plagioclimaxes whose cycle of change is unchanging and which has incorporated a certain frequency of disturbance.

Today, the pressure of human numbers and activity has accelerated the pace of disturbance, hastening loss of plant cover in these areas and prompting widespread soil erosion and degradation. Just as Plato described the process occurring in Ancient Greece, so many of these areas are now showing signs that key plant species are being lost – with potentially disastrous consequences. Some ecologists think mediterranean communities are now so

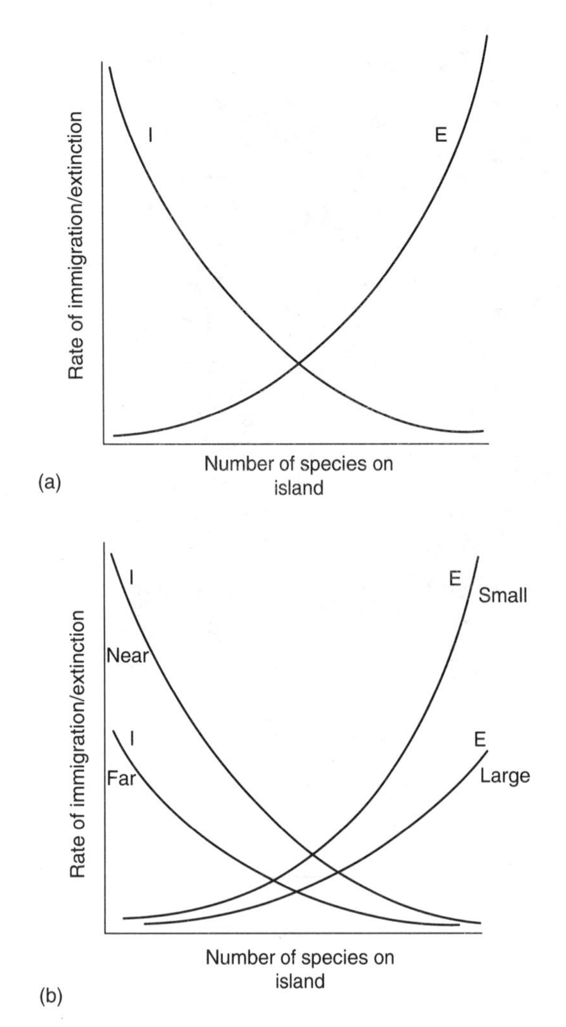

Figure 7.9 Island succession.

severely damaged that we are close to exceeding the limits of even their ecological elasticity. Despite our long history together, the pressures on the landscape from tourism, from agriculture and from industry may mean that the scale of human disturbance is just too great for the partnership to continue. Along with the threat of increased aridity due to the effects of global warming (section 8.3), some are predicting a major wave of extinctions in the near future, as many of the endemic plants of these regions are lost.

Human disturbance has taken a different form in recent years. Through our travels we have connected the five regions to each other, shuffling species between them. Not only do we now cultivate the vine in all mediterranean communities, we have also introduced some less desirable species into each region. *Carpobrotus*, a large succulent plant from South Africa's Cape Province, has now become an invasive weed in the Mediterranean basin and California. Introduced as an exotic garden ornamental, it escaped from cultivation and spread into a number of habitats. A good colonizer, it forms large mats of vegetation (Plate 31). Whilst it can stabilize sand dunes, it also swamps native plants, potentially changing the course of their succession.

Another introduced species displacing native species in mediterranean communities is *Eucalyptus* from Australia (Plate 32). Adapted to the nutrient-poor soils of the mallee, its fast growth makes it a formidable competitor as it scavenges nutrients from richer Mediterranean soils, resulting in an impoverished ground flora within *Eucalyptus* plantations. In addition, the tree itself is unpalatable to the resident invertebrate community, and its high content of essential oils both sours soil and increases the occurrence of fires. As a consequence, communities naturally controlled by periodic fires may become rapidly destabilized by the increased frequency and intensity of the disturbance and fail to recover.

Our capacity to effect change through disturbance is particularly apparent in mediterranean communities. Today cultures collide as global travel brings people – and other species – together. When viewed against industrial processes, tourism may seem benign, but it too has its negative impacts and the popularity of mediterranean climates means that these regions are in serious risk of being 'loved to death'. Even amongst the most well-meaning of visitors, the very act of visiting a place involves the use of its resources (land, water and so on), adds effluents and demands facilities which, taken together, can add up to a potent form of disturbance.

Armed with knowledge of communities and the processes which construct and sustain them, we are more able to manage natural and semi-natural areas for conservation. By examining similarities and differences, we can begin to piece together past influences and predict future trends. Such knowledge will bring with it an appreciation of conservation requirements of some of the planet's most beautiful and threatened places, along with clues as to how they might be protected.

Box 7.4

THE CLIMAX CONTROVERSY

The first three decades of this century produced a lively debate about the nature of the climax community. This has shaped the way ecologists have thought about and studied communities ever since.

In 1916 botanist Frederic Clements argued that succession led to a single common climax for its climatic region: the **monoclimax**. He saw succession as a highly directional, self-determining process which produced a closed or discrete community. For Clements, the community had an existence of its own above and beyond the sum of its constituent parts. It might be regarded as a functional 'superorganism' in its own right, tightly integrated like a living body. These ideas led to the holism of today's 'deep ecology' and have been extended beyond species and populations to include the non-biotic elements of the biosphere, as with the Gaia hypothesis discussed in the Prologue.

Others saw things very differently. Henry Gleason pointed out that we rarely find the distinct boundaries between communities that we should expect if each community consisted of close associations between species. Instead he presented a view of succession as being something of an ecological free-for-all. His **polyclimax** theory of succession was an individualistic model in which 'every plant species is a law unto itself', leading to a number of possible climax communities. Taken to this extreme, a succession was largely a series of chance events ultimately leading to a group of similarly adapted species coinciding within a given area. For Gleason, the path of succession was far from fixed and in the resulting open community any one of a number of combinations could arise.

The entrenchment behind each argument lasted decades. Nowadays succession is no longer viewed as a linear sequence of stages moving to a fixed endpoint. In a world of diversity there are many possible outcomes. As we saw with sand dunes in mediterranean climates (Figure 7.6, Box 7.2), there are at least three main endpoints for the oldest dunes. Those least disturbed and having favourable conditions might develop into dune forest. On drier, nutrient-poor soils the result might be dune heathland, whilst the harshest, most disturbed sites might never get past a mosaic composed of grass and heath. Of course, there are many intermediate forms so we rarely see distinct boundaries but instead **transitions** between these communities. There are, however, regular patterns which reflect the species interactions that develop during a succession. It is these we classify as types that are characteristic types of certain communities (Box 7.5).

SUMMARY

A species assemblage or community represents an association of species adapted to the abiotic conditions of a region, especially climate, and which also reflects the interactions between the species themselves. The mediterranean-type community is found in five regions of the world sharing a climate very similar to that of the Mediterranean basin. Some also have similar geology, topographies and geologies, though closer inspection shows that there are clear differences in the species composition of each community.

We can recognize two main clusters. One is the California–Chile complex, where coastal mountain ranges and a high latitudinal extent are characteristic.

Box 7.5

CLASSIFYING COMMUNITIES

Communities form distinct patterns within the landscape – say, where grassland gives way abruptly to woodland. The simplest classification of these vegetational communities is based on characteristic species, such as scrub oak maquis. Or we simply refer to the community itself, such as chaparral or fen. We may also make reference to some key physical factor, as in chalk grassland and acid heaths. The most comprehensive classification uses all three categories and also distinguishes variations (in the form of subcommunities).

The use of species to classify plant communities, known as **phytosociology**, has its origins in the eighteenth century. At that time scientists saw patterns within vegetation and recognized this as evidence of interactions between the plants themselves and with the environment at large. In 1825, Dureau de la Malle carried out a series of felling and regeneration experiments in his own Normandy woodlands, concluding that plant species lived socially and that this was 'a condition essential to their conservation and to their development'.

In the early years of this century, the study of phytosociology was a peculiarly European pursuit, pioneered by research groups working on plant associations in Zurich in Switzerland and Montpellier in France (the Zurich–Montpellier School) and a rival but nonetheless related Scandinavian system in Upsalla. The Zurich–Montpellier approach was descriptive and its rival was largely numerical, but both classifed communities according to predictable species associations. Since then phytosociology has developed its own terminology and conventions, which groups communities in a nested hierarchy reminiscent of the Linnaean system (Figure 7.10).

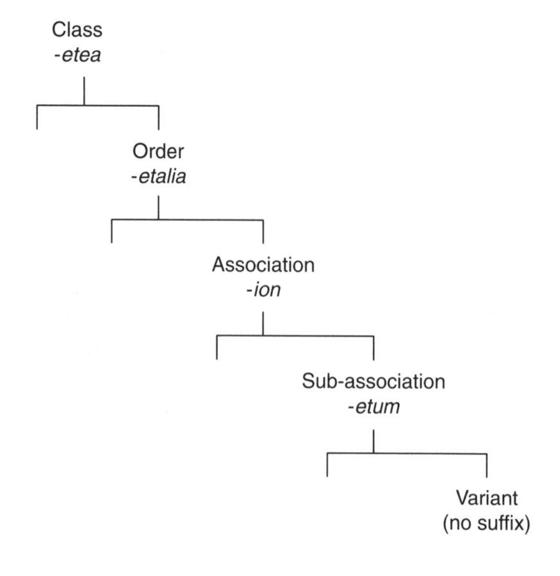

Figure 7.10 Units of continental phytosociology. These are named according to their position in a nested hierarchy. Suffixes are used from the large-scale classes right down to highly localized variants of communities. The system derives the root of a community's name from the most characteristic species and uses the Association as the fundamental unit (if compared with species classification, Association is to Alliance as species is to genus). For example, an Association within a heathland community characterized by the heather *Calluna* would be known as *Callunetum*.

Other community classification schemes have developed from these origins including Britain's National Vegetation Classification scheme (NVC). The scheme, developed by a team led by John Rodwell and Andrew Malloch, is based on 20 person-years of field work. Altogether 33 000 samples were recorded, covering 14 different ecosystems, classifying Britain's vegetation into 300 distinct communities with around 750 subcommunities.

The standard sampling unit of the NVC consists of a quadrat sample taken within a 'stand', an area uniform in both composition and structure and representative of the community. The presence of species and their abundance is measured by the five-point **Domin scale**. Their occurrence is categorized according to how strongly they are associated with a particular community or its subcommunity. There are four possible groupings of species in any community or subcommunity. First are the **constants** – species that are most likely to occur and best characterize the community (and invariably give their name to it). Next are the **differentials** – those used to pick out the different subcommunities associated with the community. **Preferentials** are rather like differentials but have a weaker association and may occur in more than one subcommunity. Finally, there are the **associates** – the 'also-rans' of the NVC which may be found in the community but have no affiliation with either the community or subcommunities. These can either be universally common or might be rare species with a patchy distribution.

The NVC has many uses. At its most basic level it standardizes terminology

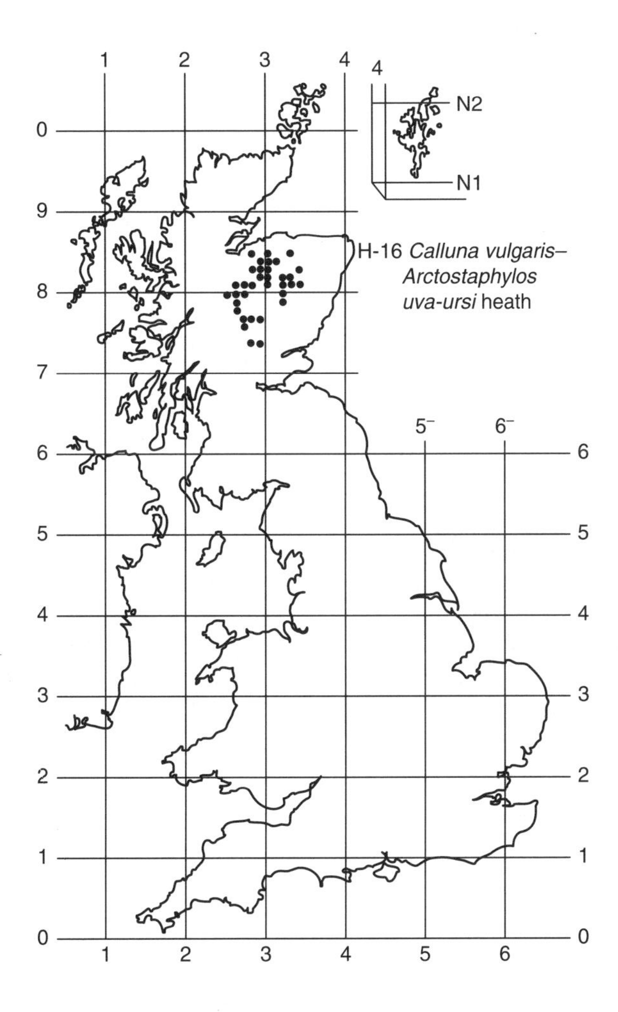

Figure 7.11 NVC distribution map of *Calluna vulgaris–Arctostaphylos uva-ursi* heath.

and survey techniques, allowing comparison between sites along with mapping both local and national distributions of different communities (Figures 7.11, 7.12). Sites can also be included in **geographical information systems (GIS)**. This computer-based technology uses aerial and satellite imaging to map large areas, and is a powerful tool in conservation ecology. The descriptions also detail threats and community transitions which predict future successional trends, valuable in conservation management. Similarly, it can provide information on a site's past communities, which is particularly useful in projects that aim to restore lost or damaged habitats.

MG1a *Arrhenatheretum, Festuca* sub-community on verge bank.
MG1b *Arrhenatheretum, Urtica* sub-community on disturbed verge.
MG1c Arrhenatheretum, Filipendula sub-community in verge ditch.
MG6a *Lolio-Cynosuretum* typical sub-community on frequently mown verge edge.
MG6b *Lolio-Cynosuretum, Anthoxanthum* sub-community with avoidance mosaic.
MG7e *Lolio-Plantaginetum* towards a gateway.
MG7f *Poo-Lolietum* in gateway.
MG10a *Holco-Junecetum* typical sub-community in ill-drained field hollow.
W24b *Rubus-Holcus* underscrub, *Arrenatherum-Heracleum* sub-community invading around field margin.

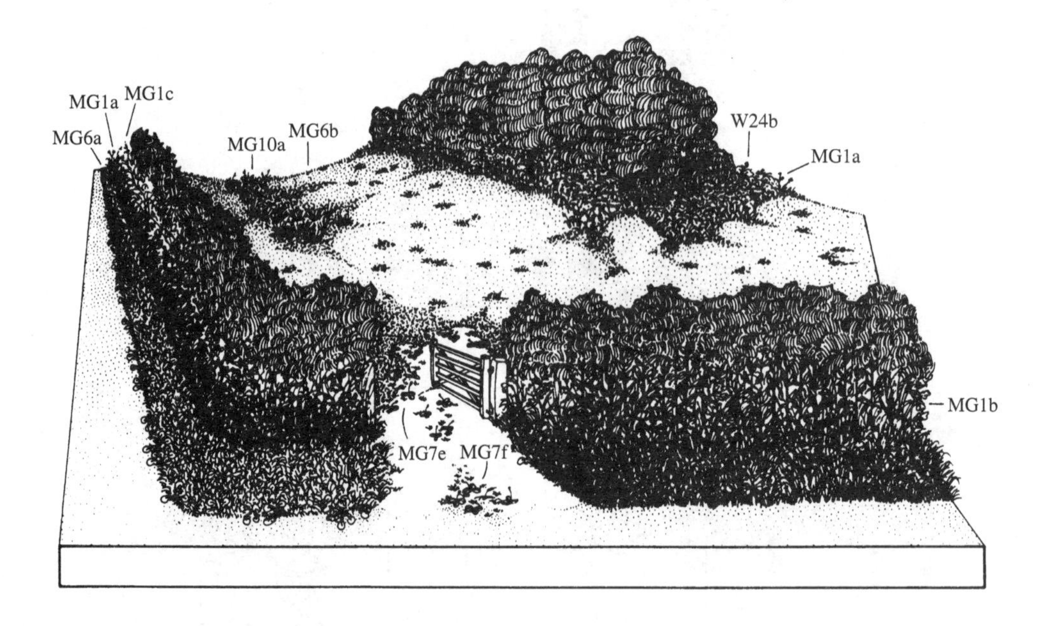

Figure 7.12 An illustrated NVC map of a mesotrophic grassland.

The second group, South Africa and Australia, have flatter landscapes and more ancient, more impoverished soils. The Mediterranean basin itself has a wider range of habitats, because of its greater extent and varied geology. All show similar plant physiognomies and some equivalence in their animal (particularly vertebrate) faunas. Disturbance, in particular fire, plays an important role in these communities and, as a result, many mediterranean-type species are fire-adapted.

However, the details of their different histories, land-use patterns and the species pool from which they have been colonized since the Pleistocene together conspire to create different communities and this suggests that a community does not always develop towards a fixed community type.

Communities change by a succession of stages in which species colonize and go extinct. In primary succession, a community starts without previous occupiers and organic matter. Secondary succession occurs in disturbed areas recolonizing from the remains of the previous community. Autogenic succession is where the community composition is largely determined by species interactions. Disturbance, particularly its frequency and scale, can also control the development of the community in allogenic succession. Species will themselves govern the succession through facilitation, inhibition and tolerance, with a general shift from rapidly dispersed and quick-reproducing pioneer species to longer-lived competitive species that eventually dominate.

When species turnover is at a minimum the community is said to be at climax. The number of species at equilibrium depends on the area and group of plants and animals concerned. Long timescales make succession difficult to investigate and the climax should not be viewed as being an inevitable result of a linear succession.

Despite having developed alongside humans for much of their history, and having incorporated disturbance into many of their communties, the mediterranean regions are today threatened by excessive disruption from tourism, increasing industrialization and changes in land management practices. We have also introduced non-native species that can shift community structure in some areas, threatening many endemic species. Conservation depends on an ability to recognize and describe communities, along with an understanding of processes, such as succession and species turnover, which control them.

FURTHER READING

Polunin, O. and Huxley, A. (1965) *Flowers of the Mediterranean*, Chatto and Windus, London.

Smith, R.L. (1992) *Elements of Ecology* (3rd edn), Harper Collins, New York.

EXERCISES

1. Complete the following paragraph by inserting the appropriate words, using the list below (some words may be used more than once, some not at all).

   Mediterranean communities are found in __________ regions throughout the world and occur in a band between __________ and __________ degrees either side of the __________. They are restricted to the __________ edges of continents and have a climate with long, __________ summers and cooler, __________ winters. Plants of these communities are therefore __________ -resistant and display __________ characteristics including thick, tough leaves with a reduced leaf area and a low growth habit. __________ is also an important factor in these communities and is a source of disturbance, occurring either naturally or artificially. __________ activity in the form of __________ helped to shape these communities in the past, but modern development in the form of industry and tourism can have serious negative impacts on mediterranean communities world-wide.

   agriculture, arid, drought, dry, Equator, exposure, flood, fire, five, human, species, western, wet, world, xerophytic, 30, 40, 60.

2. List the factors that have contributed to the high endemism amongst the species of mediterranean-type communities around the globe. In each case, indicate the scale (Global, Continental, Regional, Local) of the processes that have allowed speciation.
3. What are the two basic forms of succession and the processes which underlie them?
4. Select the correct answer to complete the following statement. Keystone species ...
   (a) are always the most dominant species within a community.
   (b) are essential for the continued existence of a community.
   (c) are always the largest predator in the community.
   (d) cannot be herbivores.
5. What are the characteristics of a good colonist?

Tutorial or seminar topic

6. It has been said that the ecology of the Mediterranean basin evolved with early humanity. Di Castri argues that there is also a Mediterranean cultural tradition, born out of the nature of the region and its landscape: 'a Mediterranean system of values ... based precisely on the co-existence of a range of different languages and cultures'.
   Are cultures influenced by their ecology today?

Landscape ecology, hierarchy theory and large-scale change – the relationship between spatial and temporal scales in ecology – climate and the major ecosystems – tropical, temperate and boreal biomes, forests and grasslands – the carbon balance of the atmosphere and global climate change.

8 Scales

Travel, it is said, broadens the mind. It certainly helped ecological thought to move forward. Like other well-travelled naturalists of the 1800s, Alfred Russel Wallace and Charles Darwin were struck by the immense variety of species found in the tropics. Ever since, ecologists have been asking why there are so many more species in the tropics than in the temperate regions.

Climate is the most obvious difference between the two zones. The major terrestrial ecosystems trace lines of latitude, to form broad bands that follow the climatic zones between the equator and the poles. But the pattern is far from a perfect match and is interrupted by distinct communities that respond to a particular landform or local climate. Nor are the bands in fixed positions: communities advance and retreat with fluctuations in the global climate, across land masses that slide around the planet in geological time.

Space and time are linked: ecological processes that operate over large areas also operate over long time-scales. For example, the ecology of the northern latitudes begins to make sense only when we understand the effects of glaciations over the last 2.5 million years. Yet most modern ecological study has been rooted firmly below the ecosystem level, confined to local communities and relatively short time periods. These were taken to be the limits of our scientific resolution, the scales within which we could answer questions with some degree of confidence.

Now ecologists are being asked about long-term change, about the history and the future of life on Earth. Today, ecology uses data from ice cores, the fossil record and satellites. The spatial and temporal horizons of the science are being stretched. In this chapter we consider the effect of scale on ecological processes and introduce some of the principles behind large-scale ecology.

We start by examining the relatively new area of landscape ecology and the connection between spatial and temporal scales in ecological change. We then go on to describe the major terrestrial ecosystems and show how their distributions reflect the prevailing global climatic pattern. Finally we explore the intricate relationship between the biosphere and the atmosphere. Our understanding of this interaction is crucial if we are to make sensible predictions and policies that anticipate climate change over the next 50 years.

8.1 LANDSCAPE ECOLOGY

Size and timing are everything in ecology. Ecologists have to be sensitive to the implications of scale for different species. The distances and time-scales over which we humans operate are inappropriate to the mayfly confined to a small stretch of river and with just 12 hours to mate and lay its eggs. On the other hand, biomes that straddle the globe move only slowly, taking decades or even centuries to shift their position significantly.

Ecology has to encompass a considerable range of scales. The chances of a bird surviving a migration from one hemisphere to another may depend on the insect community in its nest and the passengers it now carries with it. The processes of nutrient transfer and species interactions, represented here by the bird and its parasites flying south, spans scales from a few centimetres to thousands of kilometres.

Much ecological research has historically been concentrated on local and well-defined ecosystems. However, ecologists have always recognized that a bird's nest or a woodland is far from being a closed system. Exchanges of genes or species or nutrients occur across the boundaries between ecosystems and these inputs and losses have to be part of our accounting of the balances within each ecosystem. Some ecosystems are more closely linked to their neighbours than others. The greatest frequency of exchanges is likely to be between adjacent ecosystems within a naturally defined landscape.

Landscape ecology is a fusion of geography and ecology, a study of the spatial arrangement of ecosystems and the large-scale processes that unite them. For example, the slope of a watershed governs its rate of drainage, soil depth and movement, nutrient supply and the stability of the surface. Its altitude and its forest cover will influence how much rainfall it receives. The chemistry of the water in its stream will depend essentially on the soil from which it has drained. The quality of its soils may reflect the nature of the bedrock, and will also be influenced by its vegetation cover and that of adjacent ecosystems higher up the valley. This intermeshing of factors means that we will fail to understand fully how a ecosystem functions until we consider the landscape of which it is a part (Figure 8.1).

The archetypal landscape of lowland England is a patchwork of fields sewn together with hedgerows and streams, dotted with occasional patches of woodland (Plate 35). The vineyards of the Mediterranean are bordered by bramble and scrub, interrupted by rocky outcrops with woodland confined to steep slopes and upland areas. All landscapes, natural or otherwise, have these discontinuities and form a mosaic of different ecosystems, each of which is defined by a distinct plant community.

Landscapes range over kilometres and are most easily delineated by obvious discontinuities – like a

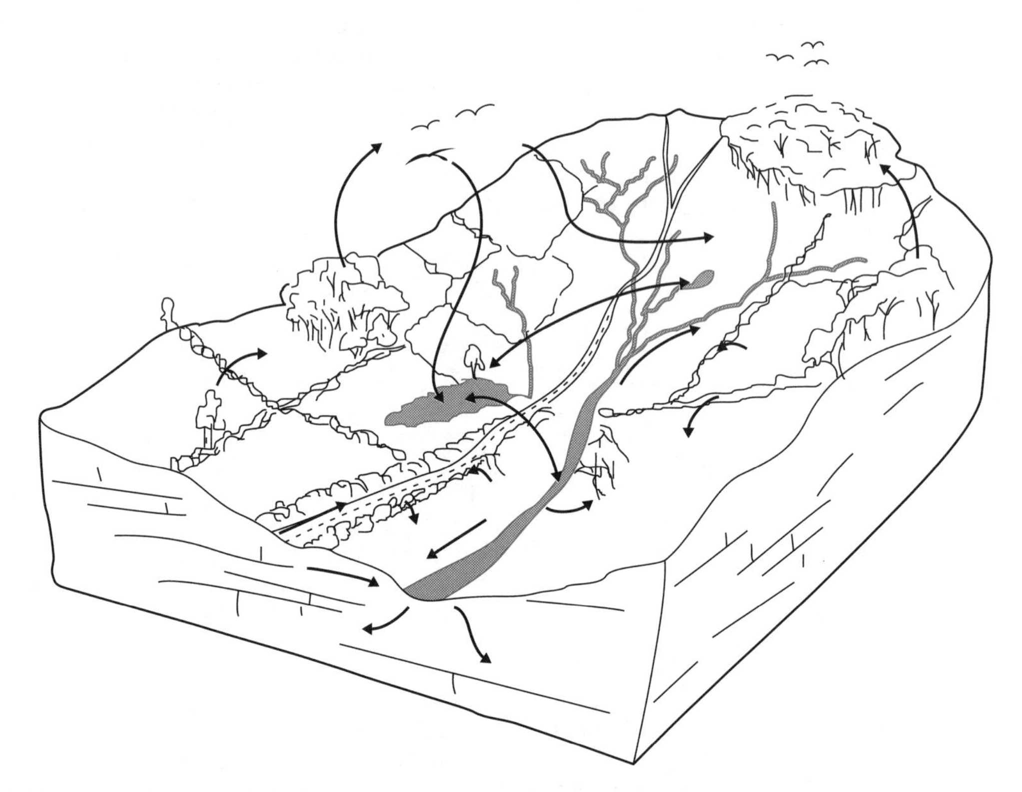

Figure 8.1 Large-scale processes in landscape ecology, in this case for a temperate agricultural area. Genes, energy and nutrients move using wind, water or the mobility of the biotic components. Networks of corridors – rivers, hedgerows or roadside verges – provide habitats down which species can disperse across the landscape. All the living organisms will be affected by large-scale effects, such as the climate within the valley, or the movement of soils and water down the valley.

watershed or a significant change in geology or land use. Its component ecosystems share the same past and the same climate and are subject to similar geomorphological processes or management practices. A series of landscapes may be collected together in a region.

At the scale of the landscape, ecologists recognize three basic spatial elements: the patch (say, an area of woodland), the matrix (the background ecosystem that dominates the landscape – such as the fields or vineyards) and the corridors that traverse the matrix, connecting the patches (the hedgerows or streambanks) (Figure 8.1). Between these elements are more or less distinct boundaries or **ecotones**. The length of the boundaries, the size of the patches, the distance between them and the extent to which they are connected by corridors can all be measured and used to quantify different landscapes. A highly heterogeneous landscape will be fractured into a mosaic of patches, but highly interconnected if there is a network of corridors. The vast fields used in intensive wheat production form a very uniform, very homogeneous landscape with short (straight-line) boundaries (Plate 36).

The patchiness or 'grain' of a landscape will vary with scale. A small patch of woodland may itself be divided into areas of closed canopy mixed with glades where sunlight reaches the ground and where a very different plant and invertebrate community can be found. Some of its invertebrates may be confined to the glade for the whole of their life cycle. In contrast, a bird flying over the landscape may distinguish only between woodlands and the open fields and therefore responds to a much larger 'grain size'.

The patchiness of the landscape at different scales has important implications for a range of ecological processes, from population dynamics to species diversity. The size of any one population is likely to be closely tied to the size of its patch (section 3.2). We have already seen how a metapopulation divided between a number of patches may be able to sustain itself even though local populations in some patches may go extinct (section 3.6). We can measure rates of colonization of unoccupied patches and relate this to the overall population size. Such movements also determine the flow of genes, so the spatial pattern of a landscape also sets the degree of isolation between populations and thus may determine rates of speciation (section 2.5). We saw earlier how the high endemism of some mediterranean-type communities has been attributed to their landscape heterogeneity (section 7.1).

Species differ in their capacity to cross the intervening matrix (or use corridors). Not only is the distance between patches important, but so is the matrix that separates them. Aquatic birds may fly readily from one pond to another, but the fish do not make the journey quite so easily. Isolated patches might be considered as habitat islands, so that their complement of species will depend on patch size and distance to the next patch (Box 7.3). In addition, distinct communities are often found at ecotones, where particular species benefit from the transitional conditions of these 'edge' habitats. For example, agricultural landscapes typically have a variety of predators and herbivores which exploit the open fields but return to the field margins and woodlands for cover.

The spatial arrangment of ecosystems within a landscape is also important when the area suffers a major disruption, such as a forest fire or the spread of an infectious disease. The heterogeneity and connectivity of a landscape then has implications for the survival or extinction of a species. A very disconnected landscape may mean the disturbance is confined to a small number of adjacent patches. On the other hand, following a large-scale disruption, a high degree of connectivity may ensure that colonization is rapid following a disturbance.

Our problem is to establish the nature of the significant interactions at different scales, to predict the overall impact and the prospect of long-term change. For example, a loss of fish or the poisoning of its waters will have more impact if a water body is a large feature of the landscape, a main tributary rather than a small, isolated pond.

One dramatic case is the pollution and overexploitation of the Aral Sea (Plate 37) in Kazakstan and Uzbekistan. It was once the fourth largest lake in the world. In the last 60 years its feeder rivers have been diverted to irrigate cotton fields and since 1960 the Aral has shrunk by nearly half – around 30 000 km^2. As its freshwater supply has dwindled, so have its waters become increasingly saline, leading to a massive loss of its unique (endemic) fish and invertebrate species. The fishery has collapsed and large steamers that once used to work the lake are today stranded in desert sands, far from any water.

The impact of the shrinking lake today extends many hundreds of kilometres from its original shoreline. Changes in the local climate have followed as it has shrunk. Now dry winds blow salt deposits on to the surrounding soil, scorching the vegetation and making the land almost useless.

Increasing atmospheric pollution and the poor quality of the Aral's waters have created major health problems for the local people. Infant mortality rates are among the highest in the world and life expectancy among the shortest. International efforts are now being made to restore water levels, but it there is no prospect of restoring the ecosystem.

One might suggest, quite reasonably, that it did not require a detailed understanding of landscape ecology to predict some of these changes, even at these larger scales. Plans to divert or dam rivers in various parts of the world, from the Danube in Europe to the Mekong in South East Asia, have also raised fears about their climatic, ecological and economic implications. Even now, we are still prepared to fund large-scale engineering of landscapes despite the warnings of ecologists, hydrologists and geomorphologists who have seen previous schemes cause environmental havoc, never paying back their promise.

To make reasoned decisions and quantified predictions requires some way of organizing such information to describe the important interactions in the landscape. One approach looks at the relationship between processes operating at different scales to understand how they interact to impart stability on a landscape. These ideas have been formulated in more rigorous scientific terms as hierarchy theory.

8.1.1 Hierarchy theory

While a landscape may appear unchanging, its component ecosystems are often much more variable. A single ecosystem may consist of a mosaic of patches, each with a different inventory of species, each undergoing change. At these smaller scales, differences between areas or between samples taken over short periods of time are marked. But averaged over the whole landscape, species composition changes little, even over long periods. The variation we observe in an ecosystem depends upon the scale over which we measure it, in both time and space.

Hierarchy theory suggests that this is a general property of many complex systems: it sees the system organized into various functional levels, differentiated by the rate at which some process operates. Most importantly, it proposes that these interactions between levels regulate the system.

Consider the example given in Figure 8.2, where the process is carbon fixation in plant tissues. Within an individual leaf, fixation fluctuates rapidly as carbon dioxide concentrations near the stomata change from one moment to another. At higher levels in the hierarchy (say, the leaves of the whole tree) average rates of fixation vary much less. In the canopy of the forest, over a larger spatial scale, rates are increasingly uniform. As we move up through the hierarchy, the variation in whatever we measure decreases.

To put it at its simplest, biomes appear unchanging because they are big. At the regional level, it may take years, perhaps decades, to detect a significant shift in carbon fixation rates, following a large-scale change in, say, the species composition of the forest. The general level of fixation for the whole forest ultimately depends on rates occurring in individual leaves.

However, constraints work in the opposite direction too. Photosynthesis in individual leaves depends on local humidity, which itself depends on the density of the tree cover. Forests change the climate of the immediate area, inducing rainfall and raising humidity: about half of the rainfall in the Amazon basin originates from the evapotranspiration of the forest itself. The speed with which water cycles through the system depends on rates of photosynthesis and the opening of stomata to fix carbon. Thus a higher level attribute (forest cover) governs a process (carbon fixation) operating at the lower level (the leaves) of this hierarchy.

Other limits will be imposed by interactions between species, or with animals, or with microor-

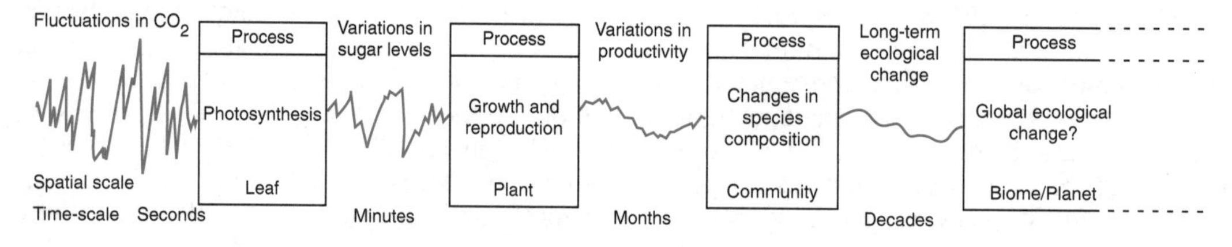

Figure 8.2 A simple example of the effects of a hierarchy of levels governing an ecological process, in this case carbon fixation by photosynthesis.

(a)

(b)

Plate 26 (a) Well-developed maquis, dominated by oaks (*Quercus coccifera* and *Q. ilex*) in the Languedoc, southern France. This is a relatively undisturbed area, too high and rocky for cultivation and long abandoned for sheep grazing. (b) *Quercus coccifera* garrigue dominating abandoned grazing in the Corbière Hills of the Languedoc.

Plate 27 Chaparral in California.

Plate 28 Phrygana on Kalymnos, a Greek island off the Turkish coast. Note the low, drought-resistant deciduous tussocks of the characteristic chamaephytes found in these very arid conditions.

Plate 29 Mallee, Western Australia.

Plate 30 Plants colonizing larva, Samoa. Pioneering plants begin the process of primary succession on cooled larva floes.

Plate 31 *Carpobrotus* invading Mediteranean sand dunes.

Plate 32 *Eucalyptus* displacing native Mediterranean oak/pine forest.

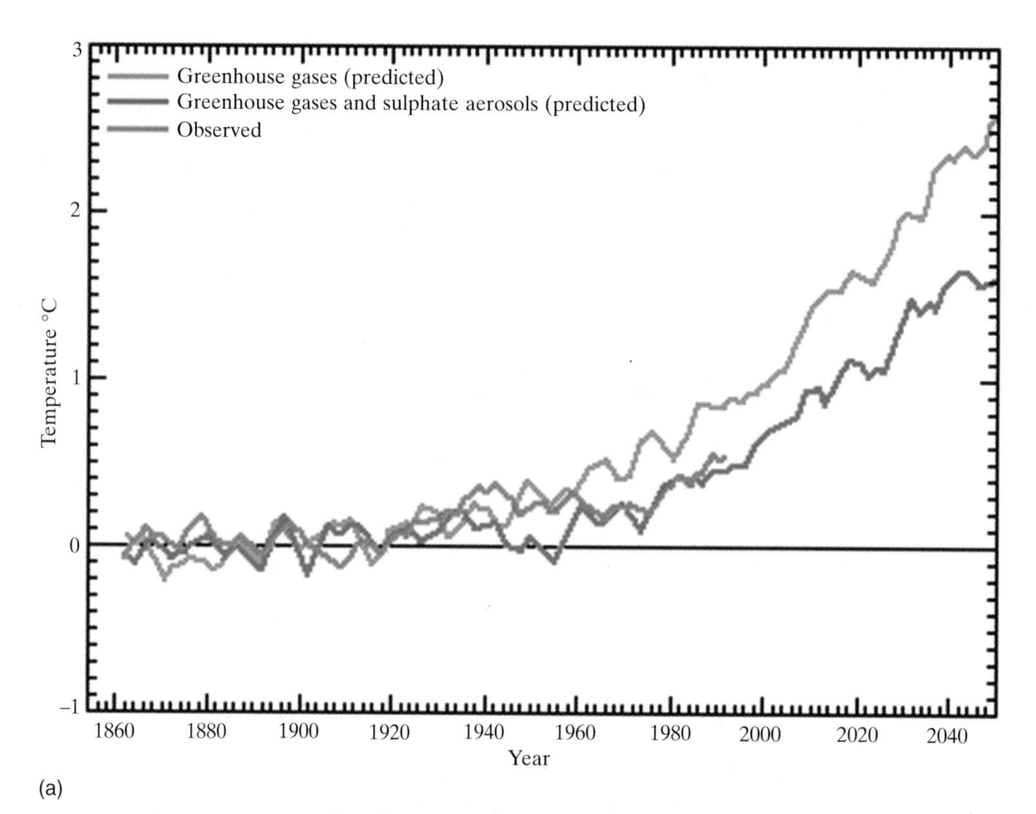

(a)

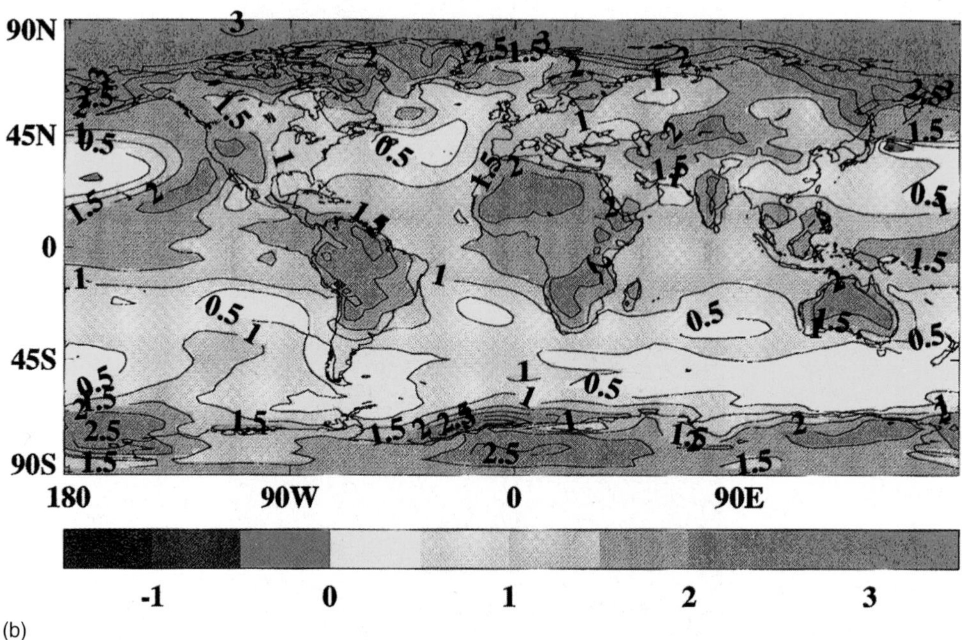

(b)

Plate 33 Modelling the temperature of the Earth in the past and in the future. The 1995 run of the GCM of the Hadley Centre of the UK Meteorological Office included the effect of sulphate aerosols. (a) For the first time a GCM accurately modelled the 0.5°C rise over the last century. (b) Even allowing for the cooling effects of the localized sulphate aerosols, mean surface air temperatures are predicted to rise over most of the globe in the next 50 years, though in some places rather more than others.

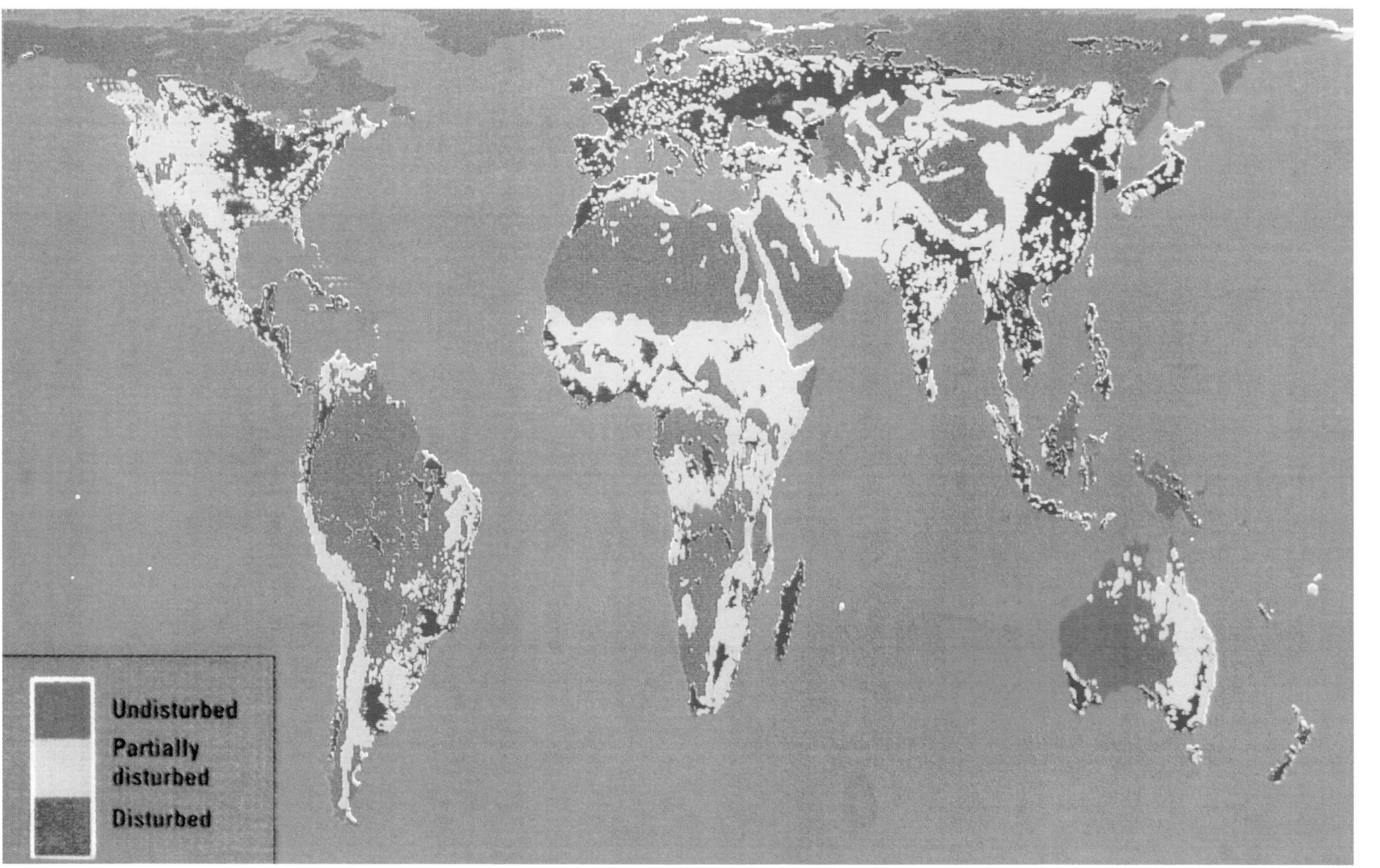

Plate 34 The pattern of disturbed habitats across the globe. Lee Hannah and his co-workers at Conservation International used a variety of sources, including satellite remote sensing data, to classify land areas into either (i) undisturbed (mainly primary vegetation with little human presence), (ii) partially disturbed (shifting agriculture or other evidence of intermittent human activity) or (iii) human dominated, with permanent agriculture or settlement.

Large areas of undisturbed land are found in the taiga and tundra and in the deserts of Africa, Australia and Central Asia and also in Amazonia. South America and Australasia have about two-thirds in this category while Europe has just one-sixth. The British Isles has 0.2% undisturbed, while the Kamchatka peninsula, of comparable size and at the same latitude but on the other side of the world, is virtually untouched.

Plate 35 The familiar irregular patchwork of the British countryside consists of an interconnecting network of hedgerows.

Plate 36 The uniform appearance of intensive agriculture.

ganisms and fungi which may thrive only in the damp and protected soils beneath a forest canopy. Together with the induced abiotic conditions (such as higher precipitation) these determine rates of decomposition and of nutrient flow between the soil and the canopy.

The community is constrained by these interactions operating up and down the hierarchy – the feedback between different functional levels. As a result, an ecosystem organizes itself: it tends to form itself into a stable configuration set by these interactions. Similar hierarchical effects may organize non-living systems, perhaps as a necessary prelude to life (the droplets discussed in section 1.5) and the hierarchy in the cell.

Because of this two-way communication up and down the hierarchy, we need to understand both the large-scale and the small-scale processes which organize ecosystems. According to hierarchy theory, our understanding of the population dynamics of a particular species may only become complete in the context of landscape or regional or sometimes global processes. These ideas have their echoes in the Gaia hypothesis, which proposes that such interactions maintain the Earth's atmosphere (Prologue).

An example within a landscape is forest regeneration following fire or a tree fall. At the local scale, regeneration is constrained by the rates of nutrient transfer and the scope for colonizing trees to form mutualistic associations with soil fungi (or mycorrhizae – section 6.2). Over a landscape, regeneration depends on the distribution and abundance of potential colonizers as well as the scale of the disturbance.

We have learnt much about the role of fire outbreaks and their scale in natural ecosystems in dramatic fashion in recent years. Like most of the extensive forested areas in North America, the Yellowstone National Park had been subject to fire-suppression patrols and prior to 1970 a vigilant forestry service had quickly extinguished the smallest fires before they could spread. Much of the coniferous forest here has extensive tracts of lodgepole pine (*Pinus contorta*), a species which becomes more flammable with age. Without a sequence of small fires to open gaps and remove some of the older trees, the population in the park became dominated by large and older trees. At the same time fuel, as litter, accumulated on the forest floor.

It needed only the extended drought of 1988 and an uncontrolled logging fire to set off the inevitable. Around 70% of the park was burnt. The scale of the fires and the dramatic pictures of its destruction led to it being reported as a national disaster. Yet the burn was not uniform. Massive though it was, different patches burnt to different degrees and temperatures high enough to kill seeds were achieved in only relatively small pockets. Trees were killed in about 20% of the park while less than 1% of its elk population was lost to the flames. Since then there has been a dramatic change in the park with a rapid regeneration of the trees, the ground flora and the animals. The nutrients and space released by the conflagration allowed opportunist plants and seedlings to flourish. The removal of old, unproductive wood was followed by a flowering of the forest floor and an increase in the resources for its herbivorous animals.

The simple lesson is that frequent but small-scale burns are part of the normal economy of coniferous forests. They do not produce dramatic change, but they do create the gaps that prevent some species from becoming dominant (section 7.1), a frequency of disturbance that ensures a high turnover of species.

Change is part of the dynamics of all ecosystems, and communities will accommodate changes to which their members are adapted. However, a disturbance that exceeds their individual capacity for recovery will cause more drastic change at the community level (section 9.2). Species disappear and the interactions within the community shift, so that its organization switches to a different configuration. A large-scale disturbance may lead to the loss or replacement of many species, perhaps reducing the capacity of the community itself to recover its former state.

At the higher levels in the hierarchy, the northern coniferous forest is showing few signs of the dramatic fire in Yellowstone in 1988, but its boundaries would move if there were long-term changes in the patterns of rainfall or temperature over its latitudinal range. According to hierarchy theory, large-scale systems are more susceptible to long-term, gradual but persistent change. The change is slow but because it operates at the higher levels in the hierarchy it leads to major shifts at the lower levels, and changes are not quickly reversed.

Diverting the rivers Amu Darya and Syr Darya was the start of just such a process of change. It took 20 years for the Aral Sea to show signs of ecological decline. Now the impact covers many thousands of square kilometres. In contrast, a short-term change may well be absorbed by the system: the fires in the Kuwaiti oil fields (Plate 22) after the Gulf War of 1991 were portrayed as major ecological disasters in the press at the time. Despite the large amount of carbon dioxide released (240 million tonnes) in the few months they burned, their effect has hardly been detectable in the global atmosphere.

Any shift in atmospheric carbon would need to last much longer to change the composition or distribution of the planet's forests. Yet, as we shall see later in this chapter, climates can change surprisingly quickly.

The major ecological regions follow behind. Moving from the landscape to the regional and global scale, we have to map the distribution of plants and animals on time-scales of millennia.

8.2 THE BIOGEOGRAPHY OF EARTH

Based on the distribution of large terrestrial vertebrates, the Earth can be divided into six distinct realms (Figure 8.3). First identified by Wallace, the boundaries of these zones represent long-standing barriers to animal dispersal, reflecting in large part the drifting of the continents across the face of the Earth. Although many mammal groups are found in all realms, some have discontinuous distributions. For example, there are no native primates (apes, monkeys and their relatives) in the Nearctic, while marsupials (the pouched mammals) are found only in the Americas and Australasia. The separation of Australia from South America and Antarctica 65 million years ago isolated its marsupials, and here they evolved into very diverse forms – the kangaroos, the koalas and the wombats – without competition from placental mammals.

Some distributions reflect more recent change. At the end of the last ice age, rapid climatic shifts led to the extinction of the mammoths and mastodonts, the northerly relatives of elephants. Reductions in the availability of water and the quality of forage stressed the herds at certain times of the year, making them easy prey. Gary Haynes (Box 3.3) suggests that the expansion of the human

Figure 8.3 The main zoogeographical realms of the planet, primarily defined by their vertebrate fauna. The mammal fauna is particularly characteristic of an area: before the advent of humans, some orders were entirely excluded from certain realms (e.g. primates in the Nearctic).

The dotted line marks Wallace's line, running through Malaysia and separating the two realms according to their native birds. The line actually follows a deep oceanic trench, marking the junction between two crustal plates. The flora and fauna of the two realms blend across the whole of this archipelago so the demarcation is not well defined, even for the marsupials.

species northwards and into North America followed as we learnt where to find the mammoths at their weakest. It seems likely that about 11 000 years ago our ancestors delivered the final blow to species struggling to adapt to rapid climate change.

Within each realm (and sometimes straddling them) are distinct plant communities, termed **biomes**. Their distribution is governed primarily by climate and, to a lesser extent, by soil type. The most important factor is the availability of water. Forests dominate where there is sufficient moisture to support them; grassland or low scrub where it is drier. Within a biome many plants have a characteristic growth form, adapted to the prevailing conditions, especially their means of conserving water.

We can distinguish four major terrestrial biomes – forest, grassland, tundra and desert – by their regimes of precipitation and temperature (Figure 8.4). For example, deserts are found where evaporation exceeds rainfall. These conditions prevail at 30° either side of the Equator, reflecting the pattern of atmospheric circulation. At this latitude, dry air is descending, having shed its rain in the tropics (Figure 8.5). Deserts are also a feature of the western

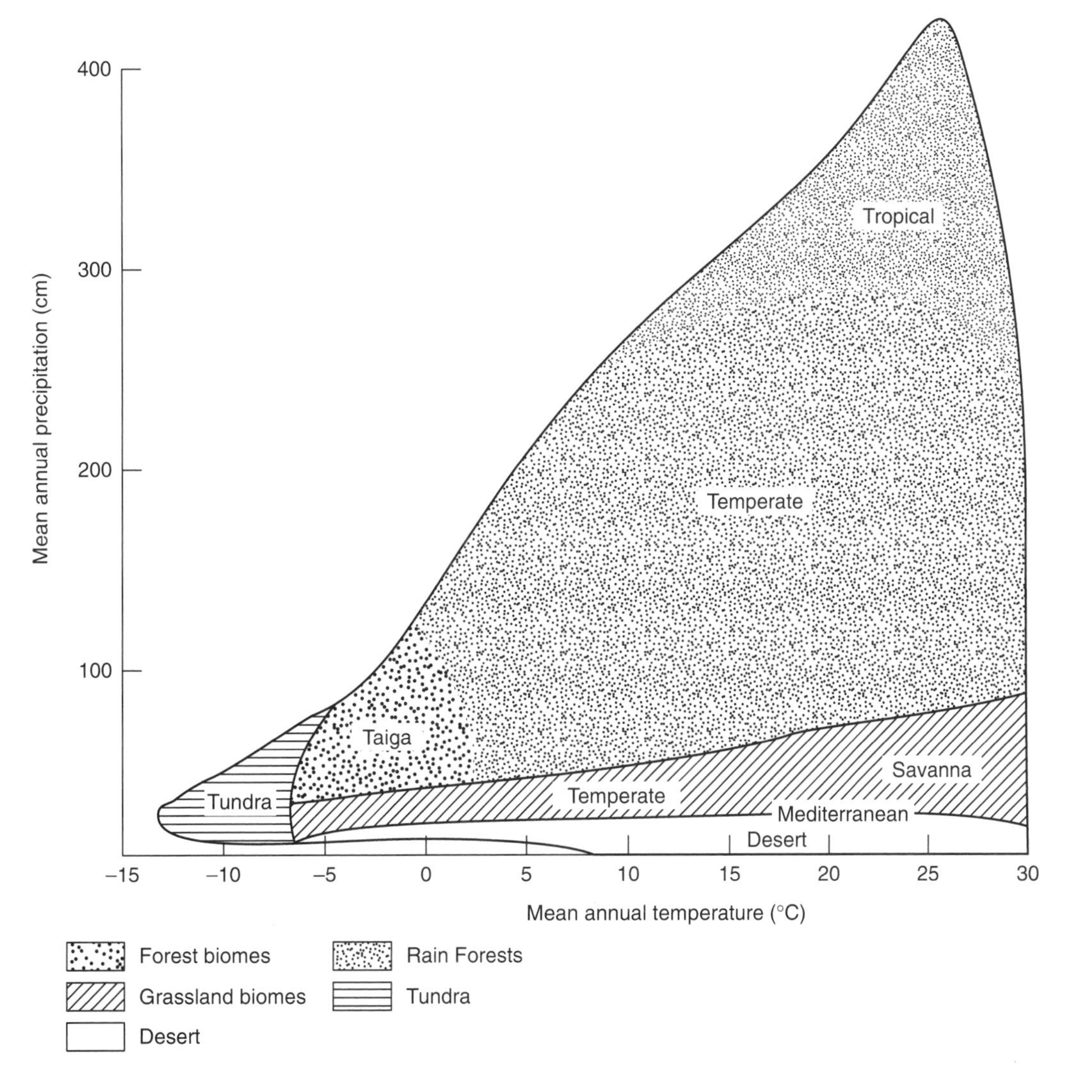

Figure 8.4 The range of the major biomes defined by temperature and moisture. Forests are only found where precipitation is above about 30 cm per year (in the coldest regions) and 120 cm per year (in the warmest regions). Grasslands develop where water is in short supply.

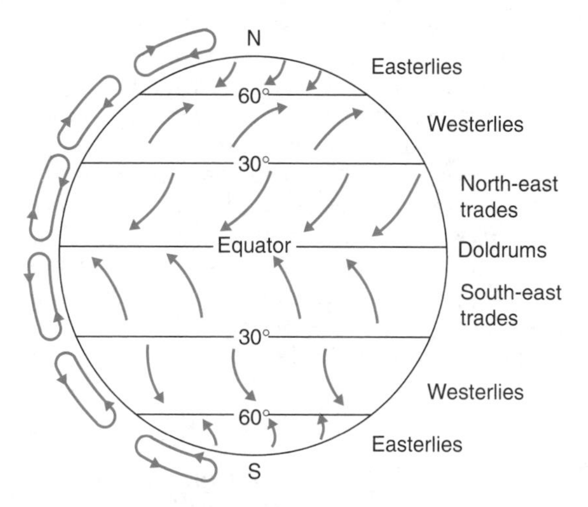

Figure 8.5 A simplified map of atmospheric circulation. Air warmed at the Earth's surface rises and then cools at altitude. Heating is greatest where the sun is directly overhead, which occurs somewhere between the tropics at various times throughout the year. Air rises and falls at particular latitudes, generating winds which are then deflected by the spin of the Earth. This creates the characteristic wind patterns of the planet and, combined with seasonal factors, zones of precipitation.

edge of several continents, where cold ocean currents induce coastal fogs that deprive inland areas of precipitation (Figure 8.6).

Such localized conditions confound the simple latitudinal pattern of the biomes. Proximity to oceans or a mountain range can create local climatic conditions and give rise to isolated, often unique plant communities (Plate 38).

Here we limit our task by concentrating on the major terrestrial biomes. This is not to deny that there are characteristic aquatic communities – far from it. Most aquatic habitats are defined by their animal community and its adaptations to depth, rates of water flow, salinity and nutrient source. As we see in the next chapter, many tropical marine communities show the same latitudinal pattern in diversity as found on land.

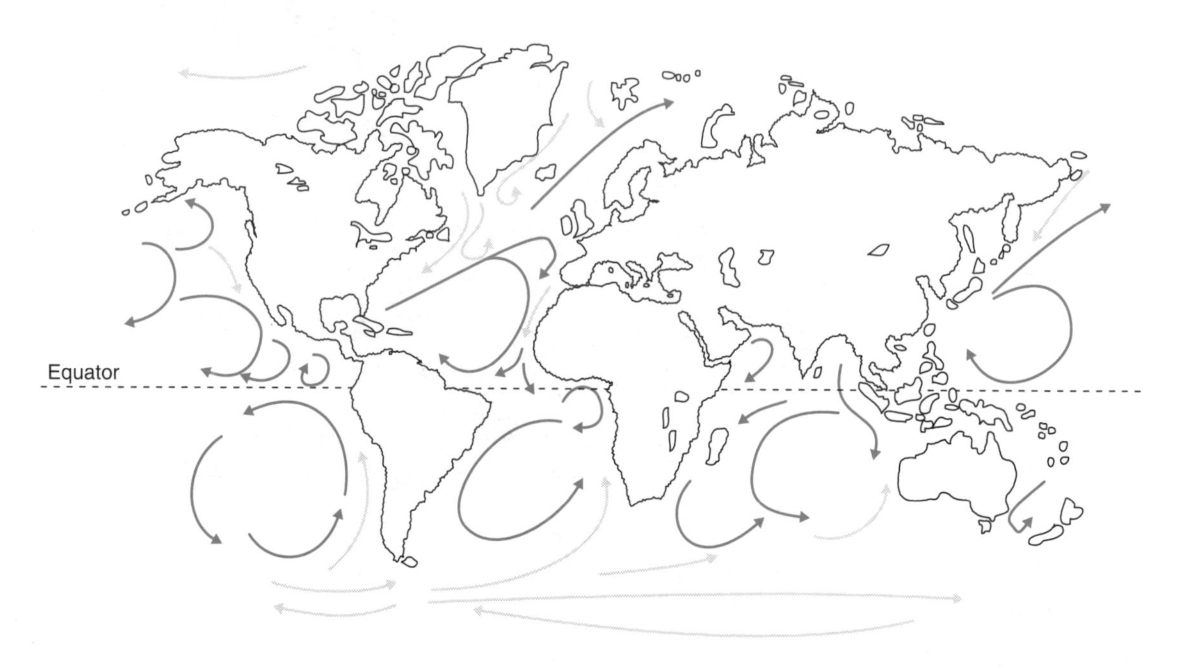

Figure 8.6 Simplified surface currents of the Earth. Warm currents (dark blue) develop either side of the Equator and flow clockwise in the northern hemisphere and counter-clockwise in the southern. This movement draws in cold polar currents (pale blue) which can deprive continental margins of rainfall.

8.2.1 Tropical biomes

Tropical rain forests (Plate 39) straddle the Equator, roughly within the tropics of Cancer and Capricorn (Figure 8.8). These are amongst the oldest communities on the planet, pockets of which went largely undisturbed even during the last glaciations. The diversity of these communities is astounding: in Amazonia there can be around 300 species of tree in just two or three square kilometres. Tropical forests have the highest net primary productivity of all terrestrial ecosystems (section 5.1; Figure 8.16), yet they grow on some of the most infertile soils.

The soils are poor because they are old and have been leached of their key minerals. Dead and decaying organic matter arriving at the forest floor is quickly scavenged by a highly competitive decomposer community. This makes for extremely rapid decomposition, promoted by the high temperatures and abundant moisture. Because its organic matter has such a brief existence, the soil has a poor structure with little capacity for chemically binding nutrients. As a result the soil is only a small reservoir of plant nutrients and competition for what is available is intense.

There is also intense competition for light. Like all forests, the plant community is stratified into

Box 8.1

THE GLOBAL CLIMATE

Consider what the atmosphere does. The main source of heat on the planet is energy arriving as radiation from the sun. Because the Earth is a sphere, spinning on a tilted axis and with an uneven distribution of sea and land, it warms at different rates in different places. Heat disperses to even out these differences. Conduction is slow through the crustal rocks compared with the convection of water, while the atmospheric circulation (also convection) is fastest (Figures 8.5, 8.6). Weather is simply the movement of air and water vapour caused by heat imbalances within and between the atmosphere and the Earth's surface.

Heat moves in the oceans as currents: warm water rises as it becomes lighter, and colder waters sink to replace it. This creates persistent circulation patterns between the cold polar and warm tropical regions (Figures 8.6, 8.7). Heat is transferred to the air mainly through the evaporation of water, producing clouds and eventually precipitation. In the atmosphere, winds blow from high pressure areas with relatively high temperatures to low pressures areas. The simple circulation between the poles and the equatorial regions is complicated by the forces created by the spin of the Earth, the seasons created by its tilt and the configuration of land and sea. The result is a well-defined if intricate pattern of wind and ocean circulation distributing heat around the planet.

To understand long-term climate change, we need to add in the solar flux (the amount of energy reaching the Earth's surface) and changes in the chemistry of the atmosphere. The solar flux varies over thousands of years because of variations in eccentricity of the Earth's orbit and its relative tilt, as well as variations in the energy output of the Sun. This can account for some climatic change in the recent past. However, other factors are important, including geological activity and the composition of the atmosphere.

Figure 8.7 The thermohaline is a major distributor of heat energy across the planet. It is a continuous flow of heat and salt in deep and massive ocean currents. Cold and saline water moves at depth (pale blue), replacing the warm and less saline waters toward the ocean surface (dark blue). The heat transfer of the thermohaline is closely associated with the global climate and changes in its circulation mark different phases in an ice age.

layers, with groups of species adapted to different degrees of shade (Figure 8.9). The highest layer is the canopy, perhaps 30–40 m above the ground and formed of the upper branches of mature trees jostling for uninterrupted light. Beneath this may be a subcanopy where sufficient light is able to penetrate the upper layer. Further down is an understorey of different species, often with large leaves, such as palms. At the forest floor, among the buttress roots supporting the taller trees, are herbaceous plants, particularly ferns, which thrive in the dark, dank conditions. Each layer contains seedlings or young trees waiting for sufficient light to join the layer above.

Where the canopy is almost continuous the interior is relatively open because light levels are too low for much undergrowth. A dense 'jungle' of young trees and climbers can only develop where sunlight penetrates well below the canopy – because of a tree fall, say, or at the edge of a river. Climbers, creepers and 'hangers-on' (**epiphytes**) all use the main trees to get access to the sunlight, avoiding the costs of building large trunks. Epiphytes attach themselves to the sides of trees, relying on the nutrients in rainfall and the water draining off their host, or which leak from the trees themselves.

Some tropical forests (for example, in South East Asia) are seasonal and may have cycles of leaf-fall prompted by a dry season. Most are dominated by broadleaved trees which maintain a canopy throughout the year. The majority of plants rely on animals for pollination and seed dispersal, producing flowers and fruits to attract insects, reptiles, birds and mammals. The structure and diversity of the plant community creates a variety of niches for the animals, which themselves are formed into distinct assemblages associated with each layer. Insects and several vertebrates – monkeys, sloths, reptiles and amphibians – confine themselves to the canopy or subcanopy, rarely descending to the ground. A range of carnivores, including birds, feed here as well. Small pools of water held in the leaves

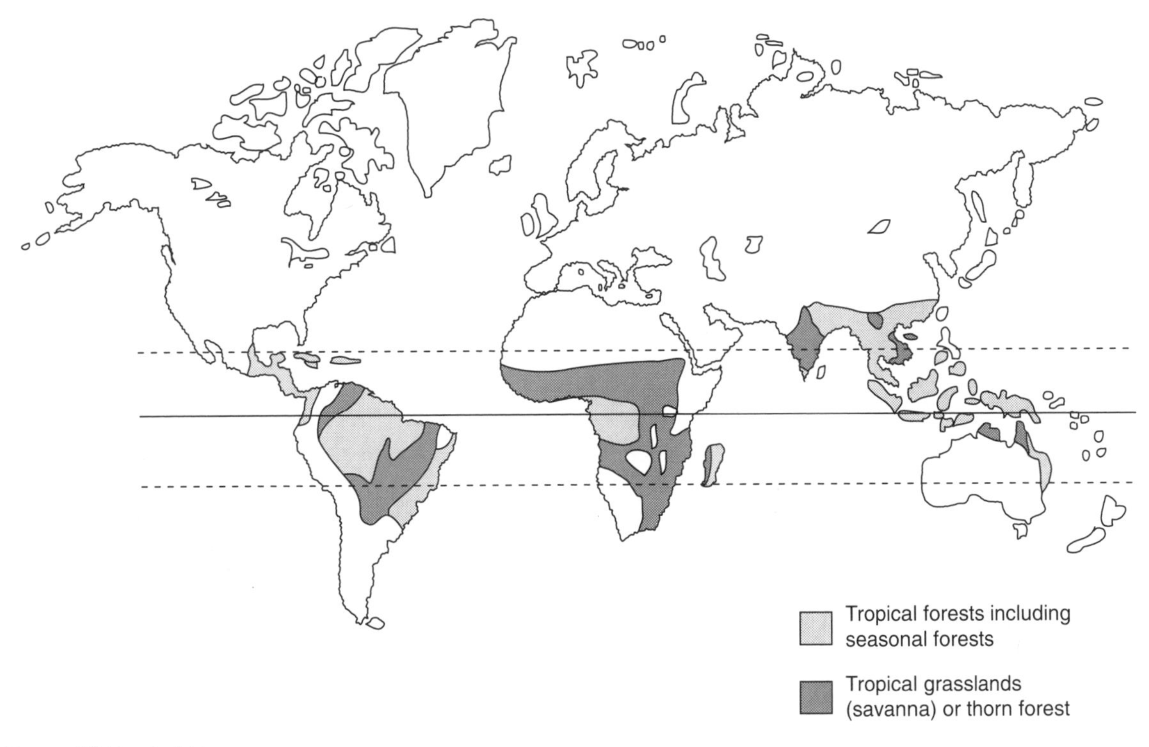

Figure 8.8 Tropical biomes.

of the upper layers and in epiphytic bromeliads support insect and amphibian communities – entire aquatic ecosystems – in the centre of a plant, perhaps 30 m above the ground. Equally diverse animal, fungal and microbial communities wait on the forest floor for material arriving from above.

Rainforest is typical of areas where precipitation exceeds 200 cm per year and where no month has much less than 12 cm. If there is a prolonged dry season, forest gives way to open woodland, thorn scrub and tropical grassland (**savanna**). The dry conditions favour trees and scrub with small leaves and with protection from browsers, such as the thorns of the acacia trees of the African plains (Plate 40). Even so, browsing of these trees serves to limit the extent of the tree and scrub zones, and elephant and giraffe are thus especially important in maintaining an open grassland.

Grazers are less selective, cropping the aboveground parts of the herbaceous (non-woody) plants. Without grazing, the grasses of the savanna may grow high – up to 3 m. Usually tussocked, they die back in the dry season, keeping their living tissues close to the ground, well protected from grazers who crop closest. Periodic fires, set by electrical storms, sweep through the savanna. These, together with the intense grazing, maintain its open structure and limit scrub growth. Many flowers produce fire-resistant seed that germinates quickly when the rains come, while others sprout from deeply hidden tubers.

The soils are again ancient and infertile, with relatively little organic matter, so that the animals grazing above are crucial for nutrient cycling. Savannas are typical of low, well-eroded flatlands, with depressions that become vital watering holes during the dry season. Their pronounced seasonality means the grass is grazed by successive waves of large mammals, cropping to different levels: in Africa, zebra is followed by wildebeest and thereafter gazelle. In attendance too are the carnivores – the big cats and hyaenas – and also the scavengers who come after: jackals and vultures.

The biomass represented by this hoof and claw testifies to the productivity of this biome. Yet in terms of biomass consumed, the most important

Figure 8.9 The edge of a rainforest, showing the multilayered structure of the forest vegetation.

herbivores are insects, primarily grasshoppers, locusts and ants. Equally important are the inconspicuous termites, obvious only by their sentinel mounds, yet whose activities are crucial in maintaining soil fertility.

8.2.2 Temperate biomes

The stark contrast in temperatures between winter and summer have the most profound effect on temperate ecosystems (Figure 8.10). Cycles of biological activity anticipate and respond to the seasons. The flowering and fruiting of plants is timed by this clock, and animals feed, breed, migrate or hibernate in tempo with the plants.

The different temperate biomes reflect the abundance of water during the year. Forests dominate where it is plentiful, grasslands where there is a dry season. We can distinguish several types of **temperate forest** (Plate 41) according to their regimes of temperature and precipitation, from broadleaved evergreen forest in the warmer south to the conifer–broadleaf mixed forest of the north.

At very low temperatures, photosynthesis ceases and respiration is reduced. Average winter temperatures make the balance uneconomic for deciduous trees and so they reduce their metabolic costs by shedding their leaves. Conifers, which can photosynthesize at well below 0°C, retain their needles (Figure 5.5), but have a series of adaptations to reduce water loss – this is important when soil water is frozen. Some herbaceous plants overwinter as seeds; others store food that will fuel growth early in the new season in corms or tubers. Animals, too, match their life cycles and activity to the times of plenty. Some overwinter as eggs or larvae, reducing their metabolic costs when food is scarce. Others hibernate or simply migrate to exploit more abundant resources.

The annual cycle of leaf-fall makes the soil an important repository of the nutrient wealth of the biome. The soils are deep and rich, with decomposer communities stratified down their profile. Much of western agriculture lives off the organic capital remaining in these soils following the clearance of the forest.

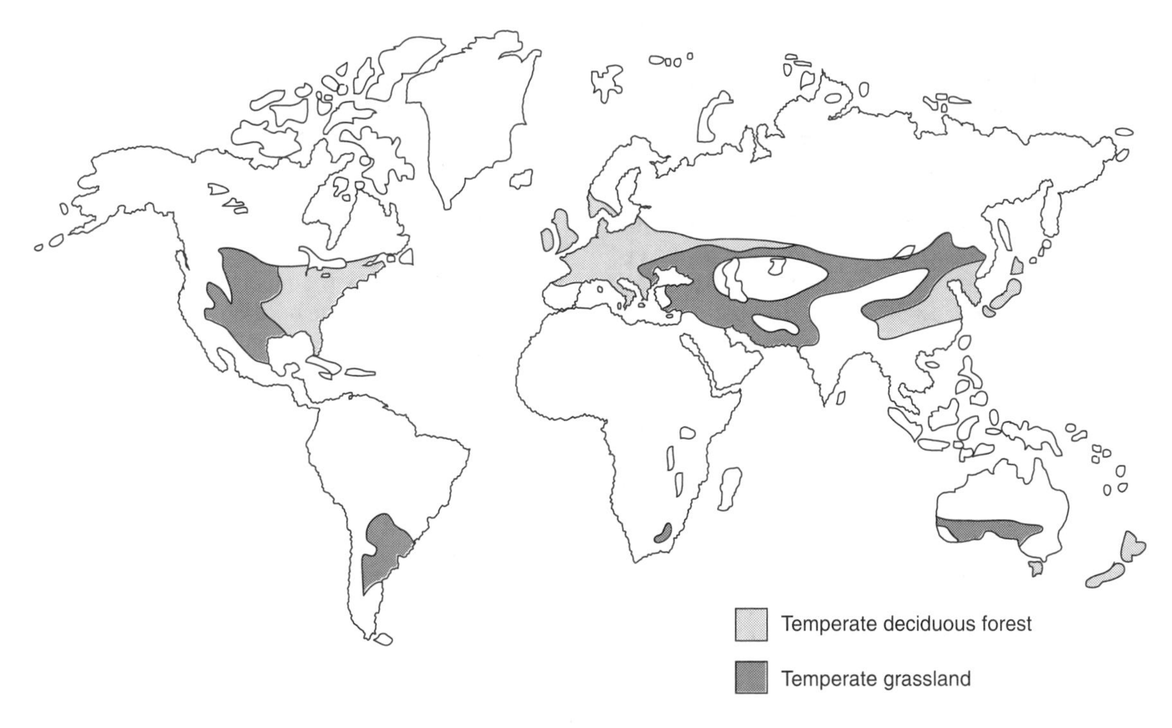

Figure 8.10 Temperate biomes.

Temperate forests have considerably fewer tree species than a tropical forest and are typically dominated by just three or four species, primarily determined by climate and soil. Often there is a mix of coniferous and broadleaved trees, with the balance shifting in favour of conifers, especially pine, where soils are nutrient-poor and there is a higher risk of forest fires. Many temperate coastal regions, with abundant rainfall or fogs throughout the year, support hemlock, fir and redwood, conifers growing in almost continuous stands.

The seasonal contrast in temperatures becomes more marked with distance from the Equator. Where moisture is available all year round, deciduous forest develops. Its composition depends on soil depth and drainage. Beech dominates on drier, shallow soils and produces a dense shade with little understorey development. Elsewhere, the forest floor beneath oak is comparatively well lit. All deciduous forest undergoes a sequence of herbaceous development each spring before the main canopy develops. As well as a distinct understorey of saplings and low bushes there are climbers (e.g. ivy, *Hedera helix* , and honeysuckle, *Lonicera periclymenum*).

Again, insects are the dominant herbivores and these, in turn, are exploited by various migratory birds. Mammals include squirrels, wild pigs and badgers with deer as the principal browsers. Before humans became the major influence on this biome, wolves, bears and several species of cat were the larger carnivores.

Where a dry season lasts for most of the summer the forest gives way to **temperate grassland** (Plate 42). In the centres of large continental masses, away from moist coastal air streams, annual precipitation falls below 60 cm at temperate latitudes. Australia, South America (pampas), South Africa (veld), Eurasia (steppes) and North America (prairies) all have such zones. The grasslands are typically flat, gently rolling landscapes, which in the northern hemisphere are associated with recent glacial activity.

Growth and flowering are distinctly seasonal, with grasses bearing their seed towards the end of the dry summer. Like the savanna, fires and grazing maintain the dominance of the grasses and the legumes. Thorn scrub and trees develop where grazing pressure is lifted. In the prairies of North America the dominance of the grasses and the variety of flowering plants were maintained by the bison. In Europe and Asia the horse and antelope were the principal vertebrate grazers, but insects, above and below ground, are major herbivores in all temperate grasslands. Each region also has a range of burrowing animals, both vertebrate and invertebrate.

Soils are deep, often with a thatch of undecomposed vegetation overlying a thick humus layer. This is important protection against wind erosion when the soil is dry. Light grazing helps to speed decompositon processes and adds moisture. Today, most grasslands around the world have few wild grazers, but instead are used for meat and cereal production (section 5.5).

Where temperatures range higher, and rainfall is confined to the winter months, mediterranean-type communities can develop (Figure 8.11, Plate 43). These have a very restricted distribution (section 7.1) and are now greatly altered from the evergreen forests that dominated before man arrived. Where rainfall is very low, true desert will form (Figure 8.11, Plate 44).

8.2.3 Boreal biomes

Towards the Poles, average temperatures decline further and so does available water (Figure 8.12). Here plants are adapted to long periods when little photosynthesis or nutrient uptake is possible.

Again we can identify a forested region, the northern coniferous forest or taiga, and a low-lying plant community, the tundra, where water is unavailable (because it is frozen for much of the year). These communities encircle the northern hemisphere where the land mass is almost continuous and together comprise the boreal (northern) biomes. Equivalent communities occur at lower latitudes but at high altitudes, in montane areas where similarly severe climates exist (Figure 8.12, Plate 45).

Where the temperate forest gradually gives way to unbroken conifer forests the **taiga** starts (Plate 46). Vast tracts of this dense forest are dominated by just two or three species – primarily pine, fir, spruce or hemlock – beneath which is a herbaceous layer of mosses, lichens and grasses. In some areas decidous trees are found, mainly birch and aspen. The fine

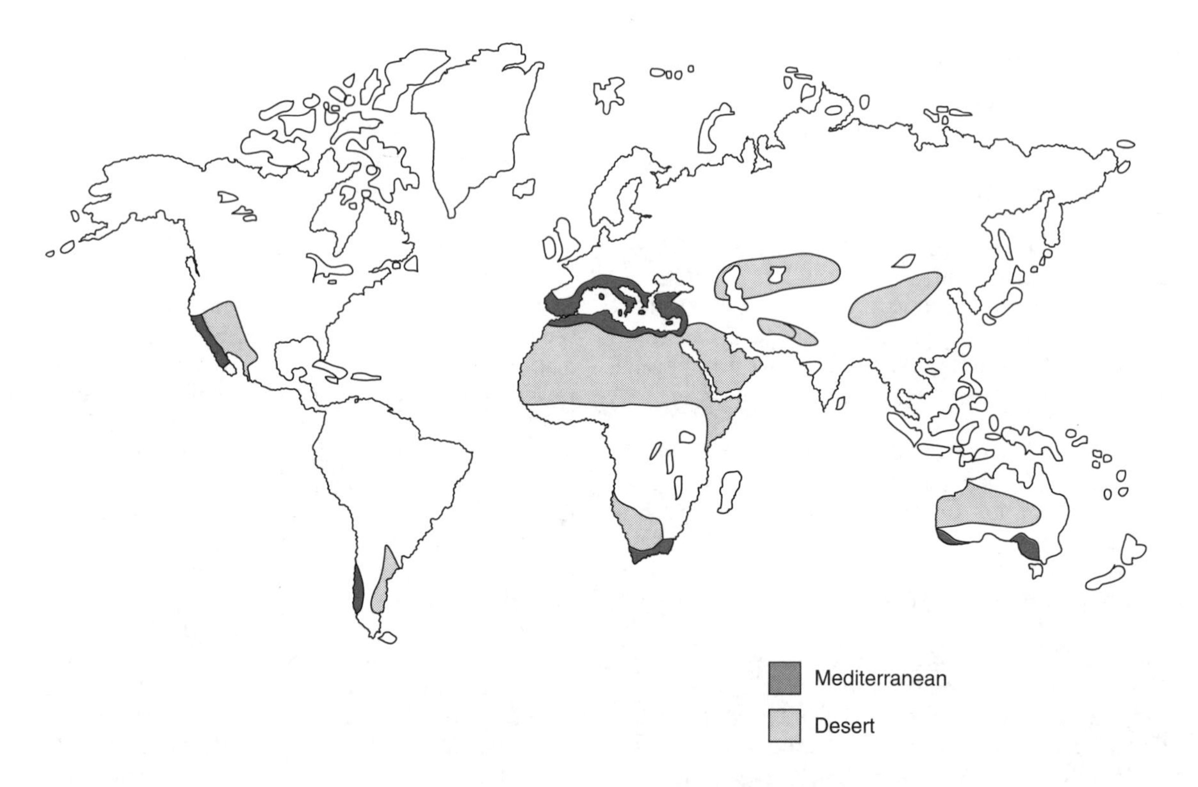

Figure 8.11 Mediterranean and desert biomes.

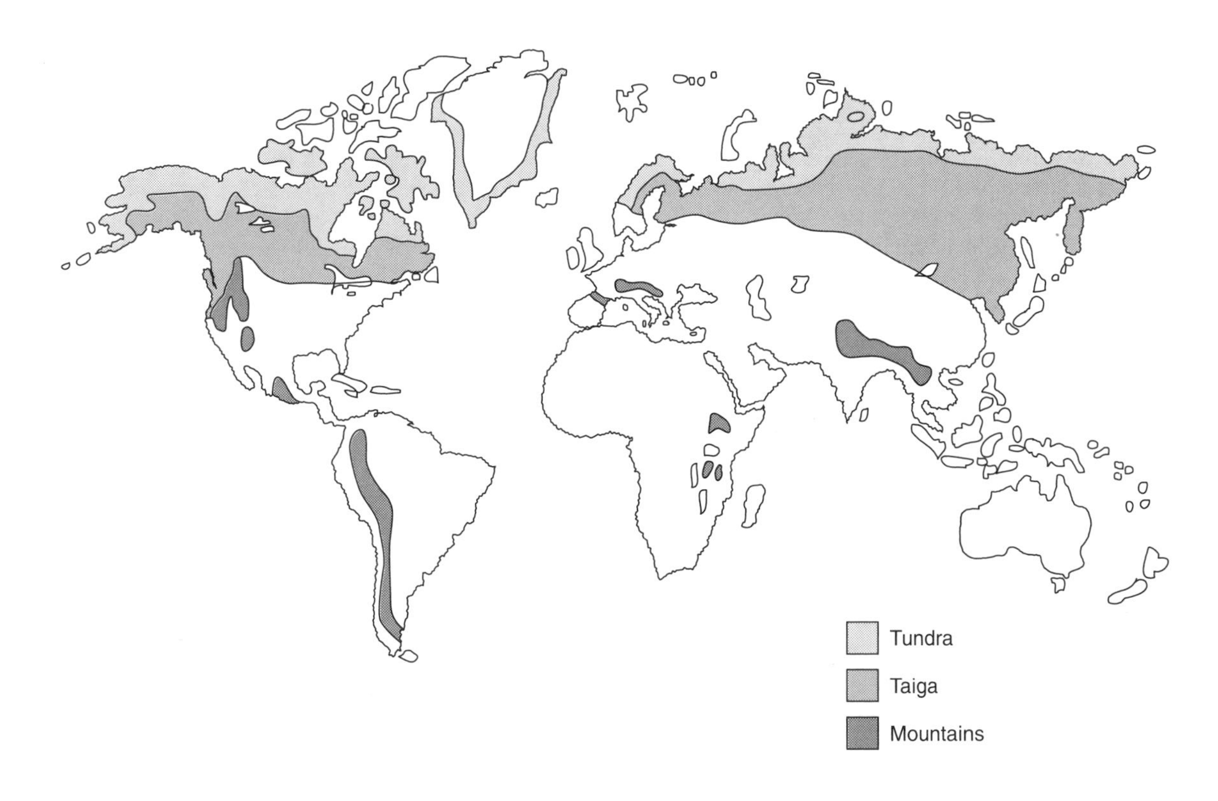

Figure 8.12 High latitude and high altitude biomes.

leaves (needles) of conifers reduce water loss and are efficient for photosynthesis at low temperatures. By retaining their leaves they can make full use of the early spring sunlight, but they also have to support or shed heavy loads of winter snow. The trees are shaped accordingly with a conical growth habit.

In more northerly areas, the soils are permanently frozen at depth. All are thin, infertile and covered with a thick layer of partially rotted needles. The nature of this litter and its slow decomposition under cold, semi-waterlogged conditions means that the soil is acidic and favours fungi as the prime decomposers. A build-up of litter can promote forest fires and these are important to taiga ecology: fires create gaps which invaders exploit. Some species (jack pine, for example) take advantage of the gaps created by these disturbances and shed their seeds only after a fire.

Relying solely on wind for pollination, conifers have no direct need for insects and defend themselves against insect attack, primarily by using resins and their characteristically pungent terpenes. Nevertheless, a variety of moths, sawflies and beetles burrow into the needles, buds and bark. Several migratory birds feed on these insects, while resident birds and squirrels take the conifer seeds.

Other herbivores, including elk, caribou and deer, feed on the low-growing vegetation. Some remain active even when there is deep snow cover. Canids (foxes and wolves), mustelids (martens, weasels, wolverines), birds (owls, eagles), bears and cats form the carnivore community. Many mammals hibernate for the winter.

Some of these herbivores and carnivores also move into the **tundra** (Plate 47) in the summer to feed. This is the zone at the edge of the ice sheet, an area once heavily glaciated and still with a short growing season of only two or three months. At this time temperatures are high enough to melt the upper layers of the soil, down perhaps a metre or so. (At greater depth the soil never thaws: this is the **permafrost**). Decomposition under these conditions is slow and organic matter accumulates as peat.

When the thaw comes, water and the nutrients dissolved in it become available. Plants bloom and grow – grasses, sedges and low shrubs, such as heathers, dwarf willows and birches – and produce seed in quick succession. Many will self-pollinate and spread by vegetative growth. Others flower to exploit the burst of insect activity. Flies (especially mosquitoes) and butterflies that have spent the winter as larvae or pupae emerge to feed and mate. Birds migrate to exploit this short period of plenty. Elk, reindeer and other herbivores move to graze further north on exposed mosses and lichens. Some predators, from wolves to hawks, follow and feed especially on lemmings and voles.

Across much of North America and Russia the tundra presents a single unbroken expanse of flat, windswept and waterlogged landscape. Its flat monotony results from its repeated glacial scouring by ice sheets that have advanced and retreated as the global climate has changed.

8.3 CLIMATE CHANGE

The positions of these biomes are far from fixed. Alongside human remains in Britain we find parts of rhino on which *Homo* fed 500 000 years ago. As one deposit succeeds another, pollen, shells and beetle wing-cases tell of communities very different from today.

The ice ages are still fresh in the memory of the planet, so that we can still find well-preserved carcasses of mammoths and occasionally frozen human beings. The fossil record describes a series of ice advances and retreats, as cold-loving and warm-loving species replaced each other in one climatic switch after another. When the last glaciation ended, it ended abruptly. Temperatures rose extremely rapidly – averages in Europe climbed by about 7°C in just 50 years and the Earth shifted from ice age to warm interglacial in the space of a human lifetime (Figure 8.13).

It is likely that this followed an increase in the energy received from the sun, perhaps as the relative position of the Earth changed because of variations in its orbit. Although there have been minor shifts in average air temperatures since then (including a 'mini ice age' in the seventeenth century), overall the last 10 000 years have been relatively mild. Today we enjoy a climate substantially warmer than the average for the last two million years, though the difference between the ice age and the present interglacial is only 6°C.

Now things seem to be changing again. Since industrialization started 200 years ago, humanity has been altering the chemistry of the atmosphere. Climatologists are trying to model the consequences for the global climate and ecologists are being asked to predict the implications for the hierarchy of ecological processes.

8.3.1 Atmospheric composition and mean temperature

Orbital variations may have started the oscillations of the ice ages (though see Chapter 9), but the amplitudes of these temperature changes – the scale of their fluctuations – have been magnified by differences in the composition of the atmosphere (Figure 8.13). Samples of the ancient atmospheres trapped in the air bubbles of ice cores dating back 160 000 years show that carbon dioxide levels have followed shifts in mean local temperature (measured by changes in ice volume). The same is true of another carbon-rich gas, methane. Methane is generated by microorganisms, and so this indicates that biological activity increased with the temperature changes.

Because of the role of living systems, the problem is to decide the extent to which this is cause or effect: has atmospheric carbon dioxide increased because it has been warmer, a consequence of increased biological activity? Or has increased carbon dioxide caused the warming? In fact, it is a combination of both. Along with water vapour, atmospheric carbon induces heating of the atmosphere via the greenhouse effect (Box 8.2). Fluctuations are 'amplified' because of the positive effect that temperature has on biological activity. If nothing else changes, a warm climate increases atmospheric carbon by increasing respiration and decomposition and that, in turn, promotes further warming.

Carbon-rich gases also originate from non-biological sources, such as volcanic activity, and can be lost through absorption by the oceans. However, the carbon content of the atmosphere is primarily regulated by biological activity – that is, it is fixed through photosynthesis and released through respiration and decomposition. But the interaction of

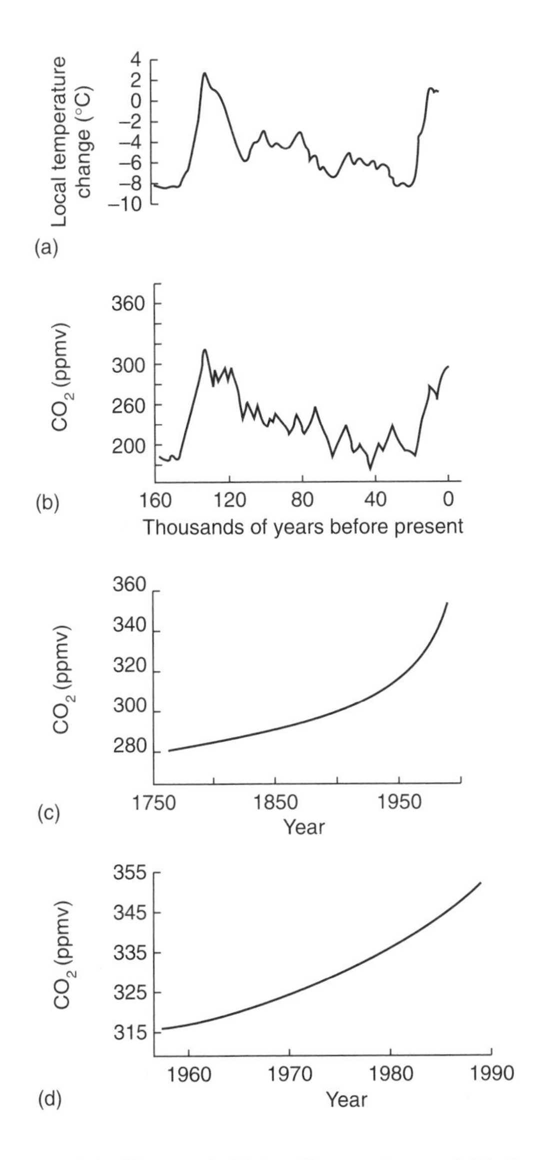

Figure 8.13 Changes in (a) local temperature and (b) atmospheric carbon dioxide levels in the last 160 000 years. Atmospheric methane levels show a similar pattern, indicating that the rise carbon-rich gases was due to an increase biological activity as the global climate warmed. A rapid rise in temperatures marks the end of the last glaciation around 10 000 years ago. (c) Carbon dioxide levels have risen consistently since the onset of industrialization in the eighteenth century; (d) this has been most rapid with the increasing fossil fuel consumption and forest clearance in the last 40 years. (Note the change of scale in b, c and d).

these processes with temperature is not simple. In many terrestrial plants, long-term rates of photosynthesis do not increase greatly with temperature, mainly because leaf surfaces are cooled by evaporation. In contrast, respiration and decomposition do rise, so net carbon release increases.

Carbon fixed in living tissues need not be released again directly, but may instead become incorporated into sediments (Figure 8.15). Large-scale shifts in the global climate can be prompted by oceanic sedimentation. One important regulator is marine phytoplankton, whose photosynthesis and productivity do rise with temperature. They will also remove significant amounts of carbon rapidly from the system if allowed to settle out, as do those marine invertebrates who use carbon dioxide to build shells.

The scale of this sedimentation can be seen in the vast formations of limestone rocks over the planet. More significant perhaps, at least as far as our economic activity is concerned, are the coal and oil deposits derived primarily from terrestrial vegetation of the geological past.

8.3.2 Anthropogenic sources of atmospheric carbon

In the last 200 years we have been releasing this fossilized carbon at an accelerating rate. On an annual basis, our inputs are relatively small (Figure 8.15), but significantly we have been burning coal and oil for a long time now. There is also an important contribution (perhaps 20%) from forest clearance (Plate 48), and which also reduces a principal route by which carbon can be removed from the air (Box 8.3). Overall, anthropogenic (human-derived) sources have led to a 26% increase in carbon dioxide over background levels since industrialization began.

Forest clearance releases carbon through burning and causes an increase in decomposition in the exposed soil. Replacing forests with agriculture does little to ameliorate this impact: 50% of the carbon fixed in the organic matter of a soil is lost under intensive cultivation. In addition, methane concentrations have increased 140%, primarily from rice paddies and from the various emissions associated with the massive increase in cattle rearing (Box 5.4). Developed nations also use colossal amounts of fuel to sustain their agricultural production. Fossil fuels drive the economic and industrial activity of these societies and they are also used to maintain the ecosystems they create, from the wheatfield to the air-conditioned office.

We have also released a range of new gases, some with very powerful 'greenhouse' properties.

Box 8.2

THE GREENHOUSE EFFECT

The greenhouse effect is a natural and vital feature of our atmosphere. Without it the mean atmospheric temperature of the Earth would be –17°C, rather than its current average of 15°C. The presence of water vapour, nitrogen, oxygen and the carbon-rich gases (primarily carbon dioxide) all absorb the heat reflected off the Earth's surface. While the comparison with a greenhouse is inaccurate (there is no physical barrier, like the glass, to the transfer of heat), the name is at least evocative.

Figure 8.14 shows how this works. Energy arrives from the sun primarily as short-wave radiation warming any surface it strikes. This is re-radiated at longer wavelengths, in the infra-red region of the spectrum, which is absorbed by different atmospheric constituents (Table 8.1). Besides water vapour, the principal greenhouse gases are CO_2, CH_4, N_2O and CFCs. The contribution that each makes to global warming depends upon several factors – their concentration, the length of time they remain in the atmosphere and the wavelength at which they absorb re-radiated energy. CFCs are particularly potent because they absorb in part of the spectrum where the atmosphere was previously transparent.

On balance, an average of about 4 watts per square metre is retained, serving to warm the atmosphere and the planet's surfaces. Over the long term, the Earth will be in some sort of balance with the incoming energy, but the stabilization can occur at different temperatures; thus the greater the greenhouse effect, the higher is the equilibrium temperature for the planet.

Chlorofluorocarbons (CFCs) and their relatives were unknown before 1930 but until very recently have been used extensively as aerosol propellants, in refrigerators and for a variety of industrial processes. CFCs absorb infra-red radiation in the one part of the spectrum where other greenhouse gases do not. They are also long-lived (Table 8.1) and contribute significantly to ozone depletion (Box 8.4).

Some of our emissions can help to ameliorate the greenhouse effect. Sulphate aerosols, themselves a product of fossil fuel combustion, can reduce short-term warming. They increase the reflectivity of the atmosphere, both by their presence and by inducing cloud formation. Since they are short-lived, sulphate aerosols have only a local cooling effect. The sulphate pollution in industrial areas may be the main reason why the northern hemisphere has warmed less rapidly in the last 200 years than the southern hemisphere. Persistent sulphate emissions from industrial areas are a major cause of acid rain (Box 6.1).

Short-term cooling also follows major volcanic eruptions. The millions of tons of ash released by the eruption of Mount Pinatubo in the Philippines in 1991 seem to have led to the cool weather, globally, the following year. Although mean temperatures were lower by 0.5°C, there appears to have been no lasting effect on the long-term trend of global warming.

8.3.3 Predicting climate change

Of all the greenhouse gases, carbon dioxide is responsible for about half the greenhouse effect. The Intergovernmental Panel on Climate Change (IPCC), commissioned by the United Nations and the World Meteorological Organization, concluded in 1992 that carbon dioxide levels would now need to fall by 60% to stabilize at current atmospheric levels. If we carry on like this (the IPCC 'business as usual' scenario) average temperatures would rise by 1°C by 2025, with carbon dioxide at about 450 ppm by 2050 compared with the current level of 350 ppm (Figure 8.13).

The original predictions of the panel, first reported in 1990, have since been revised using

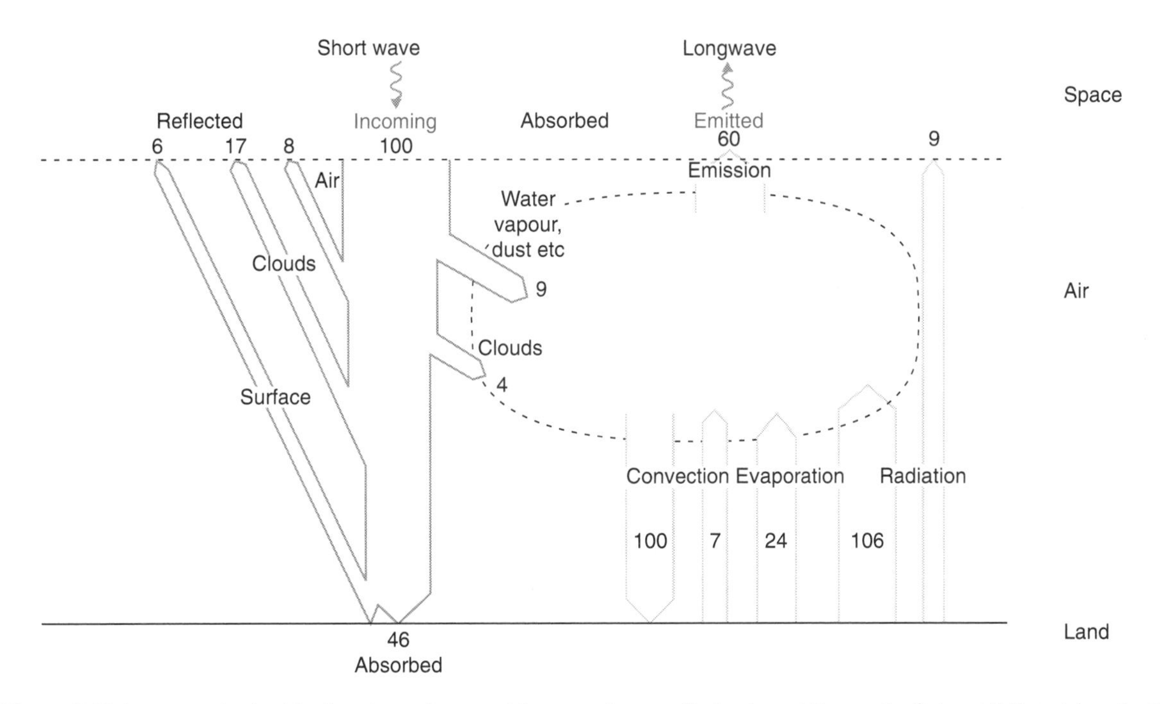

Figure 8.14 An energy budget for the atmosphere and the greenhouse effect using arbitary units. Solar radiation, primarily at the shorter wavelengths, enters the upper atmosphere and is variously reflected or absorbed by the different surfaces it strikes. The warming of the Earth's surface causes it to give off long-wave radiation (infra-red), which is absorbed by the water vapour and carbon-rich gases of the atmosphere. This is the main mechanism by which the temperature of the atmosphere is raised. Increasing the carbon and water content of the atmosphere increases its average temperature, a change in its energy balance.

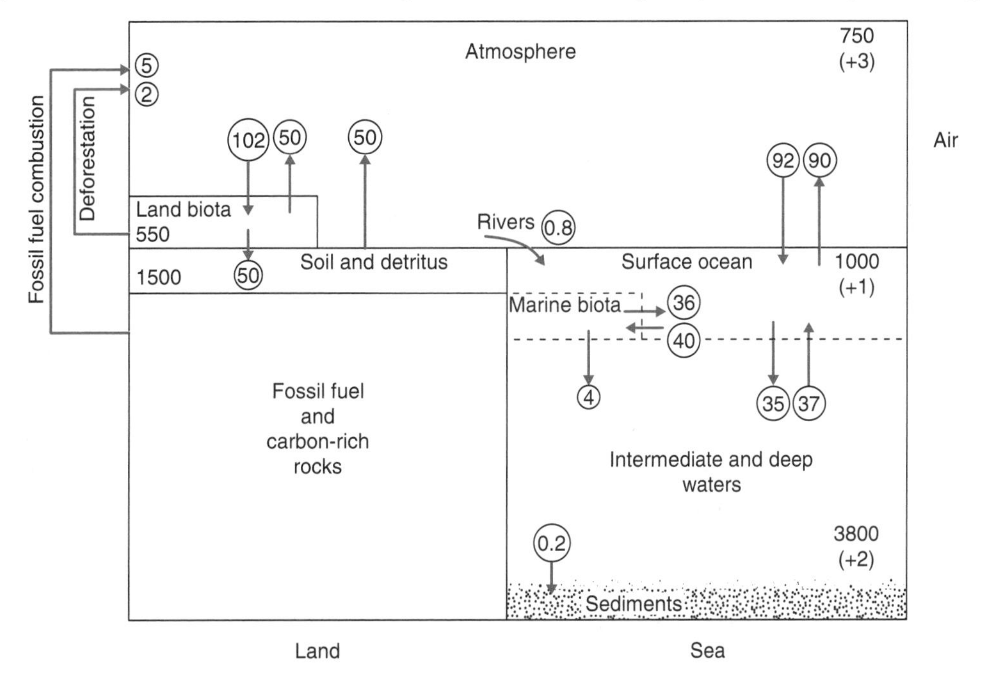

Figure 8.15 The global circulation of carbon. Notice that the most important fluxes (circled) are between the land and air, and that the sea and air are largely in balance. What has changed in the last two decades is the amounts moving through the atmosphere and a reduction in the principal sinks on land – the forests. Fluxes are in 10^{12} kg per year and reservoirs in 10^{12} kg.

Box 8.3

THE CARBON BALANCE AND THE MISSING SINK

Without anthropogenic inputs, carbon added to the atmosphere is balanced by that removed. We add an additional 7 gigatonnes (7×10^9 tonnes) each year, yet the atmosphere is accumulating only 3.4 Gt each year and the oceans an additional 2 Gt. So where does the missing 1.6 Gt go?

One possibility is that our estimates for the oceans have been too low. However, since there is little evidence of major changes in phytoplankton productivity over the last century, this is perhaps unlikely. Most studies suggest that the global forests are most likely to account for the deficit (Figure 8.16). About 31% of the Earth's land area is under forest and that supports 86% of total above-ground carbon. Around 73% of global soil carbon is in forest soils.

Some suggest that the higher rates of fixation in the tropical forests are the most likely sink, primarily because of the effect of higher temperatures on photosynthesis and the longer residence times for carbon in their biomass. These forests would respond fastest to the atmospheric changes of the last 100 years.

Others point to the considerable shrinkage in the area of tropical forest over the last 50 years and argue that these latitudes are more likely to be net contributors to the carbon budget of the atmosphere. The temperate forests, in contrast, especially those in the north, have been expanding. Commerical forestry and the abandonment of farmland has led to a slow expansion of both the deciduous and boreal forests throughout the century. In addition, the production of the northern forests may have been heightened because of the fertilizing effect of nitrogen pollution from the industrial activity of the region and the use of agricultural fertilizers. Sulphate deposition and the overall warming of the atmosphere may have accelerated the growth of established woodlands.

Roger Sedjo calculates that the increase in biomass in the northern forests represents 0.7 Gt each year of carbon, accounting for a sizeable proportion of the missing carbon. Although the region has been subject to considerable climatic fluctuation, it seems (ironically, perhaps) that forest growth in the north is the missing carbon sink.

models that take greater account of the effect of cloud cover, deep ocean currents and, more recently, sulphate aerosols. The trend is still for a substantial rise – at least 1.5°C by 2060 (Plate 33). Oceans may buffer changes over the longer term as higher temperatures lead to greater evaporation, but how long stabilization will take, when the atmosphere reaches a new constant temperature, is still uncertain.

The movement of the deep ocean currents will lead to differential heating and the IPCC predicts twice as much warming in the waters of the northern hemisphere as in the southern hemisphere. Sea levels will rise as a result, primarily because of thermal expansion (2–4 cm per decade) and only partially because of melting ice (1.5 cm per decade). A number of countries face inundation during storms, including many low-lying Pacific islands, large

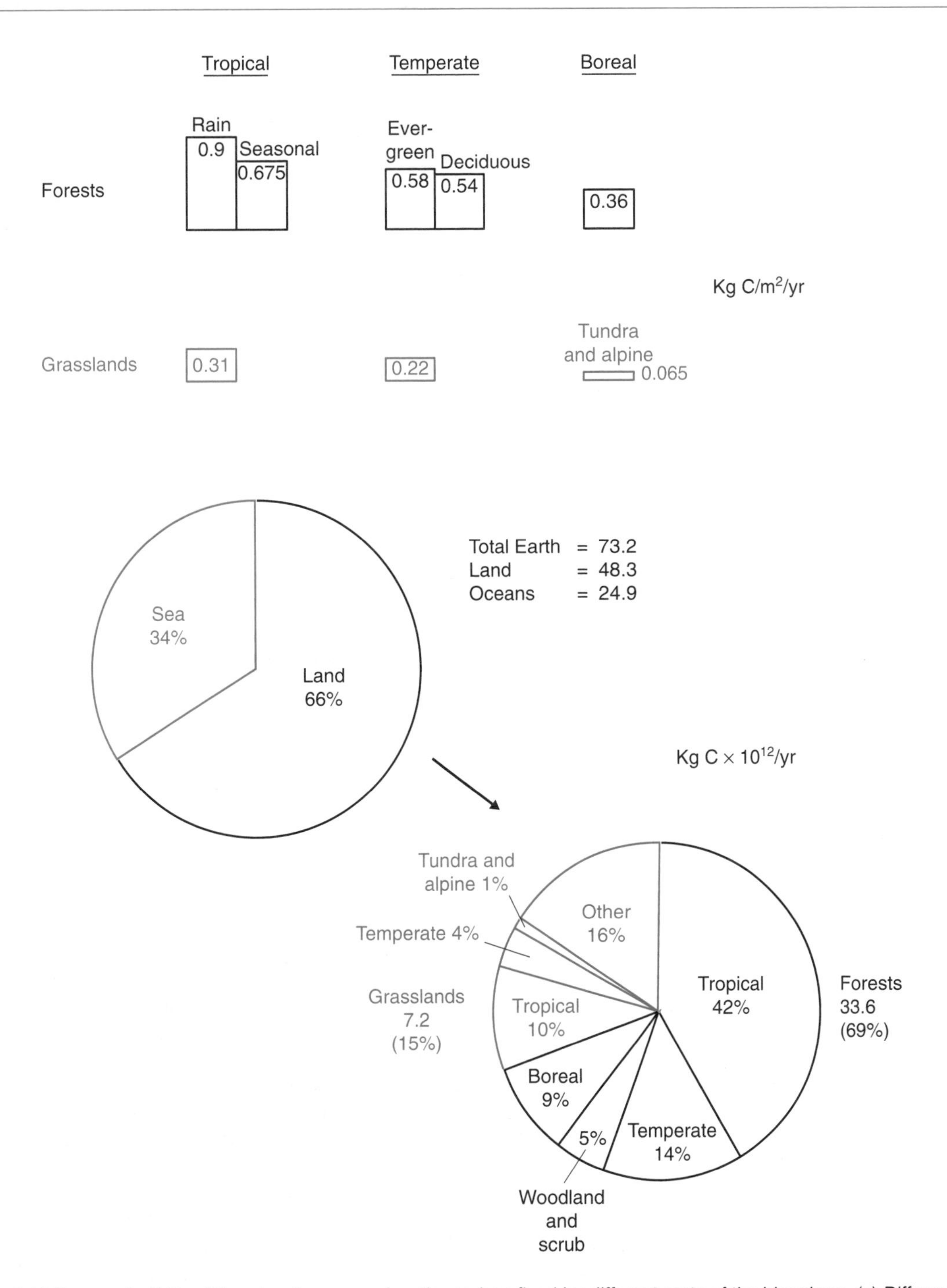

Figure 8.16 The productivity of the planet measured as the carbon fixed by different parts of the biosphere. (a) Differences in mean net primary productivity per square metre (section 5.1) between forests and grasslands and between the three major latitudinal biomes. (b) Overall, terrestrial biomes are responsible for about two-thirds of the carbon fixed in the biosphere each year, of which tropical forests represent the largest proportion.

Table 8.1 The principal greenhouse gases

	CO_2	CH_4	N_2O	CFC_{12}
Concentration (ppmv)				
pre-Industrial	280	0.8	0.29	0
now	353	1.72	0.31	0.00048
Rate of increase (% per year)	0.5	0.9	0.25	4
Lifetime (years)	50–200	10	150	130
Greenhouse effect over 100 years relative to CO_2	1.0	11	270	7100
Relative contribution over 100 years*	72	18	4	(Highly variable)
Reduction in emissions needed to stabilize	> 60%	15–20%	70–80%	75–85%

* These estimates also allow for the indirect effects of different gases on CO_2 production.

Box 8.4

OZONE DEPLETION

Changes in the chemical composition of the atmosphere can all be related to the radiation balance of the atmosphere – what is being let in and what is not being allowed to escape. Ozone (O_3) concentrations in the atmosphere are important, depending on their altitude.

In the lowest layer of the atmosphere, the troposphere, ozone levels have been increasingly locally, as pockets of pollution around cities. These are generated through photochemical reactions involving car exhaust gases. Forest dieback around some major conurbations has been exacerbated by such pollution. However, low-level ozone is also an important source of hydroxyl radicals (OH^-), important for reacting with methane, carbon monoxide, sulphur and nitrogen oxides, and helping to reduce their impact on the atmospheric energy budget.

In the next atmospheric layer, the stratosphere, ozone is being depleted. Here, the gas absorbs harmful incoming ultraviolet radiation (UV-B), protecting living systems from energy at wavelengths that are strongly absorbed by a range of organic chemicals, including DNA. Skin cancers and depressions in plant productivity follow from the damage caused by increased UV-B reaching the Earth's surface.

The depletion of stratospheric ozone is not uniform over the globe. It is most pronounced over the polar regions in the spring when low temperatures help to produce a high-altitude luminescent cloud within discrete circulation patterns. The ozone 'holes' recorded over Antarctica and the Arctic since 1977 follow from chemical reactions with sunlight and a number of gases, especially CFCs. Ultraviolet radiation slowly breaks down these compounds to release chlorine monoxide which, in turn, reacts with ozone to produce molecular oxygen.

In addition, CFCs themselves account for perhaps 15% of the greenhouse effect. Because of their longevity and impact on UV levels, action to phase them out has been relatively swift. According to the Montreal Protocol agreed in 1987 and revised in 1992, large-scale use of all CFCs and related gases will be phased out within the next few years, though their impact will extend long into the next century.

areas of Bangladesh and northern Europe. Sustained warming over the long term will almost certainly halt the thermohaline circulation as a stable stratification of the oceans establishes itself (Figure 8.7), a change known to be associated with the major climatic oscillations of the ice ages.

These predictions are based on the outputs from a number of **global circulation models** (GCMs). These are immensely complex mathematical models of the air/water system of the planet, originally derived from the weather forecasting simulations used on more local scales. There are several such models, but all require very large powerful computers running for long periods. The current UK model (the Hadley Centre model) requires four months to run on one of the fastest supercomputers.

Predictions between the models differ, primarily because of the different weights each attaches to various critical factors, especially deep ocean currents and cloud cover. These differences and the uncertainty of their predictions have been seized upon by their critics (Box 8.5). However, in 1995 the Hadley Centre model was run with due allowance for the effects of sulphate aerosols. For the first time a GCM successfully matched past observations, replicating the rise (0.5°C) of the last 100 years. This was half of the warming predicted by modelling carbon dioxide alone, suggesting that sulphate aerosols are a significant cooling agent. For the same reason, the Hadley model predicts a slower rise (compared with other models) in the future, of about 0.2°C per decade until 2050 (Plate 33).

Even so, this latest result confirms previous weather forecasts for the next century: warming will be uneven over the globe, increasing most rapidly in the northern hemisphere, perhaps more so as we reduce our sulphate emissions. Here, winters will be milder and in the middle of continents summers will be hotter and drier. Continental margins will be wetter, but although more precipitation is expected in mid and high latitudes, much less rainfall is forecast for the subtropical zones.

As time goes by we find more and more evidence that global atmospheric warming is under way. Now ecologists have to speculate on what this means for the planet's ecology (Box 8.5).

8.3.4 Ecological changes likely with atmospheric warming

Shifts in the global climate have implications for the oceans. If, as predicted by some models, the thermohaline circulation ceases, the carbon cycle and biogeochemistry of the planet will undergo profound changes (section 6.2). Nutrients released from the overturning of marine sediments will decline, limiting primary productivity in the surface waters. One consequence of global warming could thus be dramatic changes to the communities in the upper waters of the oceans.

The distribution of terrestrial biomes will shift as temperature and moisture regimes change. The extension of the monsoon rains polewards will increase the moist tropical forest cover in some areas, such as northern Australia. On the other hand, the mediterranean biome is expected to suffer drier and hotter summers, with lower soil-moisture levels and prolonged summer droughts. The rich mediterranean flora is expected to suffer major species losses and the drier top soil will be more easily eroded. Agricultural productivity in these regions is likely to fall.

The tundra will be warmer for longer and experience longer periods of biological activity. Its boundary will extend further northwards. Some more southerly plants, most probably the grasses, will encroach polewards, as will the conifers of the taiga. Current estimates suggest that the northern tree line will move 100 km northward for every 1°C rise in mean temperature. Deciduous hardwoods and grasslands will follow. Temperate mixed forests are expected to lose their conifers and become dominated by deciduous trees. Parts of Canada (northern Alberta, Saskatchewan and Manitoba) may lose their coniferous forests altogether. Some predictions see the tundra and taiga shrinking by about one third as carbon dioxide levels double and temperate forests extend their range.

Patterns of cultivation are likely to change in the same way. Growing seasons will be extended in northern latitudes, so that leaf and root vegetables could be grown in Central Alaska. Crop yields will increase in northern Europe (perhaps by a third in Denmark) but decline where water becomes limiting (down by a third in Greece). More significantly, the drying of the prairies and the steppes in the southern part of their range will reduce cereal production. The overall impact will be increased food shortages in Russia and neighbouring countries. The Sahara will extend into the Sahel and increase its range in central Africa.

These are all large-scale changes which will take place over decades, as biomes move in response to warming. In the short term, lower level changes in

Box 8.5

WHAT, ME WORRY?

The uncertainty of predictions from the different GCMs and the absence of clear signs of warming lead some commentators to caution against present action to curb carbon emissions. Fossil fuels are so central to industrial economies that emission targets equate, in their minds at least, with limiting economic growth. It would be better, they argue, to let market forces, through the costs of pollution and environmental degradation, limit fuel use. While the economic models that predict such responses are even less convincing than the GCMs, their forecasts are studied much more closely by decision-makers.

These commentators point out that the match between carbon dioxide levels and warming is not perfect. One hiccup occurred between 1940 and 1960 when, despite rising carbon dioxide levels, the atmosphere cooled slightly. This certainly demonstrated the weaknesses in atmospheric modelling and their predictive power. It was most likely caused by astronomical factors. On another occasion, warming and gas concentrations appeared to halt in mid-1991 following the eruption of Mount Pinatubo, yet both resumed their rise in the 1993. In late 1995, the UN IPCC finally accepted that global atmospheric warming was indeed a reality.

Carbon dioxide levels will continue to rise in the forseeable future if only because of the time needed for atmospheric stabilization, when inputs match outputs. Stabilization will only follow when we stop increasing fossil fuel combustion.

Setting targets for CO_2 concentrations in the future effectively means allowing a given amount of carbon to be emitted into the air between now and that date. However, there are different ways of using up that carbon allowance: we could do nothing for the first few years and then rapidly reduce our emissions, or we could opt to reduce more gradually. There may be some justification for adopting the former approach, allowing 'business as usual' for the next few years, but then seeking to reduce rapidly. In this way, our current equipment – the machinery that uses the fossil fuels and releases the carbon – needs to be replaced only slowly, allowing us time to make provision for the massive costs associated with replacement. It also gives us time to continue to research all aspects of the problem – ecological and engineering.

Of course, the target we set will depend on a range of others factors, including the scale of environmental change we expect. The lower the target, the more we shall have to reduce emissions. We need to be careful not to injure the economic prospects of many poor and malnourished peoples, wherever they are. We do need better data but we cannot afford to wait too long to confirm our predictions. The longer we delay action, the longer stabilization will take.

photosynthesis and respiration will determine carbon release and these in turn will influence our predictions about future warming.

8.3.5 Changes in primary productivity

Can we expect an increase in photosynthesis to mop up the abundant carbon dioxide and so offset the warming? Plants, especially C_3 plants (Box 5.2), should benefit from higher carbon dioxide concentrations, but photosynthesis also depends on temperature, and the availability of water and nutrients.

Within an individual leaf, carbon assimilation increases with carbon dioxide concentrations in the surrounding air. A doubling of carbon dioxide levels increases growth in C_3 plants by about 40%, and in C_4 plants by 20%, though there is considerable variation between species, in part because of the allocation of resources each makes to different functions, including root growth and the search for nutrients (section 4.1).

Experiments on a variety of crop plants have shown that photosynthesis may rise initially with rising carbon dioxide but then some species appear to acclimate, so they eventually show no overall increase in carbon fixation. One limiting factor is the capacity of the enzyme responsible for fixing the carbon (rubisco – Box 5.1) or at least the speed at which it can be regenerated. Others may include the build-up of the end-products of photosynthesis, with perhaps starch granules impairing the function of the chloroplasts.

High carbon dioxide levels cause partial closing of stomata, so reducing transpiration and making the plant more efficient in its water usage. However, in some crop plants this is offset by the plant producing more leaves, so any shortage of water may still be limiting. In others species, a lack of some nutrients (especially nitrates and phosphates) will constrain carbon fixation. Overall, the response varies between species and with the age of the plant. It also changes with environmental conditions, most especially temperature and humidity, and also with the level of ultraviolet radiation so that ozone depletion may also be significant on a regional scale.

Fixation by trees is one of the main ways in which carbon is removed from the atmosphere. We can see the significance of terrestrial vegetation in temperate regions with the regular seasonal cycle of atmospheric carbon dioxide noted in the Mauna Loa observations (Prologue: Figure 2).

8.3.6 Changes in decomposition

Higher temperatures are likely to accelerate decomposition in the colder areas of the planet and increase respiration more generally. Alone this would add to carbon release, yet experiments in the Alaskan tundra show carbon storage actually increases at elevated levels of temperature and carbon dioxide. This is probably because nutrient concentrations (especially nitrogen and phosphorus) are low in the litter of plants grown with high carbon dioxide. This litter takes longer to degrade, and its carbon remains locked up for longer. In contrast, decomposition rates for older, better quality deposits will indeed increase as mean temperatures rise.

While higher average temperatures in the tundra might promote greater plant growth, fixing more carbon, they will also allow the permafrost to thaw to a greater depth. Then the waterlogged soil will generate more methane. Wetlands produce about five times as much methane as dry areas (dry soils are an important sink for methane – particularly neutral woodland soils). Overall, microbial decomposition and respiration rates are expected to increase with high temperatures, releasing more carbon from soils rich in organic matter.

Another significant source of methane is rice paddies, which have expanded considerably in the drive to produce more food, often at the expense of tropical forests. A large proportion of their methane production is oxidized to carbon dioxide in the upper layers of the soil, but significant amounts still enter the atmosphere. Methane is also generated in large quantities by animal husbandry (the wastes, gaseous and otherwise, of cattle) and by domestic refuse in landfill sites.

Any increase in biological activity will promote a faster turnover of minerals in the biosphere. Most likely to be affected is the nitrogen and phosphorus fixation of plants with symbiotic microorganisms (Box 6.2). Their ability to supply carbon to their partner (*Rhizobium* for nitrogen fixation and the fungal mycorrhizae for phosphorus scavenging) may well be improved with carbon dioxide enrichment.

Other interactions between species – competitive, exploitative and cooperative – may change as the balances shift favouring some physiologies and metabolisms over others. As the nutrient status of their diet improves, some insect herbivores may well enjoy higher population growth rates. This, along with higher mean temperatures extending their periods of activity, may put some habitats at risk. For example, some ecologists have suggested that the heathlands of north western Europe may be threatened if a key herbivore, the heather beetle (*Lochmaea suturalis*), is able to increase its rate of reproduction under a warmer climate.

We have already seen that the composition of a community is far from fixed. As the balance shifts in favour of particular species, others may be lost and the biomes themselves change. Indeed, some of the plant assemblages recorded in the deposits of the last ice ages are unknown under today's climate.

8.3.7 Stabilizing the atmosphere: ecological methods

The main mechanism for controlling the rise in atmospheric carbon dioxide is to limit our emissions; after that, it would be to limit our deforestation programmes. Titus Bekkering suggests that together these two measures could reduce the growth in atmospheric carbon by 12% by the year 2100. Even with a massive, and perhaps unrealistic, programme of reforestation and forest regeneration extending cover by 865 million hectares, carbon levels in the atmosphere are projected to be 54% higher in 100 years time.

A variety of schemes for increasing carbon fixation have been suggested. One is based on a proposition that oceanic phytoplankton communities are limited by a shortage of iron, which primarily arrives as wind-blown dust from land. The suggestion is to add iron compounds to the open oceans to increase their primary productivity. However, significant changes in the phytoplankton community would almost certainly follow.

On land, forests are the principal means of fixing atmospheric carbon in terrestrial ecosystems, and the one component of these biomes over which we have the greatest control. Estimates in the IPCC report suggested that 370 million hectares planted at 10 million hectares per year over the next 40 years would eventually fix 80×10^{15} g carbon (80 gigatonnes), which is somewhere between 5% and 10% of the carbon derived from fossil fuel combustion. After this time, when tree death eventually matches tree growth, the forest ceases to be a sink for carbon. By then, the total area planted would be equal to about half of the Amazon basin. Unfortunately, the forecast for warmer days may mean grasslands encroach because of increased aridity at certain times of the year in many tropical areas.

Can planned reforestation be combined with a conservation strategy that protects the biological diversity of the tropics? This may only be realistic in those countries currently able to feed themselves. Bekkering suggests that this amounts to 15 countries, which between them have the potential to replant a total of 580 million hectares. The UN's Tropical Forest Action Plan (TFAP), started in 1985, attempted to introduce forest management techniques into developing nations, bringing about an eight-fold increase in reforestation in participating countries. Unfortunately the total area is small (1.9 million ha) and more carbon is released by these countries (282 million tonnes per year) than is fixed (12.4 million tonnes).

Rates of deforestation have begun to decline in Amazonia, though elsewhere in the tropics they are increasing. Recent evidence indicates that the scale of deforestation may have been over-estimated, because of a poor understanding of long-term land use patterns and a misinterpretation of the patterns on the ground today. Too often, it seems, local farmers have been blamed for poor woodland management practices, when they are actually helping to sustain the forests at their margins.

Tropical reforestation can be a realistic prospect only where the needs of local people are taken into account. Forests can be exploited without decimating them. They have an economic potential from timber and fuel production, as well as higher value goods. If exploitation is managed with a proper regard to the natural processes which determine community structure, there is every indication that this can be sustained.

The real issue is not carbon fixation by forests but carbon release by fuel consumption. The solution depends on the will of industrialized nations to curb their use of fossil fuels. Perhaps the major user nations should pay tropical countries to maintain their carbon-fixing potential – a service they rely on, but which is not part of their costs. If done on a *per capita* basis this would provide some incentive for all countries to reduce their fossil fuel consumption. Yet, as in any hierarchy, we should all remember that it is individual rates of consumption which cumulatively determine the impact humanity has on the atmosphere. We should all look to see the size of the footprint each of us is leaving on the planet.

SUMMARY

The global distribution of the major biomes reflects the climatic zones across the planet, most especially the mean temperature ranges and patterns of precipitation. Where water is abundant, forests develop; where an extended dry season occurs, grasslands are found. Between the tropics, forests are lush and productive and the savanna grasslands support a variety of game. Temperate regions have communities adapted to distinct seasonal cycles. The taiga and tundra are limited by short growing seasons and long winters.

Landscape ecology provides an approach for studying the large-scale processes that determine the nature of component ecosystems. Hierarchy theory suggests that the organization of biological systems, from the biome to the individual cell, is a product of functional interactions between different levels of organization. Some of these mechanisms become apparent when we look at the relationship between the carbon budget of the atmosphere and the climate of the Earth. The two are linked primarily through biological activity and the evidence increasingly suggests that a large-scale change is under way.

FURTHER READING

Kemp, D.D. (1994) *Global Environmental Issues: a Climatological Approach*, Routledge, London.

Pickering, K.T. and Owen, L.A. (1994) *An Introduction to Global Environmental Issues*, Routledge, London.

EXERCISES

1. Select the correct answer for each of the following.

(i) Deserts are found where ...
  (a) moist air is ascending on its way to the tropics.
  (b) 40° either side of the Equator.
  (c) rainfall is less than 10 mm each year.
  (d) dry air descends having shed its moisture in the tropics.

(ii) The global distribution of biomes is primarily a product of ...
  (a) the effects of latitude on temperature and moisture.
  (b) the effects of longitude on temperature and moisture.
  (c) the effects of altitude on temperature and moisture.
  (d) All of the above.

(iii) Grasslands are found where ...
  (a) moisture limits forest growth.
  (b) soils are thin and infertile.
  (c) in flat areas.
  (d) in the centre of continents.

(iv) Which of the following limits the supply of nutrients to tundra plants?
  (a) Waterlogging of the upper soil in summer.
  (b) The frozen soil in winter.
  (c) Grazing by vertebrates.
  (d) Lack of litter.

2. Assign the following to either tropical or temperate forests:
  (a) Deep rich soils with a high organic content.
  (b) Rich diversity of herbaceous plants.
  (c) Rapid rates of decomposition in the soil.
  (d) A distinct season when respiration exceeds primary production.
  (e) A high and continuous canopy.
  (f) An associated community of plant creepers and climbers.
  (g) Most nutrient capital is held in the living components of the ecosystem.
  (h) A resident population of frugivores.

3. Construct a diagrammatic hierarchy of the processes that govern atmospheric temperature, using the information contained in this chapter. Your final diagram should show how different components are nested into hierarchies which together determine mean atmospheric temperature.

4. Why might an increase in the concentration of carbon dioxide not lead to a corresponding increase in photosynthesis by a particular species of plant? What significance does this have for global atmospheric warming?

Tutorial or seminar topic

5. Assume that tropical reforestation is unlikely to be used to mop up carbon from the atmosphere. Speculate on what possible technological fixes might be possible within the next 50 years, either to limit carbon releases or to remove it from the atmosphere.

The problem of ecological stability and its measurement – stability and diversity – the global gradient in species diversity increasing towards the tropics – past and present rates of species extinctions – the global scale of human disturbance – human population pressure on land and resources – human origins revisited.

9 Checks

...the beauty of the cosmos derives not only from unity in variety, but also from variety in unity.

The Name of the Rose, Umberto Eco

Human beings, you will remember, originated with a series of major climatic changes that started 7 or 8 million years ago. This instability in the Earth's climate and the sequence of ice ages that followed played a large role in subsequent human evolution. Back in Chapter 1 we should, perhaps, have asked what initiated this instability, which was to have such dramatic consequences.

The short answer is we do not know. One suggestion is a change in the Earth's orbit relative to the Sun. A more controversial view is that global cooling followed when the Tibetan plateau and the Himalaya were formed, as the Indian crustal plate pushed up against the Asian continent (Plate 49). A shift in atmospheric circulation over this massive upland area changed patterns of precipitation and increased rates of erosion. The nutrients arriving in the seas prompted greater primary productivity, and this enhanced photosynthesis stripped the air of its carbon dioxide, reducing the greenhouse effect. In this case biological activity caused temperatures to fall and the polar ice sheets to expand. If this was the ultimate cause of the climatic change in East Africa and, later, the ice ages, then the birth of humanity was prompted by a collision of the continents. Literally, the Earth moved.

That new species arise following climatic change should come as no surprise to us by now, nor that some species are lost. Rates of climate change have been particularly rapid over the last 100 years, primarily because of increased atmospheric pollution and deforestation. As natural habitats disappear, so do species. Currently extinctions are running at a rate perhaps 1000 to 10 000 times higher than usual. We are at the beginning of a mass extinction event, which may eventually compare with one of the Big Five extinctions documented in the fossil record. While species may go extinct for a variety of reasons, today there is one principal cause: *Homo sapiens sapiens*.

In this final chapter we look at the ability of species to persist and of ecological communities to remain unchanged in the face of disturbance. We see how this might be related to the pattern of species diversity across the planet. We also examine the particular difficulties of describing such highly complex systems. Ecology is something of a pathfinder in its attempts to understand systems with a large number of interacting components, and ecologists are developing methods of analysis using ideas that may have a wider currency. So we finish by looking at the insights that these ideas provide into the nature of humanity itself.

9.1 ECOLOGICAL STABILITY

We expect our environment to stay more or less the same. Even as we prepare for winter, we know when spring is due and when to sow seed. Predictable change is incorporated into our biology and behaviour in the same way squirrels store caches of food for the lean times ahead. The apparent continuity of the major ecosystems suggests there must be checks and mechanisms to keep things the same from one year to the next. This stability is what most people mean when they refer to a 'balance of nature'. But how real is it?

Much of what we have to say about stability can be applied to any level in the ecological hierarchy, from individual organisms to ecosystems. To encompass these different levels we use the term 'system', meaning two or more interacting elements that remain together. This term also allows us to make allusions to other complex phenomena. Ecologists have used a variety of measures to gauge how stable ecological systems actually are, from fluctuations in biomass or rates of nutrient transfer to annual variations in the abundance of single species. We shall concentrate here on populations and communities, where the components are individuals and species, respectively.

Stability is a property of systems that change little following a disturbance, or which return quickly to their previous condition. A stable population, community or ecosystem would either resist disturbance or resume its previous pattern of change. Many disturbances cause no permanent change and may be incorporated by the ecosystem, rather like seasonal change. Occasionally the disturbance may be on a scale from which the system does not recover easily.

A good analogy is the weighted spring (Figure 9.1). At rest, undisturbed, the spring has a particular length. Pull the spring gently and it bounces around but eventually returns to its original position. Exceed its elastic limit and it will be permanently stretched, coming to rest in a lower position. It is easy to show an equivalent stability in the systems of an individual organism and to describe the mechanisms by which it maintains fairly constant internal conditions, such as its temperature or water balance. This homeostasis is necessary for the organism to maintain its metabolic efficiency.

Detecting stability in ecological systems is more problematical and depends upon the the scale over which we measure a parameter. Over a year, the seasonal cycle of a temperate woodland brings a sequence of changes to which each species is adapted: birds arrive to exploit the insects of spring and summer and plants bloom to attract them to their flowers. These seasonal variations are incorporated into the ecology of the community.

Over larger time-scales and larger areas, the forest community may cycle through various stages as trees fall over and new colonists arrive. Again, the system may incorporate such change, so over a 10-year period the woodland may show little change in its collection of trees or the abundance of a single species. But if it is sampled over a larger time period, more variability creeps in. Individuals are replaced and the species composition shifts. Some changes will come about by chance, but a persistent ('directional') change over a long period may also be detectable, perhaps as one species of tree comes to dominate its competitors.

Different ecosystems have a variety of processes operating at different rates and spatial scales. What we can detect depends upon our sampling programme (Figure 9.2). Often an ecosystem consists of a mosaic of patches, each at different stages in a cycle of change, but overall the landscape remains more or less unchanged. Our data would then show different degrees of variability with different scales of sampling. According to hierarchy theory (section 8.1) we should expect small-scale processes to operate very rapidly over short distances and large-scale processes to have a longer periodicity.

Our understanding of the link between ecosystem stability and community organization is crucial for our capacity to predict changes over the long term and large spatial scales. Experiments such as Biosphere 2 demonstrate the difficulties of mimicking some of the global-scale processes, even though such model systems may be the only way to examine some of the important interactions. Experiments in the 'Ecotron' at Silwood Park in southern England (Figure 9.3) have examined the role of community interactions and ecosystem function. These use a series of microcosms – small artificial ecosystems whose communities can be constructed following precise recipes. Although the microcosm size (around 2 m^2) limits the species that can be used, the Ecotron does allow experiments to be replicated and repeated. Data from early trials provide a timely reminder of the problems in Biosphere 2 and show how carbon fixation is related to species richness. The greater is the diversity of the microcosm, the more carbon fixed. This seems to be straight-forward enough – a result of effective light interception (and, therefore, photosynthesis) from a multilayered plant community.

So do real ecological systems show stability? It depends upon what you measure, how long you measure it for and over what spatial scale you measure it. Whether stability is a feature of populations and communities is still a matter of great debate, and an important one. We need to describe the mechanisms which keep these systems within their limits. Efforts to understand how a community might remain stable have invariably focused on the number and abundance of its component species, and the relationship between species diversity and stability.

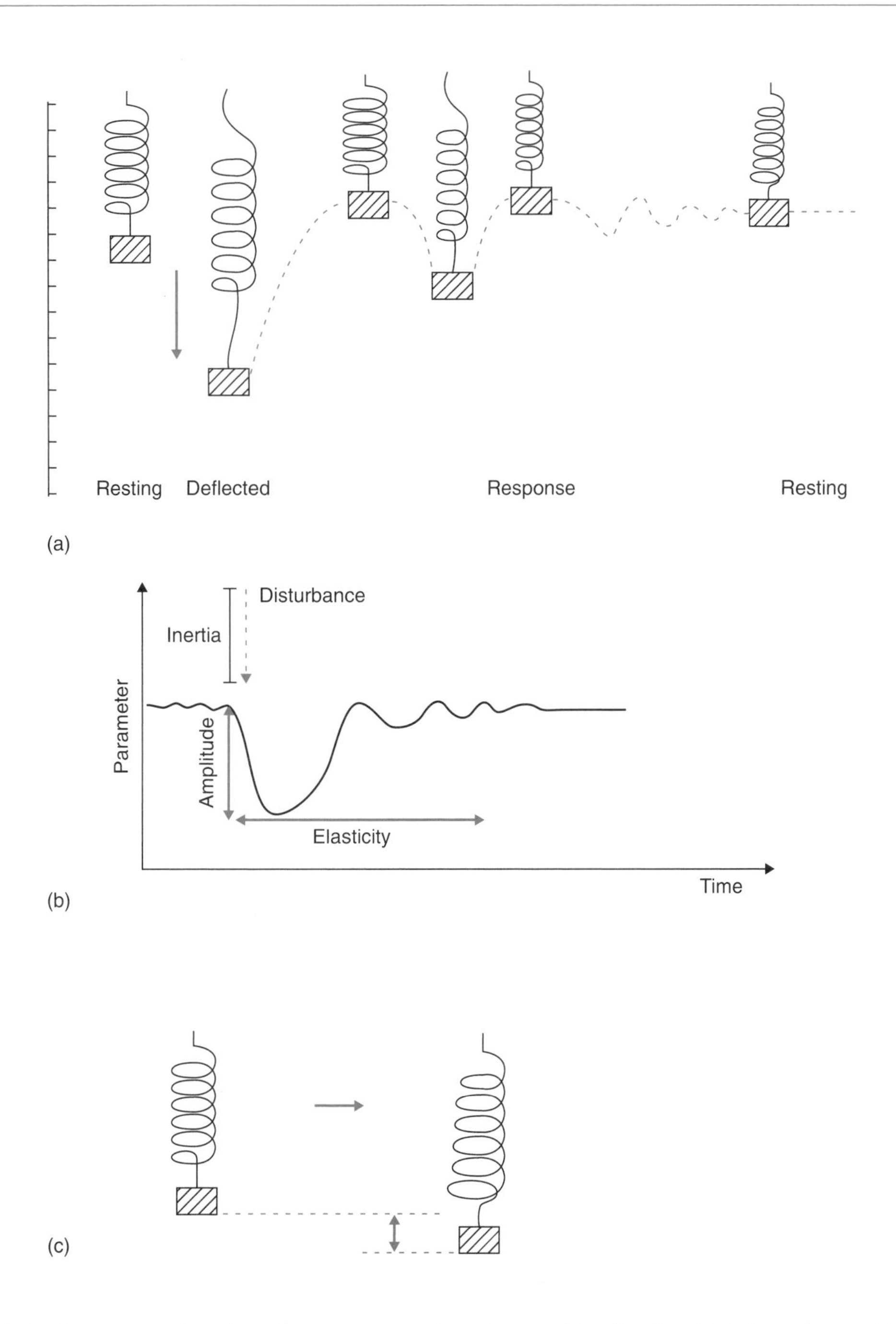

Figure 9.1 A simple representation of stability in a system, shown by a weighted spring. (a) In its resting position the spring and weight extend to a certain length. (b) If we disturb the system, the amount of force we need to deflect the spring is termed its inertial stability (or **resistance**). The scale of any deflection (its amplitude or **resilience**) is the increase in length. The time taken for the spring to settle down to its original position is its **elasticity** (or adjustment stability). (c) If we exceed the capacity of the system to return to its original position (we stretch it) then it will return to a new (slightly lower) equilibrium position. In ecological systems we might measure such dynamics as changes in population size, biomass production, species richness or a range of other measures.

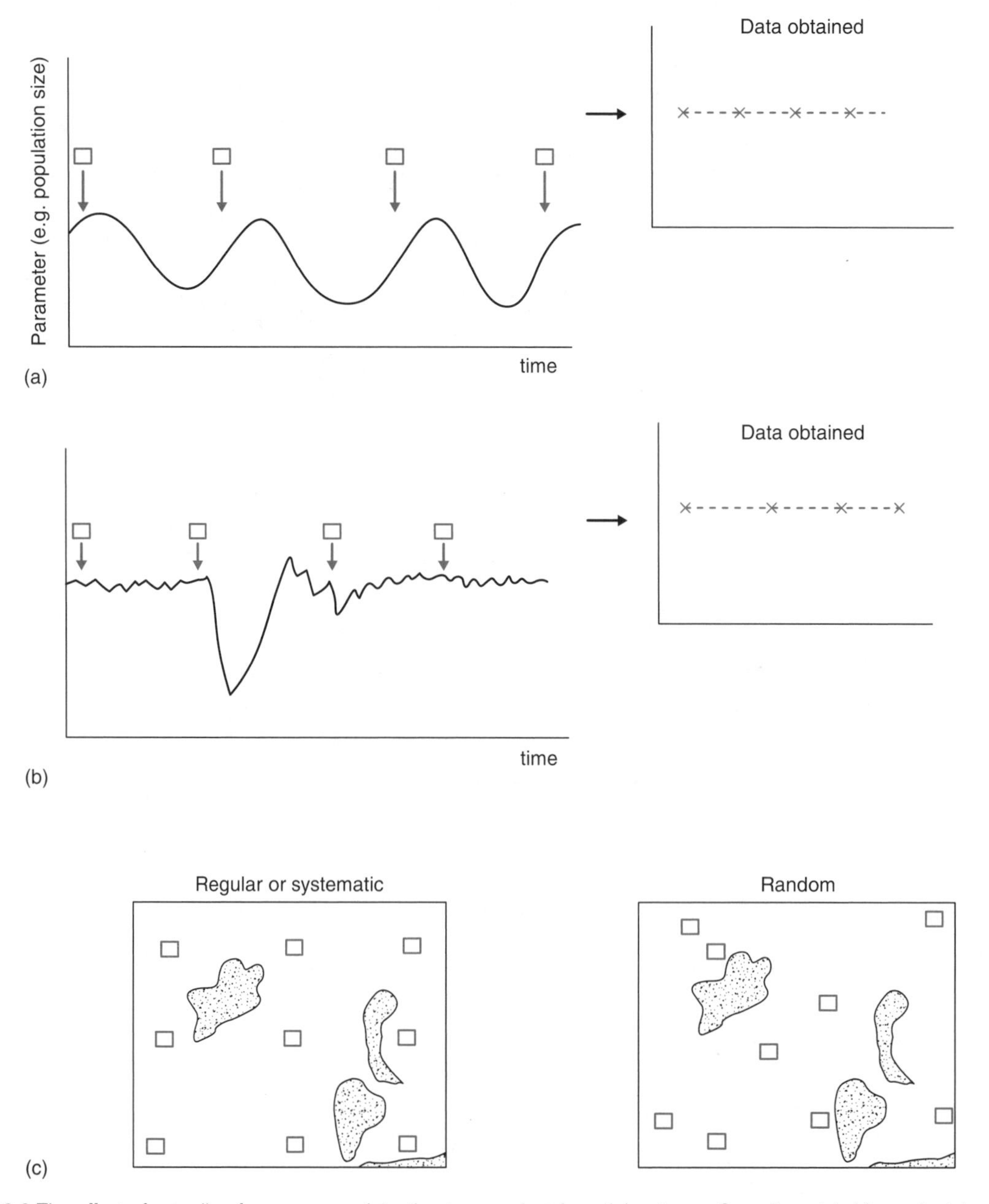

Figure 9.2 The effect of sampling frequency on detecting temporal and spatial patterns. Sampling at (a,b) regular intervals or (c) randomly in time and space can fail to detect the dynamics of a system, a disturbance or its patchiness. This makes detecting significant change and stability in variable ecological systems very difficult. We have to use preliminary surveys to match our sampling programmes to the pace of processes or the distribution of features in the real world.

Although there are problems in measuring diversity (Box 9.1), we can learn much from comparing systems, finding common features between communities and the species configurations which are most likely to persist. One example is the relatively constant ratio of predator species to prey species across very different food webs (section 5.4). Another is the similarity in the organization of

Figure 9.3 The Ecotron in operation. The Ecotron is a series of 16 microcosms that can be used in various arrangements to create small, replicated ecosystems.

mediterranean-type communities found at the five corners of the globe, even though their plants and animals have very different ancestries (section 7.1). Given such patterns, does this mean that the most persistent communities are assembled in certain ways?

9.2 STABILITY AND DIVERSITY

One of the most persistent ideas in ecology is that a diverse ecosystem is a stable ecosystem. A link between the number of species and stability was first suggested in the 1950s, based on comparisons of species-poor and species-rich communities. Pest outbreaks were thought to be more prevalent in agricultural systems compared with diverse natural ecosystems, and populations more variable in temperate and boreal biomes compared with tropical forests. While we now recognize these as oversimplifications, the idea does find support in some more recent and systematic studies (Figure 9.4).

The stability of species-rich communities was thought to come from their 'functional redundancy'; that is, more species allowed for more pathways through a food web (section 5.4). If one species was lost, others provided routes by which energy could still flow: a consumer would just switch to a different prey species. The greater the redundancy, the greater was the capacity for withstanding the

Box 9.1

MEASURING DIVERSITY

As with so many terms, the meaning of diversity depends very much on context. At different times diversity, or even biodiversity, refers to species richness (simply the number of species), to genetic diversity (the variety of genotypes), or to the structural diversity of a habitat. The shared property is the variety seen in each class, but here the term biodiversity encompasses everything from biomes to genes and nucleotides. Perhaps most important is the acknowledgement of the significance of variation at all of these levels – genetic diversity for the health of the population, or habitat diversity for global ecological processes.

Community ecologists have concentrated their efforts on measuring species diversity as a way of summarizing the structure of a community. A considerable amount of work has gone into trying to find a universally acceptable method of measuring this diversity effectively, one that allows us to compare very different ecosystems. In this context, diversity consists of two components: **species richness** (the number of species) and **species equitability** (the proportion of individuals belonging to each species).

Two communities may share the same number of species and the same total number of individuals, but the one with the more even distribution of individuals (the highest equitability) is the most diverse (Figure 9.5). High equitability implies that resources are evenly shared between species and that the system is not dominated by just one or two abundant species. For example, surveys of the deep-sea fauna of the North Atlantic show that equitability makes a significant contribution to the latitudinal gradient in the diversity of bivalve molluscs (there are few highly abundant species dominating the samples).

Various indices attempt to measure diversity and to include some measure of equitability, with varying success. In doing so, the index may make assumptions about the underlying pattern of equitability within the community, which itself can be problematical. Others treat all species as being equivalent but, again, this may not accurately reflect the significance that some species, such as keystone species, have for a community. The taxonomic difference between groups is also rarely measured: two samples may have the same species richness, but the one with four species of ant will normally be seen as less diverse than a sample with a fly, a wasp, a beetle and an ant.

Several indices depend on the size of sample taken, and this can cause problems when we are unsure about our sampling efficiency. For these reasons, no single measurement of diversity has yet been adopted as being the most effective under all circumstances.

disturbance of species loss. Populations in a diverse community would thus show less variability than a species-poor community.

In the early 1970s Robert May challenged this idea. He constructed mathematical models of food webs with different numbers of species and different numbers of connections between them, so varying the complexity of the system. A connection represented some interaction between two species (predation, competition and so on) and was of variable strength (how much the population size of one species affected the abundance of the other). May

Plate 37 An ecosystem stranded: abandoned boats at Muynak, formerly a fishing port on the Aral Sea.

Plate 38 The Teleki valley, Kenya. This montane habitat on the Equator has a unique plant community that reflects the large temperature range created by the high altitude. Here giant groundsels (*Senecio* species) grow, with a series of adaptations to survive the wide temperature range. Different species with equivalent adaptations grow on the Equator in the Andes and in Hawaii.

Plate 39 Tropical rainforest.

Plate 40 East African savanna.

Plate 41 Temperate forest: deciduous woodland in south-east England.

Plate 42 Prairie.

Plate 43 Maquis.

Plate 44 Desert.

Plate 45 The alpine ecosystem, an example of a high latitude and high altitude biome.

Plate 46 Taiga.

Plate 47 Tundra.

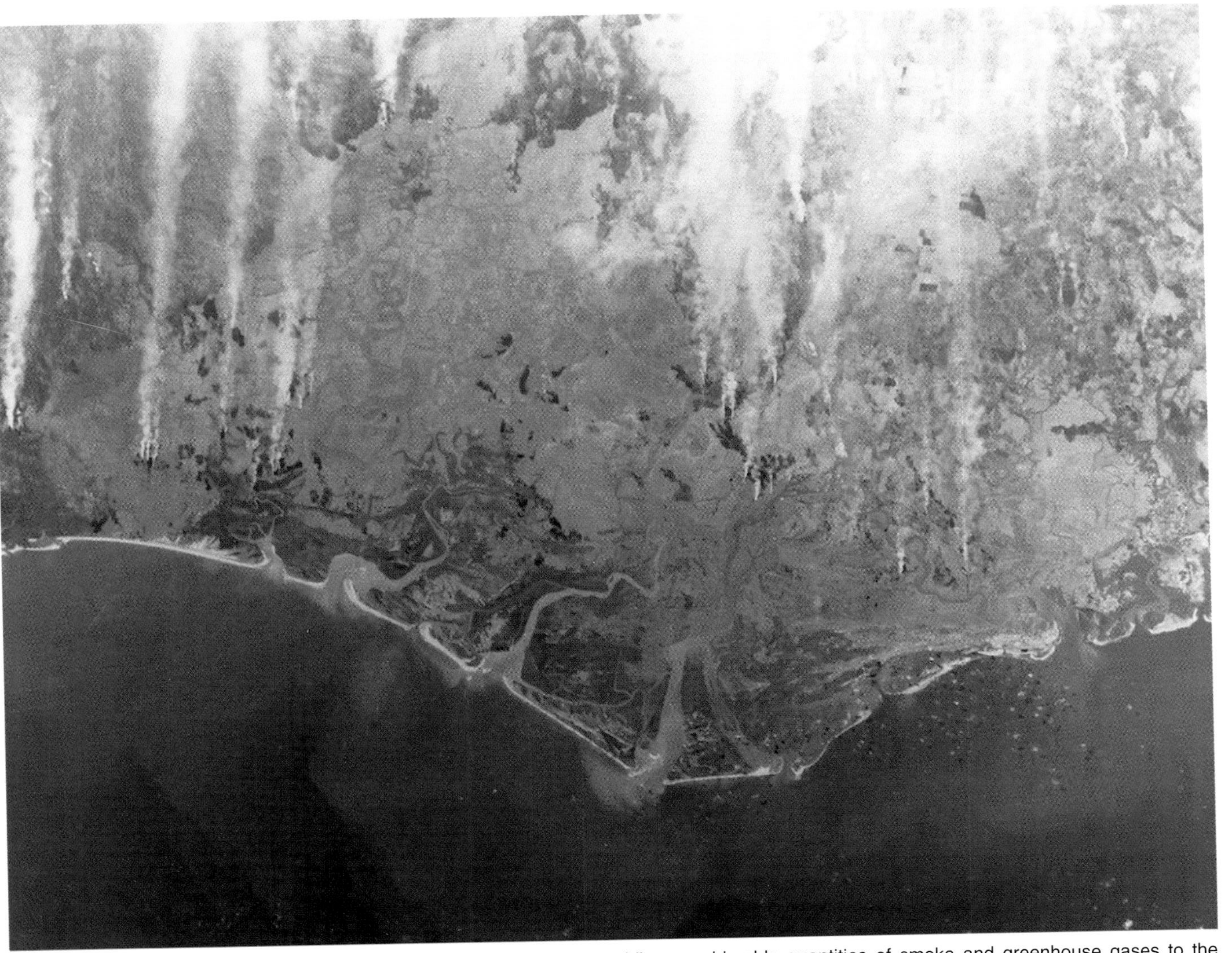

Plate 48 A near infra-red image of agricultural fires in Mozambique, adding considerable quantities of smoke and greenhouse gases to the atmosphere.

Plate 49 Anapurna south. Was the raising of the Himalaya the prime cause of human evolution? Raymo and Ruddiman propose a sequence of events that link the collision of India with Asia about 50 million years ago with the climatic changes in East Africa that gave rise to the first humans.

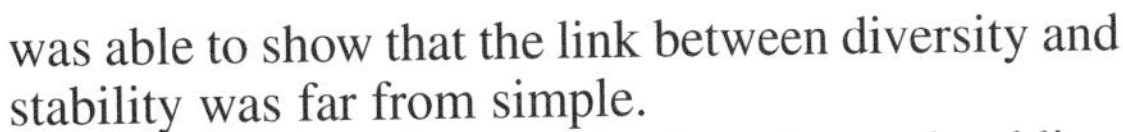

was able to show that the link between diversity and stability was far from simple.

If everything else remained unchanged, adding more species would actually reduce the stability of the populations. In May's models more species could be accommodated only if the number of connections or the strength of these interactions declined: stability in its populations followed when the system's complexity was reduced. The evidence from 40 published food webs seemed to support this – connectance within a web falls with species number. A highly diverse community could only persist if its interconnectedness was reduced.

A food web divided into a series of **compartments** might achieve this. A compartment consists of a group of species which share strong interactions with each other, but have tenuous links with the rest of the community. The abundance of one species within a compartment can have major implications for the other members but not necessarily the rest of the community. John Moore and William Hunt found evidence of such compartments in their study of nitrogen transfers in the below-ground communities of American grasslands. They describe compartments organized according to their food source, and which connect with each other only further up the food web. For

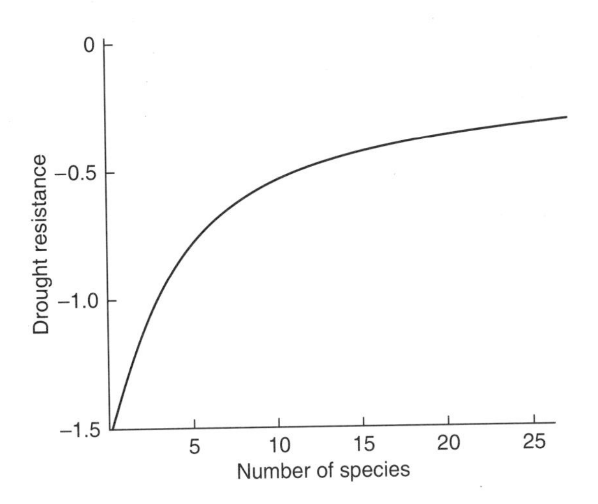

Figure 9.4 David Tilman and his co-workers have shown that temperate grassland plots with more species have a greater resistance to the effects of drought (a smaller change in total plant biomass between a drought year and a normal year). There was a limit: each additional plant species contributed less and less, so that the most diverse plots showed only marginal increases in resistance. The reason for the greater stability seems to be that species-rich plots are more likely to contain some drought-resistant plants. Beyond a certain level, however, new species are less likely to differ in this ability from established species.

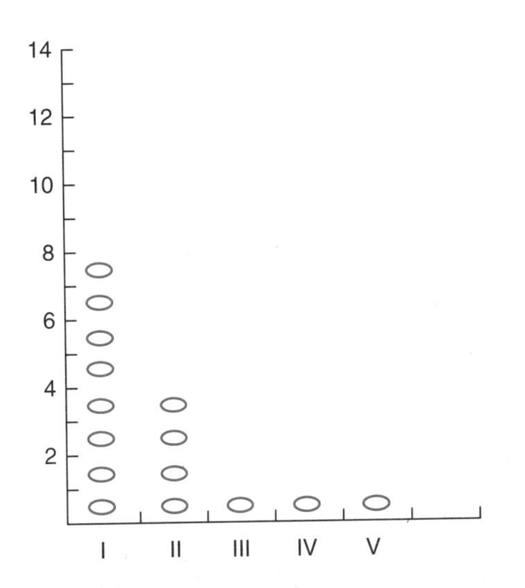

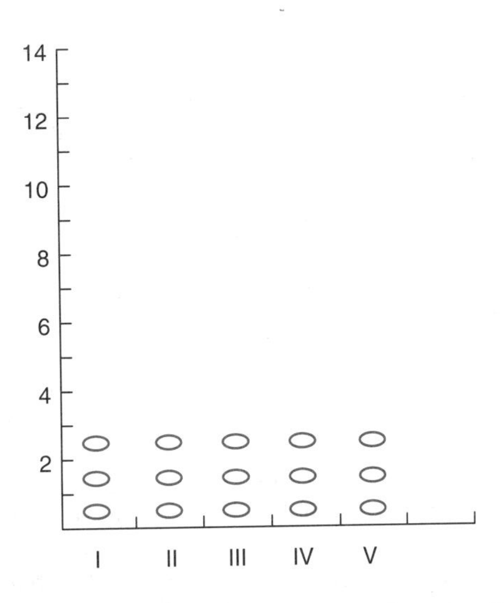

Figure 9.5 Measuring diversity. These samples from two different habitats contain the same number of species ($S = 5$) and the same total number of individuals ($N = 15$). However, we would not take them to be equally diverse because community A is dominated by just two species, whereas B show equal numbers of all species – B has high equitability. This would not be indicated by a simple ratio of S/N and so ecologists have sought more indicative measures of diversity.

example, species clusters of soil nematodes (roundworms) feeding solely on bacteria had no connection with those nematodes feeding on fungi. The two compartments were eventually linked two trophic steps away, by top predators (predatory mites). Significantly, Moore and Hunt could demonstrate the same close relationship between connectance and species richness within each compartment.

As we saw in the last chapter, a hierarchical organization of such groups might confer some stability on the community. Population ecologists, however, are not so convinced by the reality or significance of these compartments. Surveys based on population-only characteristics find compartments only between distinct habitats, and discontinuities are not obvious elsewhere. It may be that functional compartments are best revealed by examining ecological processes, like nutrient and energy flow, rather like the soil communities we have just described.

One population study, by Nigel Waltho and Jurek Kolasa on the fish communities of Jamaican coral reefs, did find evidence of compartments. Their work revealed a community structured according to habitat use, with specialist fish confined to narrow reef habitats and generalists occupying larger ranges. Not only were clusters of species (compartments) evident from the analysis, the population stability of each species depended on its ecological range. Generalists, interpreted as high up in the hierarchy, were less variable.

Robert May concluded that complex communities were found only where the environment was unchanging and large-scale disturbance was unlikely. Their complexity was possible because members of the community had had a long history of evolution in each other's company. With time, new species evolved and, through competition and other interactions, defined their role precisely (section 2.4). Long-lasting, unchanging environments would tend to accumulate more species and these would be the only circumstances where complex communities and their balancing act could develop. But such communities would also be most easily tilted from their precarious position by a disturbance.

There is evidence that these are indeed the characteristics of complex communities. As we shall see below, this may help to explain global patterns of species richness.

9.3 THE BIG QUESTIONS: LATITUDINAL GRADIENTS IN DIVERSITY

There are large and small gradients of species richness across the planet. Diversity declines with altitude up a mountainside. Down the length of a river the variety of invertebrates and plants increases, from highland stream to lowland river. Many open-water (pelagic) marine animals show a maximum diversity at some 1.5 km deep. Horizontally, across the oceans, their diversity also reflects the gyral circulation of water in each of the major oceans (Figure 8.7).

At the global scale, the number of species for many groups increases from the poles to the Equator so that the richest communities are found between the tropics. A range of benthic (sea-floor) fish and invertebrate groups (especially molluscs and crustaceans) show this pattern, at least in the North Atlantic (Figure 9.6). The gradient is less distinct in the South Atlantic and Indo-Pacific oceans, where regional patterns dominate and there are

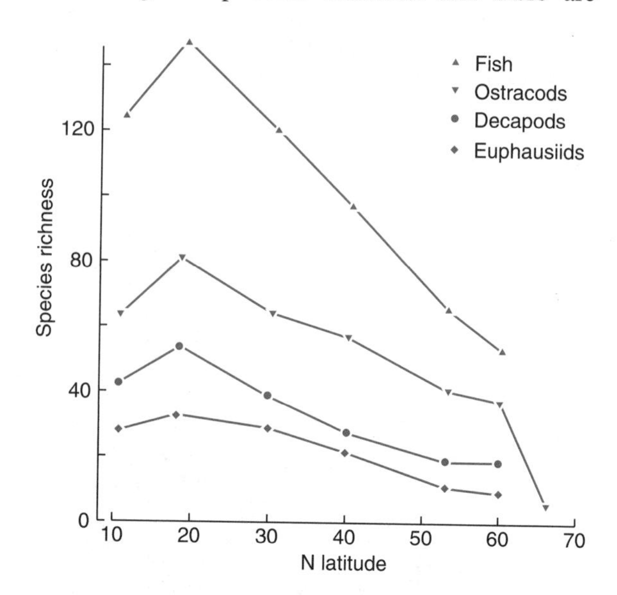

Figure 9.6 Four latitudinal gradients in the diversity of a variety of marine vertebrates and invertebrates – Euphausiids (planktonic krill), Decapods (shrimps, crabs, etc.) and Ostracods – described by Martin Angel for the North Atlantic. Mean body size of benthic invertebrates also decreases toward the Equator. The length of food chains tends to be longer and cycling of nutrients is faster closer to the Equator.

'hotspots' for some sea-floor invertebrates in the temperate latitudes. The greatest variety of marine isopods, for example, is off the Argentine coast, perhaps reflecting ancient ocean currents. The latitudinal pattern is also confounded in shallow-water communities, where local features, such as coastal configurations, are important.

Comparable processes determine the patterns found in terrestrial ecosystems – regional contrasts superimposed on the global gradient because of local climate or landform, the history of the region or its plants and animals. One interesting example is the higher diversity of mammals in the mountainous regions of North America, compared with its lowlands. This is probably caused in part by the greater range of habitats in upland areas. However, even with local hotspots, species richness of nearly all vertebrate groups rises toward the Equator in North America (Figure 9.7).

Globally, the distinction between tropical and temperate biomes could not be more dramatic: 6% of the Earth's land surface is covered with tropical forests and yet it is home to perhaps 70% of all species. Table 9.1 provides a checklist of the factors that might explain this gradient but, as we shall see, many of these are interlinked. Ecologists have looked for a prime abiotic factor that changes with latitude, one that could apply in both marine and terrestrial communities.

Temperature is the obvious candidate. Not only is it warmer near the Equator; it remains so throughout the year. One suggestion is that a consistently warm climate allows for more specialization and greater differentiation of niches (section 2.4). So, for example, tropical forests have birds and mammals that feed exclusively on fruit, possible because this resource is available all year round here. There are no equivalent fruit-eaters in temperate areas and most of their insectivorous birds have to migrate to avoid winter shortages.

However, some argue that the marked seasonality of the temperate zones should promote greater niche differentiation. We know, for example, that changes in daylength are used to cue plant growth and flowering. In the early spring of an oak forest, bluebells and a host of other flowers bloom before the closing of the tree canopy and the forthcoming competition for sunlight. The seasonal clock sequences the availability of a resource, allowing species to specialize and avoid competing with their neighbours. Even so, the temperate forests cannot match the net primary productivity of the tropical forest (Figure 8.15, Table 4.1), if only because their

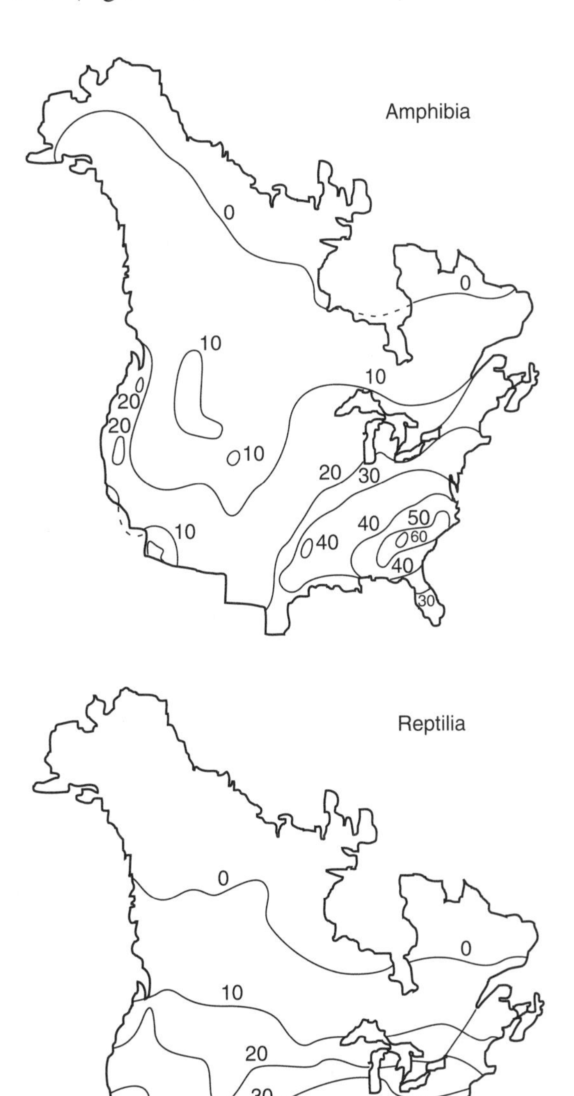

Figure 9.7 Distribution of the number of species of amphibians and reptiles in North America. David Currie attributes this primarily as a response to solar radiation and its impact on water availability and temperature, both of which change with latitude and continentality.

Table 9.1 Possible factors contributing to global patterns of species richness

Factor	*Description*
Physical conditions	Climate generally more benign and less variable in tropics, harsh and highly seasonal in higher latitudes.
Large area	Greatest area of the globe found in the tropics, allows for larger populations and therefore extinctions.
Productivity	Higher net primary productivity in tropics, more elaborate food webs.
History	Longer periods and larger areas undisturbed by glaciations in the lower latitudes.
Disturbance	Moderate levels of disturbance may create a greater range of opportunities for colonists in the tropics.
Constancy	Longer period of coevolution of species in tropics without major change in biotic or abiotic parameters.
Greater complexity	Complex plant communities promote more niche opportunities for animals in the tropics.
Greater competition and predation	High levels of competition, herbivory and predation prevent any species from becoming dominant.
Greater variety	Promotes greater variety.

carbon fixation is turned off for several months when temperatures are too low for growth.

Marine primary production shows only localized maxima around the Equator. The most productive areas are at continental margins, where sediments are stirred and their nutrients mix with the surface waters at some time during the year. The amount of sunlight arriving at the surface varies with latitude but local turbidity and turbulence also limit the radiation reaching the phytoplankton. Primary productivity is distinctly seasonal in the seas of the higher latitudes, but where nutrients and sunlight are abundant they are also highly productive. In contrast, tropical waters show less seasonality and, with little overturning of surface waters, are less productive over the year. Perhaps the greater diversity of animals on the tropical sea bed (Figure 9.6) is a consequence of the reliable year-long 'rain' of organic matter from above – consistency being more crucial than the actual quantity that arrives.

High productivity does not guarantee more species, only more biomass. Many productive plant communities are dominated by one or two species occurring in very large numbers, often at the expense of others. The addition of nutrients to soils invariably leads to a loss of plant diversity. Estuarine and wetland communities typically have just a small number of competitive but fast-growing and abundant plants (e.g. *Typha* or *Spartina*). In contrast, the nutrient-poor, semi-arid soils of the mediterranean-type biomes have a prodigious variety of flowering plants, especially the fynbos and its many endemic species (section 7.1).

Similarly, a shortage of the major nutrients may explain why tropical forests are so diverse and several studies point to a latitudinal gradient of soil fertility. The distribution of the trees often reflects the presence of key minerals in the soil. David Tilman suggests that a low level of nutrients in tropical soils allows species to coexist, preventing any one species becoming abundant and therefore dominant (section 4.3).

Much of the plant and animal richness of the tropical forests is attributable to the diversity of the trees and the opportunities they create. In turn, the variety of these species contribute themselves to the diversity of a community: adding more species adds more opportunities for other species – that is, additional niche dimensions. In species-rich assemblages these dimensions and species interactions become important in structuring the community. It may be that the diversity of tropical forests and ocean floors follows from the resources generated by their diversity. Species richness itself promotes species richness.

At higher latitudes such species interactions may have less significance. Here the seasons and the dramatic changes in temperature are the key factors to which all species must adapt. Because conditions are relatively harsh, extensive adaptations are needed – fur, feathers and, in some cases, even antifreeze in the body fluids. Any environment that demands major adaptive change will be less readily invaded. By chance alone, fewer species will evolve the adaptation. For this reason perhaps, reptilian and amphibian diversity declines in North America in close correspondence with increasing coldness and aridity (Figure 9.7).

Species in temperate areas have to adapt to a wider range of conditions – the changes from winter to summer. These favour a few adaptable generalists, rather than narrow niche or specialist species. Overall, the benign and consistent conditions of the tropics make it more likely for species to survive here. Indeed, there is evidence that a large number of plant and animal groups first evolved in the tropics.

Not only are more species likely to originate at lower latitudes; they are also less likely to go extinct there. These biomes straddle the girth of the planetary sphere, where its area is greatest. Their land area has expanded (and contracted) in geological history, at the expense of the temperate and polar zones. Larger areas should allow for larger populations which, in turn, lower the risks of extinction. In contrast, the higher latitudes have experienced a more changeable history and been subject to migrations and extinctions associated with climatic change.

The tropics might have accumulated species simply because they have been around for longer. The variety of tree species in the tropics suggests a long evolutionary history. Their diversity is attributable to different families and genera, with few species sharing the same genus. In contrast, temperate zones have more tree species per genus, a result of recent speciation. Even so, this does not explain why some of the oldest groups of higher plants, including the conifers, are adapted to life in the colder regions.

In fact, the tropics have not enjoyed an uninterrupted history. During the climatic swings of the ice ages some forests became divided into discrete patches, or refugia. This fragmentation may itself have also contributed to their diversity. Their isolation promoted local speciation within the refugia, from which new forms have since dispersed.

However, there has been a greater degree of continuity in the communities around the Equator and some large areas did survive intact. An unchanging environment causes fewer extinctions and, with time, greater specialization and niche differentiation. Certainly, the deep ocean floors have been buffered from significant change and the benthos of a tropical sea typically has a low density of individuals yet a high species richness.

No one group of animals dominates tropical forests or coral reefs (aside from the corals!) and animal species richness appears to reflect the architectural complexity of these systems. Both habitats offer a variety of niches within their superstructure and are termed three-dimensional habitats, to contrast with the limited vertical structure of, for example, grasslands or tundra. This distinction was found to be a good predictor of food web complexity by Frédéric Briand and Joel Cohen. Their survey of 113 webs from all types of ecosystems linked structural complexity and a relative constancy of the environment to longer food chains (section 5.3). It seems that, given some measure of constancy, a complex community can assemble itself, and that adding more species provides opportunities for others.

Disturbance too has its part to play in maintaining diversity. Some species rely on gaps to maintain a population and a certain frequency of disturbance to provide opportunities for them to colonize. Again, the scale of the disturbance is critical, since diversity will only be maximized when the adjustment stability of the system is not exceeded (Figure 9.1). The elasticity of a community – its capacity to restore itself after disturbance – results from the collection of adaptive strategies of its component species. We saw in our survey of mediterranean-type communities how a certain frequency of minor fires was needed to maintain the scrubland mosaic (Figure 7.2), a scale of disturbance from which these communities could recover rapidly and upon which some species relied.

Intermediate levels of disturbance remove competitive species that would otherwise exclude new arrivals (section 7.2). The intermediate disturbance hypothesis has been used to explain species richness in more variable environments, especially in coastal zones and on land. For this to create a latitudinal gradient, however, requires disturbance close to optimum in tropical regions. In fact, turnover rates for trees in temperate and tropical forest biomes do not differ.

Robert Ricklefs argues it is the nature of the gaps, not their quantity, which is key. He points out that an opening in a tropical forest creates a habitat (regeneration niche) that contrasts sharply with the forest floor under a closed canopy, with a vast increase in light. A wide range of very different niches becomes available. In temperate forests the gaps remain in partial shade because of the sun's angle at these latitudes (Figure 9.8). Ricklefs believes that this explains the astounding variety of trees in tropical areas when their understorey diversity is relatively unremarkable.

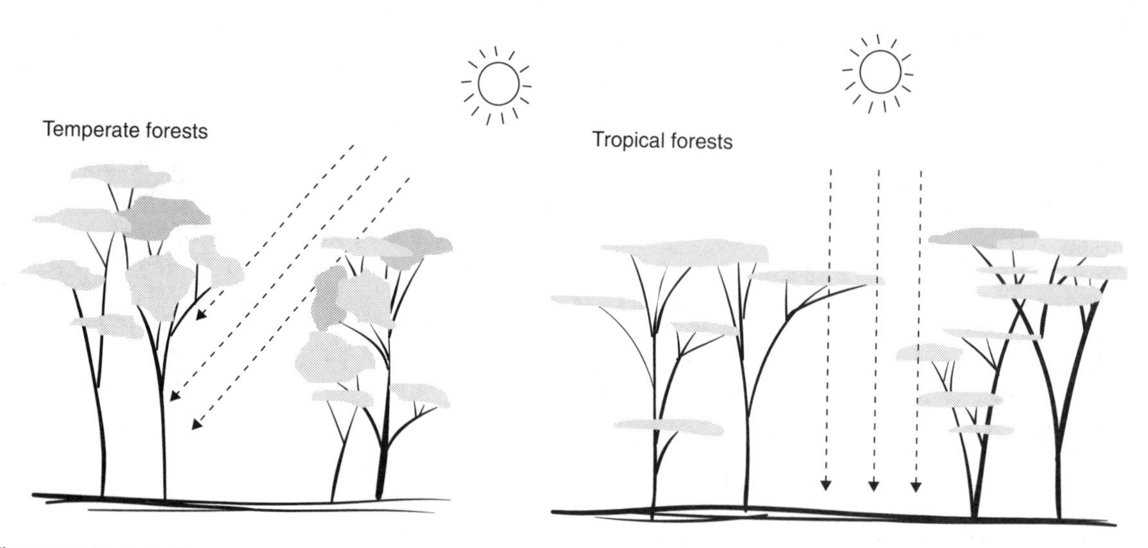

Figure 9.8 Gaps in the canopy in forests are much more fully illuminated in the tropics because of the angle of the sun, producing a much greater contrast with the closed canopy. Robert Ricklefs suggests that this creates very different niche conditions favouring a wider variety of trees, compared with temperate forests.

Predation and herbivory are probably more severe in tropical forests and this has important implications for the organization of these communities: a species consumed is a species constrained and at a high intensity, the consumed species is prevented from becoming dominant. That pressure, operating throughout the food web, itself promotes greater species diversity. Certainly the intense seed predation observed around the parent tree might explain the large distances between individuals of the same species in tropical forests (Figure 9.9).

Finally, there seems to be more sex in the tropics. Sexual reproduction may be favoured here because it promotes variability, which is essential in the continual battle to escape the attentions of the large number of pathogens, parasites and predators. At higher latitudes, there is a general tendency to resort to asexual reproduction, presumably because the selective advantage of sexual reproduction is less where there are fewer natural enemies. Again, species richness promotes species richness.

Despite its simple pattern, we are still unable to identify a single cause for the latitudinal diversity gradient. It may be that two or three abiotic factors set off a chain of biotic reactions that accentuate the differences between latitudes. Perhaps the greater energy influxes and primary productivity of the tropics provides the resources for a greater variety of herbivores, which in turn provide for more carnivores. In each case, the consumers serve to constrain species they feed on, preventing any one species from becoming dominant. In effect, as the community builds, so it serves to amplify the number of opportunities for other species. The contrast between latitudes perhaps follows from this potential for positive feedback, so that adding more species itself promotes further complexity, supported in the tropics by higher productivity.

Can we keep adding species? Highly diverse communities seem to be more fragile; yet, through natural selection, evolution will try to add new species. Perhaps only the tropics provide the stable conditions necessary for complex communities to assemble, and where, because of their long history, many 'trials' (different combinations of species) have taken place. What we see today is the current result, the current balancing act.

Can we keep removing species? Diverse ecosystems may remain stable only if there are redundant species whose removal does not cause the community to topple. We have seen how the mediterranean-type communities have, under human influence, lost species in the past and developed relatively stable configurations (section 7.1). As more are lost, however, the role played by the remainder is increasingly critical to the stability of the whole system. Lose too many and collapse becomes inevitable – a threat facing areas of several mediterranean communities today.

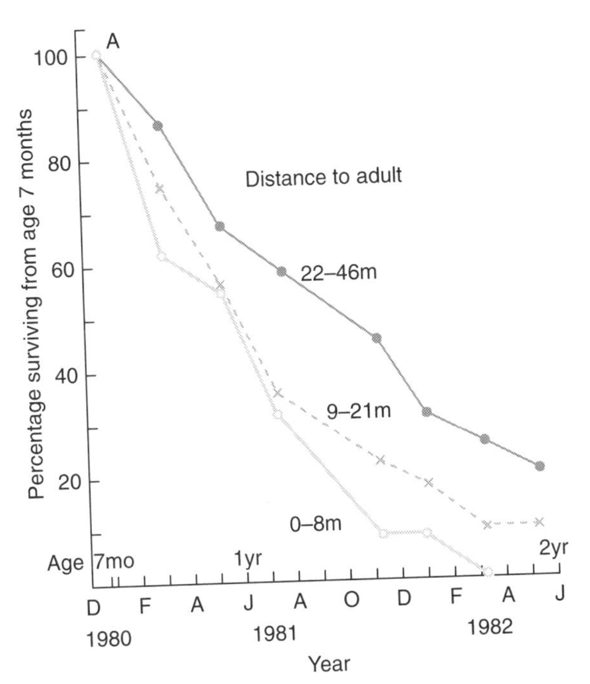

Figure 9.9 The high diversity of trees in tropical forests may be due to high mortality of juveniles around the parent tree. Janzen and Connell proposed that the pressure of herbivory on seedlings would mean that distances between the parent tree and its offspring would be very large. Clark and Clark were able to demonstrate this for one canopy tree, *Dipteryx panamensis*, by following the survival of its seedlings. Although 80% of the seeds fall within 13 m of the tree, none survived within this zone after two years. The nearest saplings from this cohort were 36 m and 42 m away from their parent. The graph shows how seedling survival decreases closer to the parent tree with time.

This provides some evidence that biotic interactions within the community, in this case herbivory or attack from pathogens, help to promote high diversity. In the same way, predation or competitive pressures may prevent a species from becoming dominant.

In the end, perhaps the most we can conclude is that communities share similar structures, with species fulfilling similar roles and with interactions that impart some measure of stability. Where they do share similar abiotic conditions (section 7.1) they may converge toward a similar response to disturbance and a corresponding set of species. As ever, what we see is a continuum, with perhaps the older communities showing the greatest complexity and integration, whose biotic interactions have had time to exert themselves.

Beyond that, we should perhaps acknowledge that communities are dynamic, living structures, responding to external pressures and internal tensions, as species come and go.

9.4 EXTINCTIONS

Species come and go at different rates according to where you look in the fossil record. There have been at least five mass animal extinction events in the planet's history: the 'Big Five', resulting from upheavals in the global environment. However, a simple count from the fossil record shows that over 90% of extinctions occur outside a single mass extinction, as part of the background rate of species turnover (Box 9.2).

The total number of extant or living species also varies in geological time. Currently, the oceans have a variety of animals about twice the average for much of the fossil record. In the past, up to 95% of oceanic species have been lost during a mass extinction, but numbers have always recovered quickly. Interestingly, their long-term richness seems to be a result of species interactions – competition, predation and so on – rather than shifts in the abiotic environment. The same is probably true of the major plant groups which have not suffered large-scale extinctions like the animals.

Corals and shelled marine invertebrates have the most complete fossil records. These show relatively few extinctions in tropical reef communities over the last two million years, despite major fluctuations in sea level. Instead, these communities have tracked the shift in the climatic regions, moving with them. Reef communities are highly complex and are easily disturbed, but as one dominant coral species is lost, another quickly replaces it. Corals are often highly localized and a species may disappear when its single population goes extinct. By the same token, these communities have shown rapid diversification after each of the Big Five, requiring only 5–10 million years to re-establish their species richness.

A more rapid evolution rate and an early restoration of species number may be a feature of all tropical ecosystems. We can see this elasticity in Caribbean mollusc communities two million years ago. As sea temperatures fell during the glacial advances, mollusc extinction rates increased markedly, but this was more than offset by an increase in speciation rates.

Box 9.2

BIOLOGICAL DIVERSITY

Professor Sir Robert May, FRS

How well do we know the world of plants, animals and microorganisms with which we share this planet? The answer, by any objective measure, must be: not very well. Firstly, estimates of the number of species that have been named and recorded (a simple, factual question, like how many books in the library) range from 1.4 million to 1.8 million. Secondly, estimates of the total number of species present on Earth today ranges over more than an order-of-magnitude, from a low of around 3 million to a high of 30 million or possibly much more. Finally, we have even less idea of the rates at which species may currently be going extinct, as a result of habitat destruction and other consequences of human population growth.

The history of life on Earth, written in the fossil record over the past 600 million years (my) since the Cambrian explosion in the diversity of multicellular organisms, is one of broadly increasing diversity, albeit with many fluctuations and punctuated by episodes of mass extinction. The average life span of a species in the fossil record, from origination to extinction, is around 5–10 my. However, some groups have lifetimes significantly longer or shorter than this. Comparing this average with the 600 my of the fossil record, we might estimate that 1–2% of all species ever to have lived are with us today. But, allowing for the overall increase in species diversity since the Cambrian, a better estimate might be 2–4%. And if we recognize that most of today's species are terrestrial invertebrates (mainly insects) whose diversification began 450 my ago and whose species' life span may typically be longer than 10 my, it could be that today's species represent more like 5–10% of those ever to have graced our planet.

Our knowledge of some groups is much better than of others. Bird and mammal species are comparatively well documented; even though three to five new bird species and around 10 new mammal species are found each life-year, such numbers are small fractions of the totals recorded in these classes (approximately 9000 species of birds and 4000 of mammals). The 270 000 recorded species of vascular plants probably represent 90% or so of the true total. But comprehensive surveys of other groups in previously unstudied places – tropical canopy insects, deep-sea benthic macrofauna, fungi – typically find 20–50% of the species are new to science.

The true total number of living species is very uncertain. My guesstimate is around 3–8 million. Dramatically higher numbers have been proposed: 30 million insects on the basis of studies of beetles in tropical canopies; tens of millions of benthic invertebrates on the basis of a deep-shelf transect off the north-eastern United States; 1.5 million fungi on the basis of scaling up the species ratio of fungi to vascular plants in Britain. I am sceptical of all of these estimates, but they could be correct. The fact that reasonable estimates vary so widely says a lot about how little we know.

Over the past century, extinctions in well-studied groups – primarily birds and mammals – have run at around one species per year. Because tropical species receive less attention, true extinction rates of birds and mammals are undoubtedly higher. But even one per year among the 13 000 species of birds and mammals translates to an expected species' lifetime of around 10 000 years; that is, about one thousand times shorter than the background average lifespan of 5–10 my seen in the fossil record. Put another way, recent extinction rates in well-documented groups have run around one thousand times faster than average background rates.

Looking toward the immediate future, three different approaches to estimating impending extinction rates suggests species' life expectancies of around 200–400 years. One of these approaches is based on species–area relations coupled with assessments of current rates of habitat loss. The other two are based in different ways on the International Union for the Conservation of Nature's current catalogue of 'endangered' or 'vulnerable' species, for better-known groups such as birds, mammals or palm trees. Such figures correspond to extinction rates 10 000 times that of the background rate over the next century or so. This will represent a sixth great wave of extinction, comparable with the Big Five mass extinctions of the geological past, except that this one results from the activities of a single species rather than from external environmental changes.

Why should we care about such impending loss of the richness of species we have inherited from the workings of evolutionary processes over geological time-scales? Why should we value biological diversity?

One argument emphasizes the benefits we already derive from natural products, as foods, medicines and so on. Currently, 25% of the drugs on the pharmacy shelves derive from a mere 120 species of plants. Yet throughout the world, the traditional medicines of native peoples make use of around 25 000 species – about 10% of all plant species. We have much to learn. As our understanding of the natural world advances, both at the level of new species and at the level of the molecular machinery from which all diversity is constructed, it is a pity to be burning the books before we can read them, let alone create wealth from the recipes on their pages.

Other arguments refer to global benefits. The interactions between biological and physical processes created and maintain the earth's biosphere as a place where life can flourish. With impending changes in climate caused by the increasing scale of human activity, we should be worried about reductions in biological diversity, at least until we understand its role in maintaining the planet's life support systems. The first rule of intelligent tinkering is to keep all the pieces.

For me, however, a third class of argument is the most compelling. It is clearly set out in the UK Government's White Paper *This Common Inheritance*:

> *The starting point for this Government is the ethical imperative of stewardship ... we have a moral duty to look after our planet and hand it on in good order to future generations.*

The devastation of manatees, turtles, jewfish and conches through hunting appears to have had little effect on Caribbean reef communities over the last 500 years. Other species, primarily invertebrates, have increased their abundance to exploit the ecological opportunities. However, human impact over the last 50 years may have tested the elastic limit of the reef ecosystems. Increased sedimentation and pollution have contributed to widespread coral death in the region and now the reef systems are showing clear signs of permanent change.

Even within larger, more continuous habitats, species with poor powers of dispersal or those adapted to a narrow range of habitats will disappear quickly as the habitat changes. The loss of specialist species many years ago may explain the present resilience of the remaining flora of the Mediterranean basin (section 7.1). In the recent past, just 0.15% of its higher plants has been lost, presumably because the survivors are less sensitive to disturbance. Over the same period, a greater proportion of species has disappeared from the other regions, especially the mallee of Western Australia (0.66% or 54 species), the area most recently exploited. Not surprisingly, the scale of the threat across the different regions with a mediterranean climate shows a close correlation with the degree of human disturbance. While current losses are not yet significantly above the background rate, all of these areas face much higher losses in the near future as aridity increases. Overall, 17% of their floral species are today threatened with extinction.

Human activities can disturb communities in more subtle ways, by adding species and creating combinations that cannot persist. A long list of birds and mammals have lost competitive battles with alien species we have introduced, deliberately or otherwise. Rats, cats, dogs, pigs, snakes and birds have devastated island faunas and floras in particular. Islands are centres of speciation, often supporting species with few competitors or predators, but which fall early victim to aggressive or competitive invaders (section 4.3).

In any inventory of biodiversity, losing the last member of a family of plant or animal must be deemed a greater loss than a subspecies or variety. Even so, every individual has a value. A species depleted to small numbers has a reduced gene pool from which future generations can be drawn. When we score diversity, we must remember to value the genetic diversity that resides in individuals and populations and not be concerned simply to conserve the species. A species needs this genetic diversity to ensure its long-term survival. As we have seen, this has economic implications in domesticated species, as a source of novel code for producing new strains (Box 2.4).

Species that are more widely dispersed are usually more abundant in each locality. They move readily between habitat patches and will re-establish populations in areas where a local extinction has occurred. Endangered species, on the other hand, tend to have a restricted distribution and occur in low numbers in each case. Ecologists use models of metapopulations (Box 3.1) and the exchange of genetic information between populations to measure the prospects for endangered species. In some cases, very detailed records are available and breeding programmes are established to avoid the problems of inbreeding depression (and outbreeding depression) by moving mates or semen between patches (section 3.6). One such success was the Arabian oryx (*Oryx leucoryx*), a victim of overhunting. Using a captive breeding programme based on a metapopulation divided amongst zoos, a vigorous population was evenutally reintroduced into the wild.

Incidentally, models of metapopulations provide an important service for human welfare. The same principles can be used to map the spread of disease and the conditions under which it becomes epidemic, treating a human being as a habitat patch that can become populated by a pathogen. Roy Anderson and Robert May have followed the population dynamics of a number of pathogens and made predictions about the spread of the AIDS virus, in the absence of any changes in human behaviour.

Today we are at the start of another mass extinction, this time caused by the increased abundance of one species – ourselves. It is too simplistic to suggest that all environmental degradation can be traced back to the explosion of the human population. Instead, it is more the number of individuals multiplied by the amount of resources we each demand, set against the fragility of the environments we exploit. What we each consume and demand of our environment varies greatly across the globe and is shaped by our culture, tradition and the political constraints on our activity. Nevertheless, the number of *Homo sapiens sapiens* is a prime factor in determining the rate of habitat and species loss. We have used our technology and ingenuity to bring us this far; now the question is whether we have the wisdom not to cut the ground from beneath our own two feet.

9.5 THE SCALE OF HUMAN DISTURBANCE

A world so altered by human activity offers the opportunity and the challenge to expand the scope of the discipline of ecology.

Ecological Society of America, launching its SBI initiative (1991)

Few would dispute that the number of human beings is set to increase, more than double, over the next 50 years. This will place a considerable strain on the capacity of the planet to feed and support us (Box 9.3). Yet satellite data indicate that of the vegetated surface of the Earth that could be cultivated, about 11% has been moderately or severely degraded. Together population growth and soil degradation will combine to reduce the amount of productive land per head by more than a half in the next 30 years.

This is not a new threat. In some form or other, the idea of a limited Earth giving way under the weight of human numbers has been around for at least 200 years, and in some cultures a lot longer. The forecasts of our imminent demise have proved premature time and time again. So should we take them more seriously now?

Certainly the scientific community is beginning to see the danger as very real (Table 9.2). Ecologists have attempted to quantify the scale of change we have wrought on the globe. Today, little of the natural environment remains untouched and that is largely confined to marginal zones, the cold and the arid parts of the planet (Plate 34). There are still significant stretches of tropical forest in South America classified as undisturbed, but the most disturbed area is the Oriental realm, where the forest has been largely replaced by rice paddies. Europe has the largest proportion of land area dominated by human activity (65%).

The global gradient of diversity also tells us something about the agricultural capacity of the different biomes. Generally, low latitudes have poor and infertile soils, largely a result of their age. The increase in plant diversity towards the Equator reflects these conditions, and implies a low agricultural potential. Without costly fertilizers and management practices, these regions are never likely to produce food at the same rate as the younger fertile soils of the recently glaciated temperate regions. Michael Huston suggests that this should give us every incentive to stop the wholesale destruction of the tropical forests, since they have a more useful role locking up carbon from the atmosphere and preserving biodiversity. Simple economics should demand that farmers do not seek to cultivate such land.

Huston describes yet another global gradient: that of declining GNP (gross national product, a measure of the economic wealth of a nation) towards the Equator, because of the predominance of agriculture in their economies. Yet intensive agriculture is probably not the best motor for the development of these nations, which occupy ecosystems unsuited to such large-scale disruption (section 5.5).

The different regions of the planet have different histories of disturbance. The Mediterranean basin has existed in something similar to its present form for close on 2500 years, yet changes in land use over the last 50 years have brought the threat of drastic change to both the terrestrial biome and the sea (section 7.3). Many of the consequences of these changes will be apparent only to future generations, as communities shift and adjust. Examples are the increased nutrient loads in groundwater sup-

Table 9.2 The Sustainable Biosphere Initiative (SBI)

The scale of the ecological problem facing humanity was the subject of a special study by the Ecological Society of America. After consulting their membership, they issued an agenda to define the most urgent priorities for research that

'hold the greatest promise for advancing our base knowledge and for improving the human condition'

and that addressed the need to

'ameliorate the rapidly deteriorating state of the environment and to enhance its capacity to sustain the needs of the world's population'.

The society's Sustainable Biosphere Initiative (SBI) identified three key areas for enhanced research effort:

1. **How ecological complexity controls global processes**
   Key factors to be considered are species and habitat diversity, the distribution of ecological assemblages, their productivities and storage capacities.
2. **The importance of biological diversity in controlling ecological processes and the role they play in shaping patterns of diversity at different scales of time and space**
   This will include studying the role of diversity in determining the behaviour of ecological systems (including responses to changes in nutrient flow, pollutants, climate change) and how biological diversity is determined.
3. **Integrated research on the sustainability of ecological systems to prescribe restoration and management strategies to enhance their sustainability**
   Such research would seek to develop management strategies for sustaining renewable natural resources (agriculture, forestry etc.) and which also look at multiple resources over large spatial scales, with integration of social, physical and biological sciences.

   Allied to this, the committee recognized the need to make ecological information accessible and comprehensible to as wide a public as possible, to address educational needs in ecology and for ecologists to bring their expertise to bear more directly on the research prioirities.

Box 9.3

THE GROWTH OF THE HUMAN POPULATION

As far as we can tell, the human population has been relatively stable for much of its recorded history. Various methods have been used to measure the size of past populations in different parts of the world and the overall picture is one of a slow rise, but with one or two noticeable declines due to disease.

The global rate of population growth showed a marked increased in the 1900s. Different nations started to grow at different times, primarily with the adoption of new medical techniques and a better understanding of the need for hygiene, both of which led to major reductions in death rates.

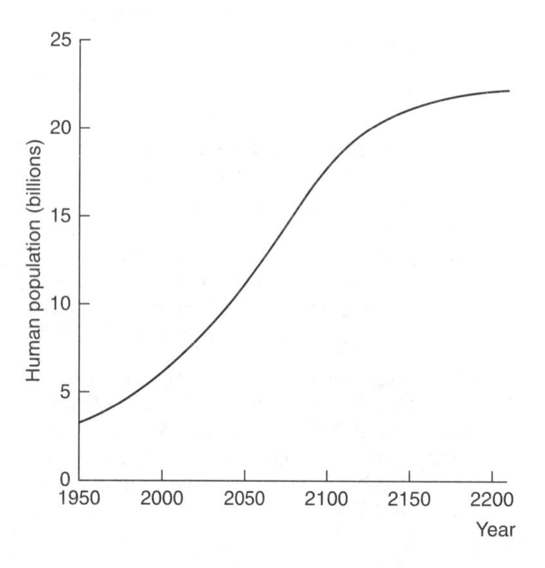

Figure 9.10 Growth in the world population, modelled according to the logistic model described in section 3.2. Henry Tuckwell and James Koziol found that this simple model gave the best fit to the data between 1950 and 1985 and that it accurately predicted the population of 5.48 billion in 1992. If it continues to be accurate, a carrying capacity of around 24 billion, four times the current size, will be reached in the year 2200.

At its current rate of growth (1.7% per year) the global population will double in the next 40 years (Figure 9.10). Today it numbers 5.5 billion people and has an age structure characteristic of rapid growth. However, the increase is not uniform across the planet. Generally, industrialized nations have little or no growth, whereas 95% of the total increase is in the non-developed nations. Overall the rate of growth has declined in recent years, but now this rate is operating on much larger populations, so the increase in absolute numbers is set to continue for some decades to come.

The United Nations has produced various projections for the global population. At the current fertility rate it will reach 7.8 billion in 2050, but this does not allow for the current age structure. A more realistic projection is for the population to double to around 10 billion in the same time. As Tuckwell and Koziol (Figure 9.10) put it, 'there are apparently logistic-type regimes which persist until some major event occurs'.

This is not news. Predictions about the consequences of large human populations have been around for centuries (and perhaps longer). Data on the scale of the increase in the 1950s and 1960s provided momentum for the early environmental movements, though forecasts of imminent environmental collapse were somewhat premature.

This is one reason why some commentators suggest we should not be overly concerned. They point out that our technologies have managed to reduce hunger and malnutrition, despite the increased population growth of this century. Fewer go hungry, fewer die at birth and most live longer. They argue that larger numbers to come will, in the same way, be sustained by future technological developments. Certainly it is true that living standards, in terms of health and wealth, have increased for most peoples over the last 100 years, despite the booming numbers. This has been especially true of the last 20 or 30 years, where technological advances have been the spur behind the rapid industrialization of many developing economies.

It seems that goods and services circulate faster. What has sustained developing countries has been an increase in economic activity around the globe. Adding individuals or at least complexity, perhaps like ecological communities or the massive ant nests we met in Chapter 4, seems to lead to increases in productivity and perhaps the provision for each individual.

Of course, this assumes that connectivity of the system remains intact. Besides the welter of social, economic and political problems which this prognosis fails to address, many ecologists point to our past experience that ecological systems are not amenable to quick technological fixes. Complexity and diversity do not have a market value, yet many of the services that ecosystems provide depend on the interactions between its species. These communities and their complex balancing act are the first things to be disrupted when we exploit an ecosystem. Similarly, the interactions between the biota and the soils which support our agriculture are not easily modified without a price. Since it is a reasonable guess that we will depend on ecological systems for our food for the forseeable future, forecasts of 'business as usual' seem recklessly optimistic.

One alternative – waiting for natural constraints to halt the growth – is, as Nathan Keyfitz says, 'to accept famine, low living standards, unemployment, political instability and ecological destruction'. At least for some.

plies, silt loads in rivers, soil erosion and salinization of large areas from some agricultural practices – effects which will have a legacy well into the next century and beyond.

Each year, more of these topsoils, which may have taken 5000 years to form, are finding their way to the bottom of the oceans or are silting up our reservoirs. A complex series of factors combine to reduce the agricultural potential of the soil so that productivities quickly decline (Figure 9.11). This is not simply a problem of developing nations. Estimates from the American Midwest suggest that, under arable cultivation, lands there have lost at least 7 cm of topsoil, equivalent to 900 tonnes per hectare. Overall, the United States was losing 3 billion tonnes of sediment in surface run-off each year in the 1930s but this had reached 4 billion tonnes in the 1970s. Somewhere between 100 and 200 tonnes per hectare are lost annually from cultivated slopes in tropical areas.

The soils beneath moist tropical forests have little long-term agricultural potential following clearance. The loss of large areas of forest leads to local climatic change, in turn causing soils to become drier and more easily eroded. The indigenous practice of slash-and-burn in small clearances, followed by cultivation for two to five years, is well suited to exploiting its short-lived fertility, without overcoming the elasticity of the forest ecosystem. Indeed, the disturbance may be important for the gaps it creates, promoting diversity through regeneration.

Today massive commercial logging and burning operations have devastated large tracts of tropical forests, especially in Amazonia. These exceed the adjustment stability of the system and consequently their recovery is slow. Many collapse to a different community structure, typically semi-arid grassland. Even where there has been commercial reforestation (for rubber or paper pulp production) the poor soil has meant that yields have been low. In the process, we can only guess how many species have been lost.

Drought and periodic famine are written into the history of many tropical and semi-arid areas, but today it seems that the scale of habitat degradation (and, with it, the rate of species loss) indicates more permanent change. Much of the developing world relies upon wood as its principal fuel. The pressure of population has stripped the land of its trees, especially in sub-Saharan Africa, India and the Himalaya. As the trees disappear, so does the soil.

Henry Kendall and David Pimental describe how, in the last 30 years, most nations have ceased to be self-sufficient in food. Despite the advances in agricultural technology – the 'green revolution' based on new crop strains, artificial fertilizers and pesticides (section 5.5) – per capita food production has slowed down. Food production in 70% of developing nations can no longer match their population growth.

The usual response to food shortages is to bring more land into production, causing more natural habitats to be lost. The land available, however, is largely marginal and cannot sustain agriculture over the long term.

Kendall and Pimental review the scope for increasing food production to meet the demand from a rising population, in a way which is broadly sustainable and without increasing losses in biodiversity and habitats. One change would be a shift to lower meat production and consumption and to increase the proportion of cereals and vegetables in the human diet. This does not mean that we should simply switch to greater use of fish stocks for protein. The evidence from nearly all fisheries is that maximum sustainable yields are being exceeded (section 3.3) and these populations also face collapse.

Much of the productivity of western agriculture has been achieved by a large energy subsidy from fossil fuels – from mechanization to the manufacture of pesticides and fertilizers. Expanding food production in the temperate zones would need an increase in this subsidy. At the same time we also need to reforest large areas to combat rising carbon levels in the atmosphere. This is our present dilemma. Limiting forest clearance in tropical areas means the area per head of the population used for food production cannot match the swelling population. Yet, the forest supplies other essential services, helping to maintain the global environment.

Forty years ago much of the central highlands of Ethiopia was covered in relatively lush forest. Today the forests of East Africa are in retreat again, this time because of an environmental change humankind has brought upon itself. The once nomadic peoples of sub-Saharan Africa have become increasingly settled, primarily around centres where medicine or water is available. In 1984, Ethiopia had its driest year since records began and the combination of a civil war, a large settled population and a severely depleted ecosystem based on unprotected soils conspired to produce one of the most severe famines of this century.

Without the pressure to produce increasing amounts of food we could, with time, allow some of these marginal ecosystems to re-establish themselves. Where chemical changes to their soils have been small, many show an elasticity that offers the promise of restoration. With sufficient rain, even the Sahel and some of the more arid or saline soils will support vegetation. Today, rain and cultivation techniques that maintain a cover of vegetation are offering hope in some East African regions. Much is being pinned upon a long-term programme of reforestation. More people does not necessarily mean more environmental degradation, but it does mean planning and regulating land use.

Much of the world population is hungry and perhaps one-fifth suffer from malnutrition. Yet even as our numbers have grown, we have made significant reductions in these proportions throughout the twentieth century. That encourages some to argue that human ingenuity can solve shortages of food and other resources. Despite the spherical nature of the Earth, they suggest our ingenuity effectively makes resources infinite. In the same way, they see the pricing of these commodities helping to safeguard the biosphere and its diversity. However, predictions of our ability to improve the lot of all peoples, feeding those here and those about to arrive, need to describe a more transparent mechanism by which this will be achieved, as well as its impact on ecological processes and the global climate.

We will inevitably look for a technological fix. Perhaps genetic manipulation might improve the disease resistance of key crops, or equip cereals with the capacity to fix their own nitrogen. Even assuming that this technology is freely shared, and that genetically modified organisms pose no environmental risks, we shall still need to use sustainable agricultural practices to preserve the complex ecological interactions that makes food production possible. Carl Jordan argues that tropical areas need to avoid adopting western methods and their reliance on energy subsidies, either directly or indirectly, to control local conditions. It would be better, he suggests, to use practices that harness the natural processes, employing more imaginative cultivation practices. This was recognized by most

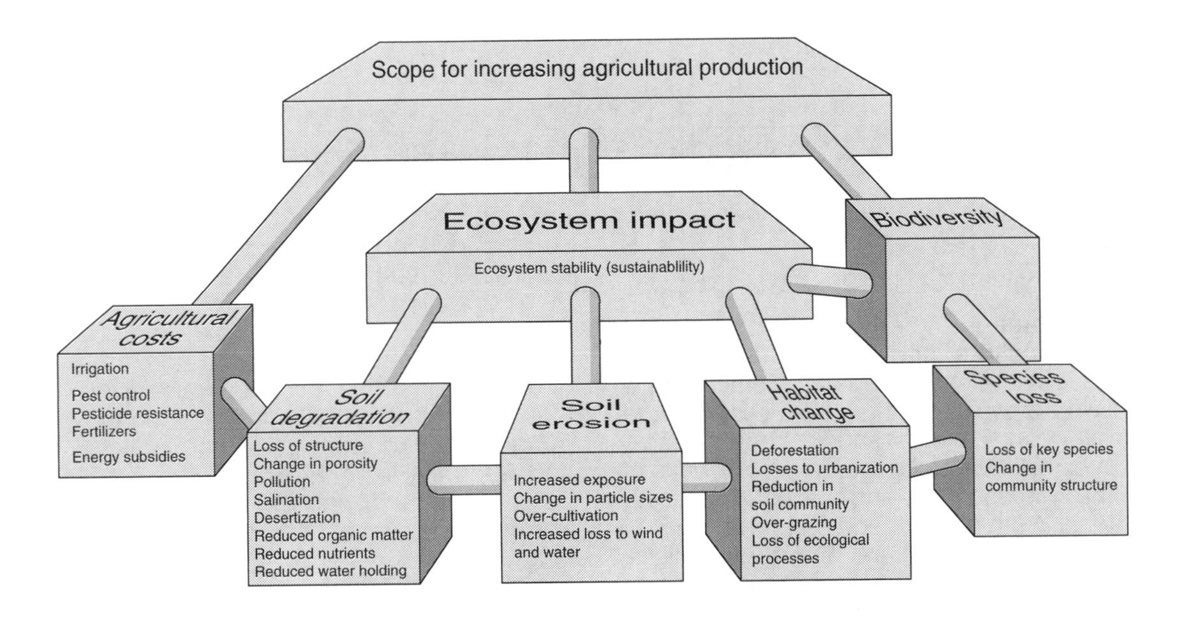

Figure 9.11 The hierarchy of factors which determine the agricultural potential for a region.

Figure 9.12 Humankind walked this way. Litter left by tourists trekking in the Himalaya.

Box 9.4

AN ECOLOGY OF LANGUAGE

A language, like a species, when extinct, never ... reappears.

Charles Darwin, *The Descent of Man*

Language is the embodiment of a people's culture and, as Charles Darwin observed, it has the characteristics of a living species. Languages evolve (and merge, unlike most species) and can also become extinct. Languages carry ways of thinking and of seeing the world, passing ideas from one generation to the next.

Indeed, the cultural and genetic histories of peoples are closely associated. In recent years, there have been attempts to trace the origins of our languages, to find an ancestral stock language, as an indication of our past. The most widely accepted is Proto-Indo-European which provides a root for languages as dispersed as Sanskrit and Gaelic. This can be traced back to a single tribe, which some locate originally in Europe and southern Russia and others put in Anatolia. The latter is favoured by genetic evidence from populations in Turkey today. One suggestion is that the spread of their language was possible because they had a successful agricultural technology which was quickly adopted by other peoples. If this is true, human genes and a language spread on the back of a set of ideas.

When cultures clash, languages often suffer. As Steven Pinker says, languages disappear for the same reason that species disappear: by loss of habitat. Their speakers are assimilated into other cultures, or in some cases laws are passed to suppress a language and its culture. Currently there are around 6500 languages, of which up to 90% are threatened with extinction. There may be no more than 300 extant languages in a hundred years time. Michael Kraus notes that 150 (out of 250) North American languages are under threat. California alone had 50 native languages and those still used are spoken only by older citizens. The last speaker of the Northern Pomo language died in 1995 and, with her, that living tongue. Michael Kraus estimates that a language needs a minimum viable population of several hundred thousand people to persist.

A universal language may be convenient but, as with all losses of diversity, we lose so much more. A language is a view of the world and its grammar shows how its users see the relation between things. It often embodies valuable information, including indigenous knowledge about a habitat, what is good to eat, what has curative properties – important information when we consider the variety of tropical plants.

Languages can be reborn from a written record. Cornish died out 200 years ago but today there are 150 speakers. Electronic media and the speed of global communications have contributed to this demise, but also provide the means to preserve languages. Rather like frozen embryos or stored DNA, we have the prospect of preserving something of their diversity.

world leaders at the Rio Summit of 1992 with its Agenda 21 and the commitment to find sustainable strategies for all economic development. Even though the financing of the initiative remains unresolved, it does at least recognize the link between human welfare and environmental quality.

In the end, although the resolution of the situation will have to be ecological, the problem is far larger. Change will follow when we address the cultural, political, social and economic factors that underlie environmental degradation. We have to be careful not to blame local peoples for their predicament.

Box 9.5

AN ECOLOGY OF THE MIND?

Consciousness, the mental narrative inside our heads that gives us our sense of being, has been described by philosopher Daniel Dennett as the last great mystery and the one character that most people choose as our distinguishing human characteristic. The relationship between mind and body has taxed philosophers since we first sat down to think. Dennett now offers an explanation which will sound familiar to ecologists and which explains consciousness as an evolutionary phenomenon.

His 'multiple drafts theory' sees consciousness as the sum of higher mental events, entirely attributable to physical processes in the cortex (outer layer) of the brain. The brain functions rather like a sophisticated parallel computer. Different areas of the brain process information simultaneously, communicating with each other to detect salient features, editing and interpreting it in line with previous experience and sequencing it according to its significance. In doing so, the condition of the whole cortex is changed and the mental environment of higher functions, including decision-making, is altered. For example, an angry cortex colours judgements differently from a happy one.

Events in the mind propagate themselves by inducing other events elsewhere in the brain. The brain does not have a fixed version of any sequence of mental events, but several – multiple drafts – and reaches a judgement about appropriate action or behaviour according to whichever appears to be the best fit. Ideas or predispositions resident in the brain are part of the 'selective environment' of the cortex.

Consider the process of speaking. Most of the time, we begin a sentence with a desire to express a general meaning, but with no pre-planned word-sequence. According to Dennett, a choice of words offer themselves for selection at each point, but the context, the prevailing environment in the cortex, makes the selection of the best option of those available. We know what we think when we have said it.

Dennett sees parallels here with natural selection. Mental events which continue to demand activity enter the central workspace of the mind and become part of our stream of consciousness. If there are multiple drafts circulating in the brain, which one becomes operational? The one chosen by a natural selection, determined by the mental environment. Thoughts or ideas that have been useful in the past are used again in similar circumstances. They are propagated, or reinforced by their use, and perhaps evolve, changing in the light of experience. So the mind, itself a product of evolution by natural selection, operates using a form of natural selection.

The appearance of control and continuity follows from the stream of events passing through the mental workspace to become part of the narrative. What we take to be the self, the inner observer who turns out the mental light when we go to sleep, is an emergent property of this continuity.

It seems, then, that the complex systems of the mind exploit the advantages of diversity, of different approaches, and the evolution of behaviour (and ideas) that this enables. If Dennett is right, we could reasonably talk about an ecology of the mind, with the ecosystem of the cortex populated by a variety of ideas and predispositions, shaped by language and culture. Ideas survive according to their utility, the number of times we use them, and evolve by being used, exposed to the selection of experience. The baggage of attitudes, ideas, hopes and fears we each accumulate with age, overlaid on our genetic dispositions, is the overall (emergent) property we call a personality. And that makes each of us an individual.

Ecologists are naturally modest scientists, but it does seem that everything is ecology, as Wittgenstein nearly said.

The changes in Africa, as elsewhere, have followed from a meeting of cultures and an exchange of ideas. Sometimes the benefits of a new technology are so obvious that their transfer is very rapid, as with basic medicine and sanitation. Others, like contraception, may clash with traditions and may be resisted. The history of our species is of cultures meeting and changing in the process. We can perhaps discern the history of humanity from the traces these encounters have left behind in our languages (Box 9.4).

If nothing else, ecology teaches us that diversity has its own virtue, as insurance against drastic change. In the same way, different cultures represent a diversity of approaches to surviving in very different environments. Yet although economic wealth or technological expertise may ensure the dominance of one culture over another, it is no guarantee of long-term success.

We should, perhaps, learn that diversity is something to be cherished. Differences produce change: new species and new ideas. Uniformity of habitats or cultures or language make life uninteresting and ultimately unsustainable. What distinguishes peoples should be a source of wonder and enquiry rather than a flag to rally around or a badge to identify the enemy. In English we claim that variety is the spice of life, but perhaps the French say it better still with *vive la difference*!

9.6 LAST, ECOLOGY

Ecology has given us an insight into how complex systems are organized and evolution theory shows us how complex systems can arise. Ecology is the study of big systems with large numbers. It has had to develop new ways of seeing problems – ideas which are now being found to have a wider currency.

We also learn about ourselves and perhaps understand ourselves more fully. We are a product of natural selection and it seems that this may also explain that most human characteristic, consciousness (Box 9.5). Consciousness, at least to an ecologist, is a useful strategy for an omnivorous, non-specialist animal. It allows a species to adapt more readily, to learn from previous experience and previous generations. We shall need this ingenuity if we are not to cheat ourselves of our heritage.

SUMMARY

Mechanisms of homeostasis help to maintain the stability of the internal environment of an individual, as an adaptive response and the product of natural selection. It is difficult to demonstrate stability in ecological systems – populations, communities and ecosystems – or the mechanisms by which it is produced. Stability is a property of a system that resumes its original state after disturbance. It can be divided into inertial stability (the capacity to resist deflection) and adjustment stability (the capacity to return to the undisturbed state). The association between the diversity of a community and its stability is not simple; often the most diverse communities are the most fragile, and are typically a feature of long-standing environments which have suffered little disturbance. This may account for the global gradient of diversity toward the tropics, in which the oldest, least disturbed biomes around the equator show the highest diversity, both on land and in the sea. A range of other factors, including their benign climate and high productivity, may be important, and also the positive feedback of high species richness which creates further niche opportunities.

The fragile nature of the tropical communities has serious implications for economic activity and the well-being of peoples in these areas. We can trace gradients of economic activity that reflect the limited agricultural potential of these areas and which mean we need to rethink our exploitation of the tropical biomes.

FURTHER READING

Edwards, P.J., May, R.M. and Webb, N.R. (eds) (1993) *Large-scale Ecology and Conservation Ecology*, Blackwell, Oxford.

Jordan, C.F. (1995) *Conservation*, John Wiley & Sons, New York.

Lawton, J.H. and May, R.M. (eds) (1994) *Extinction Rates*, Oxford University Press, Oxford.

EXERCISES

1. Choose the correct answer for each of the following.

(i) In ecological communities functional redundancy implies...
   (a) that some species have no significant role in the organization of their community.
   (b) that there are no keystone species in that community.
   (c) that some species might be lost without significantly disturbing the organization of the community.
   (d) some function within the community does not determine its organization.

(ii) The 'connectance' between members of a community may determine the relationship between its diversity and stability. Between two species, connectance measures...
   (a) the frequency with which they exchange energy.
   (b) the scale of all interactions between species.
   (c) how dependent a consumer is upon its prey species.
   (d) how the abundance of one species determines the abundance of another.

2. Which of the following are said to help explain the great diversity of the tropical forest?
   (a) larger area
   (b) greater rainfall
   (c) greater sunlight
   (d) higher soil fertility.

3. Complete the following paragraph by inserting the appropriate words, using the list below (some words may be used more than once, some not at all).

   The latitudinal gradient of diversity in many plant and animal groups, increasing towards the __________, is the result of a number of interacting factors. For example, one suggestion is that the __________ structural diversity of tropical forests and coral reefs offer a wide variety of __________, well-defined and __________, which promotes specialization between neigbouring species __________ for space or __________. In turn the larger number of species accomodated creates __________ for other species that may derive resources from them or as a result of their activity. In effect, adding more __________ serves, through __________ feedback, to amplify the number of __________, at least while __________ and other resources are available. This has led some ecologists to suggest that this high diversity is relatively __________, when the loss of a __________ number of species can lead to __________ scale change.

   competing, energy, high, large, low, negative, niches, opportunities, partitioned, positive, resources, small, species, stable, temperate, tropics, unstable

4. Some ecologists have argued that the higher prevalence of sexual reproduction in the the tropics is an attempt to escape the attention of specialist predators and parasites. Why should sexual reproduction confer an advantage under such circumstances?

Tutorial or seminar topics

5. To meet the 'sustainability' criteria of Agenda 21 from the Rio summit, much effort is being devoted to methods of reducing the disruption to ecosystems by cutting nutrient and energy subsidies, two of the hallmarks of the agricultural revolution that has supported human population growth. Should we offer financial incentives to farmers to adopt such practices?

6. How might the use of language confer a selective advantage in a modern society?

7. Both the brain and the immune system may work by using something close to natural selection. What are the features of natural selection that make it a useful process (give it a selective advantage) in each case? What plays the part of the selective environment in each case?

Appendix A: Glossary

Box numbers are given where a full description is given in the main text.

Adaptation The collection of characters that enable an organism to survive. An organism may adapt in the short term, within its lifetime, by adjustments of its behaviour, physiology or development. A species may adapt over generations by changes to its genetic code, through natural selection (q.v.).

Age structures (age distribution: Box 3.1) The proportion of individuals in each age class of a population. A stable age distribution is dominated by the younger age classes and is characteristic of a rapidly growing population. Closer to the carrying capacity a stationary age distribution is approached and the distribution is far more even.

Allele (Box 1.2) The various forms of a gene found at a particular position (locus) on a chromosome, and which thereby code for a particular character.

Allelopathy The inhibition of one organism by another using chemical means. Some plants may inhibit the growth of others (especially those of the same species) by the chemical nature of their litter or by special secretions.

Allopatric speciation Speciation that occurs when populations become geographically isolated.

Anaerobic (anoxic) Of an environment that is devoid or has very low concentrations of oxygen. Anaerobic metabolism does not use oxygen; aerobic metabolism does.

Annual A plant species that completes its life cycle within a year.

Antibody A protein that is produced by some animals in response to a foreign substance (antigen) as part of their immune system.

Apomict (micro-species) A species which reproduces without fertilization.

Asexual reproduction Reproduction in which there is no swapping of genetic material between individuals.

Autotroph An organism that is able to obtain its own energy either from sunlight (cf. photoautotroph) or from chemicals (cf. chemoautotroph).

Benthic communities Those found on the bed of an aquatic ecosystem. Pelagic communities are found in the open water.

Biennial A plant that lives for two years, with growth and establishment in the first followed by maturation, fruiting and death in the second.

Bioaccumulation An increase in the concentration of a pollutant from water and diet by a living organism.

Bioconcentration An increase in the concentration of a pollutant from water by a living organism (cf. bioaccumulation, biomagnification).

Biodiversity (Box 9.1) Often taken to be simply the number of species (cf. species richness), but used by some ecologists to encompass biological variety at all levels, from the genetic diversity within a species to variations within ecosystems.

Biological control (Box 4.4) Pest control that makes use of a species' natural enemies.

Biological species concept Species defined as a group of individuals that breed exclusively with each other.

Biomagnification An increase in the concentration in the tissues of a living organism either from the soil (in plants) or from the diet (animals) (cf. bioaccumulation).

Biomanipulation Changing the structure of an ecological community to bring about long-term changes in ecosystem structure and processes. This technique is used especially with eutrophic freshwaters (q.v.).

Biomass The weight of living material.

Biome (Box 1.1) The characteristic vegetational community associated with a particular latitude or biogeographical region, such as tropical forests or temperate grasslands, and primarily defined by climate.

Biosphere The part of the Earth that supports living organisms; the global ecosystem.

Catchable stock Individuals that are sufficiently large to be harvested, such as fish being caught in nets with a certain mesh size.

Character displacement The divergence of trait between two or more species competing for the same resource.

Chemoautotroph Microorganisms that derive their energy from oxidizing inorganic compound and use carbon dioxide as their carbon source. These include the nitrifying bacteria.

Chromosome The main site of the genetic material in living cells, composed of DNA (q.v.) and protein.

Cohort (Box 3.2) A group of individuals of the same age in a population.

Commensalism An association between two species of benefit to one partner only but of no detriment to the other.

Community (Box 1.1) A collection of plants and animals that live in the same habitat and interact with each other.

Compartmentation The extent to which a community, or more specifically a food web, is divided into discrete units each with strong internal interactions between its members, but weaker links to others.

Compensation point Condition in plants when energy fixed in photosynthesis is balanced by energy used in respiration. The extinction of light down a water column means this is represented by a certain depth in aquatic ecosystems.

Competition The fight for a limited resource between individuals of the same species (intraspecific competition) or between different species (interspecific competition). Organisms can compete for food, water or space, and such competitive battles are one of the major driving forces behind natural selection.

Competitive exclusion principle Two species cannot coexist in the same niche at the same time in the same place. Of two species with identical resource requirements, one will eventually be ousted.

Connectance The number of direct interactions between different species in a food web expressed as a proportion of the total number of possible interactions. This is a measure of the complexity of the system.

Connectivity In landscape ecology, the extent to which habitat patches are connected by corridors.

Consumer A species that feeds on another species or organic matter – that is, herbivores, carnivores, parasites, detritivores and others.

Cultivar A distinct form of a cultivated species that has been produced as a result of artificial selection.

Decomposer An organism that feeds on dead organic matter. Detritivores feed on detritus (fragmented organic matter).

Density-dependence Populations that grow in a density-dependent way increase their numbers in relation to the density of the existing population: usually, the rate of population increase falls as the habitat becomes more and more crowded. Other organisms have density-independent growth and a population growth that has no relation to the existing population size.

Diversity (Box 9.1) Species diversity simply refers to the number of different species in a habitat, or species richness (q.v.). Other measurements of diversity include the relative proportions of different species, or equitability. Genetic diversity refers to the totality of the variation in the gene pool (q.v.).

DNA (Box 1.2) Deoxyribose nucleic acid, the molecule that carries the genetic code.

Ecological niche The totality of adaptations of a species to the biotic and abiotic factors in its environment; a species' role in its community. Niche breadth measures the range of a resource exploited by a species. Niche overlap is the range of a resource used by two species occurring together. Fundamental niche is the niche breadth that a species would use (occupy) in the absence of competitors or predators; realized niche is the niche breadth a species is restricted to in the presence of competing species.

Ecology The science that studies the relationship between living things and their environment.

Ecosystem (Box 1.1) A community of living organisms and its physical environment.

Ecotone The boundary between two adjoining communities.

Ectotherms Animals which, when active, attempt to regulate their body temperature by gaining heat from, or losing heat to, the environment, primarily by their behaviour (cf. endotherms).

Effective population size The size of the breeding population at a particular time and location, taking into account the breeding behaviour of the species, such as the number of males needed to inseminate the reproductive females.

Elaisome An oil-filled fleshy appendage of plant seeds which induces species of ants to collect the seed.

Emergent property A characteristic of a population or community that could not be detected by looking at individuals and that only emerges from seeing the system operating together. Important in deciding just how much the communities within an ecosystems are interdependent on each other.

Endemic species Native to and only found in an area.

Endotherms Animals that regulate their body temperature by generating heat internally through their metabolism (cf. ectotherms).

Epiphyte A plant that grows entirely on another plant. Typical epiphytes include the Bromeliads of tropical forests.

Eukaryotes Organisms whose cells have a discrete nucleus within a membrane. Eukaryotes include all algae, fungi and all multicellular plants and animals.

Eutrophication Changes in the structure of an ecological community as a result of nutrient enrichment, most often phosphate and nitrogen enrichment.

Evolution A gradual, directional change in the characters of an organism. The process by which one species might arise from another.

Extinction vortex A descriptive title given to the three factors (genetic, population and habitat) that can act in combination to drive a population extinct.

Fitness The adaptive success of an individual (or genotype q.v.) usually measured by its relative reproductive success compared to other members of the population. This is determined by the proportion of genes it contributes to the gene pool of the next generation.

Food web The trophic or feeding relations betwen different members of a community. A food chain consists of a series of trophic levels – primary producer, herbivore, carnivore.

Fossil record The entire catalogue of fossilized plants and animals that have been dated in geological time.

Founder effect The genetic differentiation of a small population isolated from a larger (parent) population. In this small gene pool, some alleles may have different frequencies compared with the main population, and these differences become more pronounced as the isolated population interbreeds.

Gaia Hypothesis (Prologue) The theory that the atmosphere of the Earth is purposefully maintained in its present form by the biological activity in the biosphere. Whilst the chemistry of the atmosphere is a product of living processes, many scientists question any implication of a united purpose, of organisms acting together to maintain the atmosphere.

Gamete The sex cells, eggs (ova) or sperm, which are haploid, having a single set of chromosomes. The fertilization of an ovum by a sperm produces a diploid zygote with two sets of chromosomes.

GCM (global circulation models: Box 8.5) Complex computer models used to describe and predict the behaviour of the climate under the influence of the land, the oceans and the chemistry of the atmosphere.

Gene pool/genome (Box 1.2) The totality of genetic material of a species.

Generation The individuals belonging to one age class within a population.

Genet (Box 3.1) An organism that grows from a fertilized egg and is therefore genetically distinct. In contrast, a ramet is an individual that has arisen by asexual reproduction.

Genetic drift The genetic differentiation of a small, isolated population from random changes (and independent of any selective pressure).

Genotype (Box 1.2) The genetic code of an individual, or sometimes, rather confusingly, all individuals that share that code (cf. phenotype).

Genotypic variation Variation between individuals that is wholly attributable to differences in their genetic code. Genetic diversity refers to the totality of the variation in the gene pool (qv.).

Genus A category between species and family in which a number of closely related species are grouped.

Geomorphology The study of landforms. Topography is the study of surface features of an area.

Greenhouse effect (Box 8.2) The capacity of the Earth's atmosphere to absorb long-wave (infra-red) radiation, while being largely transparent to the incoming short-wave radiation, resulting in global warming.

Habitat The place where an organism lives; sometimes used in the general sense to refer to the type of place in which it lives.

Herbivore An animal that feeds on plants.

Heterotroph An organism that obtains its energy from other species; a consumer.

Heterozygote (Box 1.2) An individual which, for a particular character, has different genes on each of its paired chromosomes. A homozygote has the same alleles (q.v.) on each chromosome.

Hierarchy theory The proposal that large and complex systems are self-organizing by virtue of the constraints and limitations imposed by the interactions between different levels. A level is defined by the spatial and temporal scale over which key processes (such as nutrient transfer) operate. Levels are nested, or formed into a hierarchy, in which process happen fastest at the smallest scales; change is slowest at the largest scales.

Homoplasy A structural resemblance arising out of the convergent evolution of two different species or groups.

Hybrid A crossbred individual derived from gametes from different populations, species or genera.

Inbreeding depression The reduced reproductive potential of a population in which individuals share much of the same genetic code. Conversely, outbreeding depression occurs when the genetic differences between individuals are too large to produce viable offspring.

Intermediate disturbance hypothesis The suggestion that maximum diversity occurs in communities subjected to disturbance that is frequent enough to prevent competitive species becoming dominant but not so frequent as to depress colonization rates by immigrant species.

K The carrying capacity (or maximum sustainable size) for a population in a limited environment.

***K*-selection** Applied to an organism whose reproductive and life history strategies are primarily adapted to life in an unchanging and limited environment where there is intense competition for resources (c.f. *r*-selection).

Keystone species An organism whose abundance or activity is central to maintaining the nature of a habitat.

Landscape ecology The study of ecological processes across several ecosystems united by a shared landform, climate and regime of disturbance. Adjacent landscapes under an equivalent climate or topographical area may be collected together as a region.

Life history strategy The allocation of time and resources that an organism makes between different stages of its life cycle so as to maximize its reproductive potential.

Life table A summary of the rates of mortality and survivorship of different age groups in a population.

Locus The site of an individual gene on a chromosome.

Maximum sustainable yield (MSY) The largest number of individuals or mass that can be harvested from a population year on year without damaging its reproductive potential.

Meiosis The division of the paired chromosomes necessary to form a gamete (q.v.). Mitosis divides the nucleus, but not the chromosomes, as a prelude to cell division.

Metapopulation A population of populations. A series of populations that may swap individuals with each other, but are usually divided into discrete patches.

Microcosm An ecosystem or part of an ecosystem isolated in the laboratory for experimental purposes (for example, an aquarium). A mesocosm is part of a real ecosystem partitioned for the same reason, sometimes called enclosures.

Mortality rate (*m*) The death rate.

Mutualism (Box 7.2) A form of symbiotic relationship between two species to their mutual benefit.

Mycorrhiza A close association between plant and fungus, where the plant root acts as a host to a network of fungal threads. The plant benefits by having an enlarged 'root' system to collect more phosphates and water and the fungus benefits from the supply of carbon from the plant.

Natality rate (*b*) The birth rate.

Natural selection (Box 1.3) The reproductive success of different individuals in the face of the constraints placed upon them by their environment. Less fit individuals fail to reproduce and their genes are lost from the gene pool (q.v.). The persistence of particular adaptive traits through the generations produces the evolutionary change that sometimes produces new species.

Neutral variation Variation within a population or species that confers no selective advantage or contributes to its speciation. It is the 'genetic noise' of a species.

Niche *See* Ecological niche.

Ombrotrophs Plants that obtain their nutrients from precipitation.

Omnivore An animal that feeds on plants and animals.

Parapatric speciation Speciation amongst adjacent populations that share a common boundary but that do not have overlapping ranges (q.v. sympatric speciation).

Parasite (Box 4.3) An organism that is metabolically dependent on another, at the expense of the host. A definitive host is one in which the parasite matures (and may reproduce sexually); an intermediate host serves as a vector in transmitting immature stages between definitive hosts.

Parasitoid Usually insects (Hymenoptera and Diptera) that are somewhere between a parasite and and predator – an egg laid inside the host feeds on and eventually kills the host.

Pathogen An organism that can cause disease.

Perennial A plant that lives for more than one year.

Permafrost The soil layer found at depth beneath tundra vegetation which remains frozen throughout the year.

Pest A species whose presence causes a nuisance and results in some economic cost.

Phenotype (Box 1.2) The physical characteristics of an individual, the product of the expression of its genetic code and its interaction with the organism's environment.

Phenotypic variation Variation between individuals that is wholly attributable to physiological or developmental (i.e. non-genetic) differences.

Pheromone A specially produced chemical used by one individual to alter the behaviour of another, usually of the same species, perhaps to attract a mate. These are especially important among social insects for communication and control.

Photoautotroph An organism that obtains its energy from sunlight through the process of photosynthesis (Box 5.1).

Phylogeny The evolutionary relationships and history of an organism.

Physiognomy The characteristic features of a plant community, reflecting their adaptation to life in that environment.

Polymorphism Variation within a species where individuals may take different forms.

Polyploidy Organisms that possess multiple copies of the entire genome.

Population (Box 3.1) Individuals of the same species living in a defined area at a defined time.

Predator An animal that kills another animal to feed.

Primary producer An autotroph (q.v.). A secondary producer derives its energy from a primary producer.

Prokaryotes Cells without a well-defined nucleus; the bacteria and the blue-green algae.

r The rate of change in a population per individual; r_{max} is the maximum possible rate of population growth per individual under ideal conditions.

***r*-selection** Organisms that are *r*-selected are primarily adapted to life in changeable, short-lived habitats, where rapid population growth is favoured (c.f. *K*-selection).

Recombination The swapping of fragments of genetic code between chromosomes during meiosis and which may therefore produce a new genotype (q.v.). Such variation is essential for the evolutionary process.

Recruitment The addition of new individuals to the catchable stock (q.v.) from growth or immigration.

Reproductive rate (R_0) The average number of offspring per individual in a given time. The net reproductive rate (R_N) is the average number alive per individual in the next generation, allowing for births, deaths and survivors.

Resource spectrum The range of a resource that is available to organisms in a habitat. Different species may use different parts of the spectrum – say, food items of different sizes.

Ruderal A species that is characteristic of disturbed and temporary habitats.

Rules of assembly The suggestion that there are certain ways in which a community can be configured, particularly regarding the structure of food webs. If there are rules of assembly, we expect to find similar food webs configurations in very different ecosystems.

Saprotrophs Organisms that derive their nutrients from dead and decaying organisms.

Sclerophyllous Used to describe individual plants or a plant community with small, tough and leathery leaves, adapted for drought. Typical of mediterranean-type communities.

Secondary plant metabolites Biologically active compounds produced by plants, which also have a defensive role against herbivores.

Species A collection of individuals able to breed with each other and produce viable (fertile) offspring. Note that this is a functional definition: the simple concept of a species as the basic unit of biological classification is now regarded as untenable, and is used largely as a convenient unit in taxonomy.

Species richness; species diversity (Box 9.1) The number of species in a habitat. Species equitability is the proportion of individuals in each species, sometimes used together to measure diversity (q.v.).

Stability The capacity of a system to resist change (inertial stability) or to return to its original position or original rate of change following a disturbance (elasticity or adjustment stability).

Succession A directional sequence of changes in the species composition of a community. A primary succession is where there has been no previous life on a site; a secondary succession re-establishes life on a site which may have some nutrients or organic matter from its previous occupants.

Symbiosis The association of two species living together. A variety of associations are possible. Mutualism (q.v.) describes where each species gains an advantage from the association.

Systematics A method of classifying organisms according to their evolutionary relationships.

Taxonomy The description, naming and classification of organisms.

Tolerator A species able to tolerate stressful conditions; or, in the case of a succession, one that can tolerate the presence of other species but that will eventually come to dominate the community.

Trophic level Position along a food chain, from primary producer to a sequence of consumers.

Variety A distinct form within a species which occurs naturally.

Appendix B: Answers to exercises

CHAPTER 1

1. The abundant oxygen of the present atmosphere would quickly combine with the organic molecules to prevent extensive polymerization.
2. Wherever extreme chemical conditions exist (hot vents, highly saline waters), some of which may recreate environments on the primeval Earth, or which have not changed since its formation. Microbial species adapted to these conditions may have undergone little adaptive change and are unlikely to have been replaced by competing colonizers. Deep oceanic trenches and the communities surrounding their volcanic vents have the means of deriving from chemical compounds. Their physiologies are also adapted to the absence of free oxygen, conditions to which all organisms would have been adapted before photosynthesis became important.
3. The DNA in the mitochondria has been inherited from the mother with the egg cell. It has not been subject to meiotic divisions or recombination, nor has it been combined with DNA from the father. Additionally the code in the mitochondria is very conservative; that is, it shows relatively few changes over time. Since all the code comes from the mother, the name Eve seems preferable to Adam.
4. A frequently encountered habitat (the host) provides more opportunities for larval parasites moving between hosts, and for parasites to become adapted to this type. A genotype that only occurs infrequently is less likely to be colonized by a parasite suitably adapted to its tissues and so, in turn, to induce selective pressures on future parasite generations. This is an example of frequency-dependent selection: the parasite adapted to the most common genotype is more likely to be succesful in finding a new host and then reproducing.
5. Any replicating system produces copies of itself. For the system to change, the copies cannot be perfect; there must be *variation* between them. Those *differences* may have no significance: unless they confer some advantage or disadvantage, the ability to make copies, any variation would be *neutral*. If, however, the *environment* favours some varieties over others, so that some are more likely to *reproduce*, then their *proportion* in the population will rise. The population will have changed and *evolution* will have occurred. Thus *replication*, *variation* and *selection* are the three elements needed for evolution to occur.

CHAPTER 2

1.

Kingdom	Plantae
Phylum	Trachaeophyta
Class	Angiospermae
Order	Ranales
Family	Ranunculaceae
Genus	*Clematis*
Species	*vitalba*

2.

Key derived from Table Ex.2.1

1. Wings present	got to 2
Wings absent	ant
2. More than 2 wings	go to 3
2 wings	fly
3. Hard wing cases present	beetle
Hard wing cases absent	go to 4
4. Large dusty wings present	butterfly
Large dusty wings absent	go to 5
5. Large hind legs for jumping	grasshopper
No large jumping legs	wasp

3. An ecological niche is the sum of a species adaptations to both the *biotic* and *abiotic* environment. It is defined by the use of *resources* such as food, water, light etc. and these can be plotted along a *resource axis*. The overall range of resources a population exploits is known as its *niche breadth*. This is *wide* for generalist and *narrow* for specialists. Where niches of two species overlap, the organisms concerned *compete* for the resources concerned. Complete niche overlap leads to competitive *exclusion* of one or other of the species.
4. (c).

CHAPTER 3

1. - Number of plant species – both woody and herbaceous. A well-developed woodland (and unmanaged woodland) will have a variety of trees and shrubs; and with age, its floral diversity will also rise.
   - Average girth of tree – a direct measure of the average age of the trees themselves.
   - Tree height – as above.
   - The distance between trees – as mature trees become large (and the canopy closes over) this will generally become larger.
   - The amount of light penetrating to the forest floor – a measure of how complete the tree canopy is and, indirectly, a measure of the competition for this resource.
2. (a) 90; (b) 45; (c) 45.

As the population increases, K is having a greater and greater effect on r, so that close to K the net increase in the population becomes very small.

A comparison of all the data (these three calculations, plus the example given in the text) shows that doubling either r or K serves to double the maximum rate of increase.

The maximum rate of increase given K can be calculated quickly as:

$$0.5 \times r \times N$$

when $N = 0.5K$.

3. (a), (b) and (d) are false; (c) is true.
4. The selective effect of this type of overfishing will be to favour those individuals able to spawn at an earlier age. The population will tend to be dominated by smaller fish which reproduce earlier. Such an effect is well documented for the cod of the North Sea.
5. The r-selected species tend to be small, rapidly reproducing and rapidly dispersed with an ability to exploit a range of resources. These are ideal characteristics to allow a species to exploit a variable environment, allowing their population to expand rapidly when resources are abundant. Many insect pests show such behaviour (but not all).

   K-selected species are the opposite: slow-growing with long generation times, poor powers of dispersal. Their slow population growth and often specialized habitat requirements mean they are readily threatened when their habitat is lost, and their numbers may take a long time to recover.

   Exceptions are plentiful. Some pest species are highly competitive, sometimes slow-growing species that simply crowd out other species (e.g. *Rhododendron*). Some rapidly reproducing species (e.g. the passenger pigeon) have been hunted out of existence in a relatively brief time. Many insects have become extinct because they have rapidly speciated and adapted to specific habitats – especially in isolated habitats – and have been lost with the habitat.

CHAPTER 4

1. Mutualism: (a).
   Commensalism: (b), though egret–insect = predation; (c).

Parasitism: (d); (g); (h).
Interspecific competition: (j); (k).
Intraspecific competition: (f).
Intraspecific competition/parasitism: (i).
Interspecific competiton/parasitism: (e).

2. A predator may respond in one of two ways when a prey species increases its abundance. It may show a functional response in which its *feeding rate* increases or it may show a *numerical* response where its population increases. Predator numbers can rise because of reduced *juvenile* mortality and by an increase in *reproductive* effort. Both have *delays* associated with them, but the response will be fastest with improved *juvenile* survival.

   Predators may *switch* from another prey species to reduce their *search costs*, but again there will be a delay. Generally predators tend to keep to *patches* where prey numbers are *high* and that means some prey may go *undiscovered*. Some *pest* species, such as aphids, outgrow their natural enemies by simply having a much higher rate of increase.
3. A generalist predator, able to switch to a variety of prey species, would be useful where the aim is to maintain a pest species, such as an insect, that undergoes periodic outbreaks and shows large fluctuations in numbers. A pest, like aphids, with highly seasonal population fluctuations may then be checked, at least to some extent, by a natural enemy that has overwintered by feeding on other insects.

   A specialist, closely tied to one prey species (such as many parasitoids), may be more appropriate where a pest is ever-present, at least in low numbers. Long-term control will mean that some pests will escape, but that should allow the coexistence of both prey and predator. Alternatively, we might use a specialist in a programme where the aim is the eradication of the pest in a very short space of time and when its reintroduction is not likely. In that case, we could allow the specialist predator eventually to go extinct (locally) as well.
4. It meets every one of them and was also cheap to produce and easy to release. It was grown on excised pads of *Opuntia* and dispersed in wax straws stuck on the spines of the cactus.

CHAPTER 5

1. Photosynthesis converts *sunlight* into *chemical* energy. It does this through the action of the light-absorbing pigment known as *chlorophyll* which is found in the *thylakoid* membranes within the chloroplast. The light reaction is the first part of photosynthesis and is used to fuel the *Calvin–Benson* cycle which is part of the dark reaction and which fixes *carbon* and manufactures *sugars*. Some plants, known as C_4 plants have evolved alternative mechanisms to help them fix carbon more efficiently to avoid *water* loss. Others have developed this still further using *Crassulacean acid* metabolism so that they do not have to open their *stomata* in the heat of the day.
2. (d) (Table 5.1. provides the answer).
3. Food web (below) derived from Table Ex.5.1.

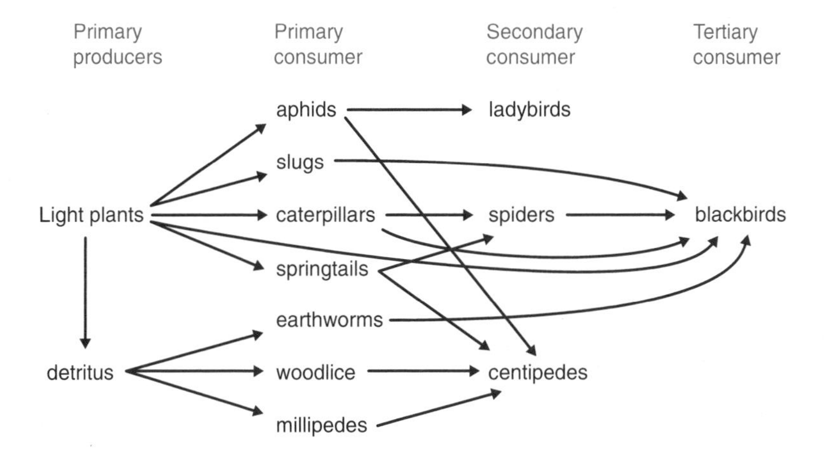

4. The crop is maize (*Zea mays*). It requires 368 g of water to produce 1 g dry weight. It is a more efficient water user as it is a C_4 plant (Box 5.2); this group of plants have evolved mechanisms which enable them to be more efficient at fixing carbon dioxide. They open their stomata for less time and therefore reduce their water loss.

CHAPTER 6

1. (b).
2. The oversupply of nutrients is known as *eutrophication*. It can affect both *aquatic* and *terrestrial* ecosystems, causing destabilization of plant and animal communities. Domestic and industrial *effluent* are the principal causes of eutrophication in aquatic environments. Agricultural sources are also a significant source of eutrophication, with nutrients coming from *effluent slurry* and *fertilizers*, from livestock and cultivated land respectively. The main nutrients responsible for eutrophication are *nitrogen* and *phosphorus*, with *phosphorus* being the most problematic because of its short supply in nature.
3. Firstly, remember that prevention is better than cure. Seek to limit the amount of oil reaching the shore and sea bed (press for quick salvage work in form of oil pumped off the ship and from containment areas on the sea surface). Use booms to contain spill; where these are breached, use materials (chalk, sawdust, wool) to absorb oil for collection or sinking.

   Divide the area under threat into sections, according to shore type, wave energy, wildlife and amenity value. Assign different clean-up contingency plans to each area (e.g. high energy beaches may be able to be left to recover naturally – provided wildlife value is not high). Wherever possible favour physical removal of oil.
   Do not forget post-spill work such as rehabilitation of injured wildlife and site monitoring.
4. Full restoration not always possible or practical. **Reclamation** seeks to restore the site to an acceptable state. Objectives here might be either ecological (habitat creation), social (amenity use) or economic (returning land to productivity – agriculture).

   Sites requiring **restoration** include natural and artificial degradation; in the latter case the substrate might be waste material from industrial processes (also landfill of refuse dumps).

   Rehabilitation is partial restoration which starts the process and allows nature to continue the development of the habitat. Replacement provides an alternative habitat – a possibility where the pre-impact site had limited ecological value (e.g. intensive agriculture).

   Note the importance of time-scale: a project can either revert to a previous habitat, or anticipate future change by restoring the site to how it might have changed without the impact. Also consider the techniques available – treatment of substrates (lack of and oversupply of nutrients), sowing of seeds, turves, planting of trees/shrubs, stabilizing land surfaces from sowing seeds. In some situations restoration is required to rescue a site, re-establishing it elsewhere. Restoration makes use of existing knowledge and can be a learning process too. It is important to consider the ultimate aim of the project, which should be to produce a sustainable site.
6. The growth rate of sycamore increases the nearer the trees are to alder. Alder is a nitrogen fixer and this shows that plants other than the nitrogen fixers themselves can benefit from the process. Consider the use of nitrogen fixers in restoration projects. Nitrogen can be a limiting factor in establishing vegetation on profoundly disturbed sites as it is mobile and easily lost through leaching. The introduction of nitrogen fixers such as legumes, with a dose of nitrogen fertilizer, can assist establishment of the early vegetation.

CHAPTER 7

1. Mediterranean communities are found in *five* regions throughout the world and occur in a band between *30* and *40* degrees either side of the *Equator*. They are restricted to the *western* edges of continents and have a climate with long, *dry* summers and cooler, *wet* winters. Plants of these communities are therefore *drought*-resistant and display *xerophytic* characteristics including thick, tough leaves with a reduced leaf area and a low growth habit. *Fire* is also an important factor in these communities and is a source of disturbance, occurring either naturally or artificially. *Human* activity in the form of *agriculture* helped to shape these communities in the past, but modern development in the form of industry and tourism can have seri-

ous negative impacts on mediterranean communities world-wide.

2. Global:
   - Five widely separated regions.
   - Distinct climatic conditions formed at the junction of tropical and temperate regions.

   Continental:
   - Geological and more recent differences between the five regions, having particularly contributed to the differences in colonizers in each region.
   - Different regional species pools for community development.

   Regional:
   - Often adjacent to very dry areas.
   - Need to adapt to extended dry periods in summer months.
   - Effects of coastal fogs and mists with moisture gradient inland.
   - Often abrupt changes in soil moisture and nutrient content when compared with surrounding areas.

   Local:
   - Often highly fractured landscapes (except for Australia and South Africa).
   - Fragmented features such as discrete valleys with different soils and microclimates.
   - Often a variety of geologies creating different soils and aridity regimes, favouring different plant communities.
   - Land use patterns associated with culture, local traditions and history.
3. Primary succession occurs when an area is colonized for the fist time, by pioneer species. Today this type of succession happens on profoundly disturbed sites and newly formed land such as mudflats and sand dunes. Secondary succession occurs in places where the vegetation has been lost either partially or totally, but may retain soil and possibly seeds. It is a recovery process associated with natural regeneration of communities.

   Disturbance is important as it is a cause of succession; it can be either natural or artificial. It can come from within (autogenic) or outside the community (allogenic): in autogenic succession the successional process is driven by the species within the community, whereas disturbance from natural causes or human activity results in allogenic succession. These two categories of disturbance can also be related to primary and secondary succession, with allogenic being linked with the former and autogenic with the latter.
4. (b). A keystone species is one whose numbers or activity determines the nature or structure of a community. Without them, the community would change to something very different.
5. Opportunist, short-lived species are good colonizers where conditions such as soil and climate are favourable. Colonists might either facilitate or inhibit other species colonizing the site. Some species take a long time to become established even though they may have been present during the initial colonization process – these are known as tolerators. Colonizers of hostile environments (such as arctic, alpine and tundra) may share some of the tolerators' features and need to be slow-growing, long-lived and persistent in order to survive.

CHAPTER 8

1. (i): (d); (ii): (d); (iii): (a); (iv): (a).
2. Tropical forest: (c), (e), (g), (h); temperate forests: (a), (b), (d); both: (f).
3. Figure Ex.8.1 (below).

   Figure Ex.8.1 is just one possible way to construct the hierarchy – you may have chosen to produce something slightly different. You might now like to use the figure overleaf as a template on which to establish the connections between various components. One possible answer is given in Figure Ex.8.2 on page 273.
4. A range of other factors limit photosynthesis, and carbon fixation. These include the availability of water and nutrients, sunlight and the build up of photosynthetic products in the tissues. Increased UV-B radiation, as ozone thinning progresses, will also affect some species. Note that the response differs between species, especially between C_3 and C_4 plants (Box 5.2), and that some plants may well be able to grow more rapidly. This means responses are difficult to predict, but it does seem unlikely that an increase in growth will mop up the excess carbon in the air predicted over the next 50 years.

CHAPTER 9

1. (i): (c); (ii): (d).

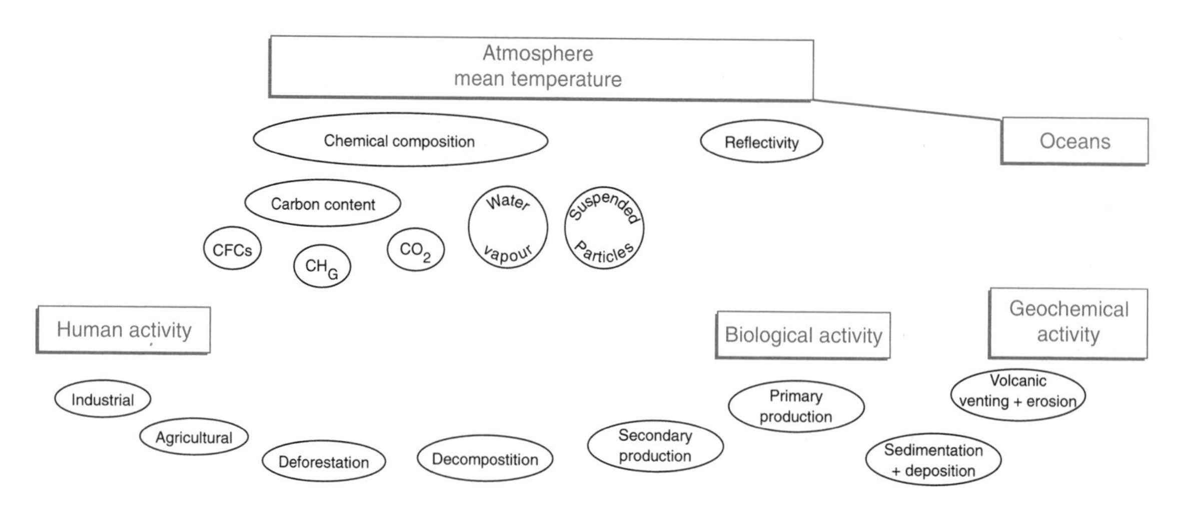

Figure Ex.8.1

2. (a), (c).
3. The latitudinal gradient of diversity in many plant and animal groups, increasing toward the *tropics*, is the result of a number of interacting factors. For example, one suggestion is that the *high* structural diversity of tropical forests and coral reefs offer a wide variety of *niches*, well-defined and *partitioned*, which promotes specialization between neigbouring species *competing* for space or *resources*. In turn the larger number of species accomodated creates *opportunities* for other species that may derive resources from them or as a result of their activity. In effect, adding more *species* serves, through *positive* feedback, to amplify the number of *niches*, at least while *energy* and other resources are available. This has led some ecologists to suggest that this high diversity is relatively *unstable*, when the loss of a *small* number of species can lead to *large* scale change.
4. Sexual reproduction generates variation between individuals, which, on occasion, may lead to sufficient change to cause a parasite or predator to fail to recognize its host or prey. Under these circumstances, sharing the most common genotype in the population makes an individual more likely to be the victim of a succesful attack since this genotype is probably the one to which the parasite or predator is adapted. Having a markedly different genotype may mean escape, at least in the short term. Under these circumstances, sexual reproduction and the variation it generates will be favoured.

   Parasitism and predation appear to have a greater intensity in tropical systems, perhaps because there greater diversity and specialization here.

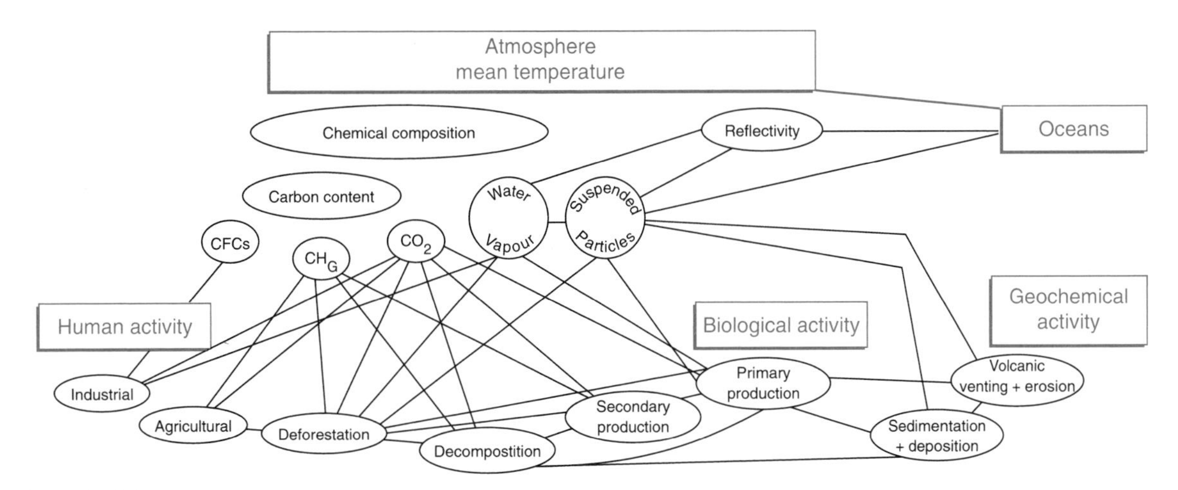
Atmosphere
mean temperature
Chemical composition
Reflectivity
Oceans
Carbon content
Water
Vapour
Suspended
Particles
CFCs
CH_G
CO_2
Human activity
Biological activity
Geochemical activity
Industrial
Primary production
Volcanic venting + erosion
Agricultural
Deforestation
Secondary production
Decompostition
Sedimentation + deposition

Figure Ex.8.2

Bibliography

PROLOGUE

Allen, J. (1991) *Biosphere 2. The Human Experiment*, Viking Press, New York.

Kasting, J.F., Toon, O.B. and Pollack, J.B. (1988) How the climate evolved on the terrestrial planets. *Scientific American* **258**, 90–97.

Kelly, K.W. (ed.) (1988) *The Home Planet*, Mir Publishers, Moscow.

Lovelock, J.E. (1987) *Gaia. A New Look at Life on Earth*, Oxford University Press, Oxford.

Sagan, D. (1990) *Biospheres. Metamorphosis of Planet Earth*, McGraw-Hill, New York.

White, F. (1987) *The Overview Effect, Space Exploration and Human Evolution*, Houghton Mifflin, Boston.

1 ORIGINS

Campbell, N.A. (1993) *Biology* (3rd edn), Benjamin/Cummings Redwood, California.

Dawkins, R. (1982) *The Extended Phenotype*, Oxford University Press, Oxford.

Dawkins, R. (1995) *River Out of Eden*, Weidenfeld & Nicholson, London.

Isaac, G. (1978) The food-sharing behaviour of protohuman hominids. *Scientific American* **238**, 90–108.

Leakey, R. (1995) *The Origin of Humankind*, Weidenfeld & Nicholson, London.

Leakey, R. and Lewin, R. (1992) *Origins Reconsidered*, Abacus, London.

Lewin, R. (1984) *Human Evolution*, Blackwell, Oxford.

Lincoln, R.J., Boxshall, G.A. and Clark, P.F. (1982) *A Dictionary of Ecology, Evolution and Systematics*, Cambridge University Press, Cambridge.

Mace, R. (1995) Why do we do what we do? *Trends in Ecology and Evolution* **10**, 4–5.

Margulis, L. (1992) Biodiversity: molecular biological domains, symbiosis and kingdom origins. *BioSystems* **27**, 39–51.

Parker, S. (1992) *The Dawn of Man*, Crescent, New York.

Trinkaus, E. and Howells, W.W. (1979) The neanderthals. *Scientific American* **241**, 118–133.

Walker, A. and Leakey, R.E.F. (1978) The hominids of east Turkana. *Scientific American* **239**, 54–66.

Wessells, N.K. and Hopson, J.L. (1988) *Biology*, Random House, New York.

Wood, B. (1992) Origin and evolution of the genus *Homo*. *Nature* **355**, 783–790.

2 SPECIES

Altschul, S.F. and Lipman, D.J. (1990) Equal animals. *Nature* **348**, 493–494.

Ammerman, A. and Cavalli-Sforza, L.L. (1971) Measuring the rate of spread of early farming in Europe. *Man* **6**, 674–688.

Antonovics, J., Bradshaw, A.D. and Turner, R.G. (1971) Heavy metal nutrient tolerance in plants. *Advances in Ecological Research* **7**, 1–85.

Begon, M., Harper, J.L. and Townsend, C.R. (1990) *Ecology: Individuals, Populations, and Communities* (2nd edn), Blackwell, Oxford.

Camp, W.H. (1951) Biosystematy. *Brittonia* **7**, 113–127.

Carr, S.M., Ballinger, S.W., Derr, J.N. *et al.* (1986) Mitochondrial DNA analysis of hybridisation between sympatric whitetailed and mule deer in West Texas. *Proceedings of the National Academy of Science of the United States of America* **83**, 9576–9580.

Chapman, G.P. (1992) *Grass Evolution and Domestication*, Cambridge University Press, Cambridge.

Dowling, T.E., Moritz, C. and Palmer, J.D. (1990) Nucleic acids II. Restriction site analysis. In *Molecular Systematics* (eds D.M. Hillis and C. Moritz), Sinauer Associates, Sunderland, Massachusetts.

Ehrendorfer, F. (1970) Evolutionary patterns and strategies in seed plants. *Taxon* **19**, 185–195.

Ford, E.B. (1940) Polymorphism and taxonomy. In *The New Systematics* (ed. J. Huxley), Clarendon Press, Oxford.

Grant, V. (1985) *Plant Speciation* (2nd edn), Columbia University Press, New York.

Gregg, K.B. (1983) Variation in floral fragrances and morphology: incipient speciation in *Cycnoches*? *Botanical Gazette* **144**, 566–576.

Grubb, P. (1977) The maintenance of species richness in plant communities: the importance of the regeneration niche. *Biological Reviews* **52**, 107–145.

Harlan, J.R. (1992) Origins and processes of domestication. In *Grass Evolution and Domestication* (ed. G.P. Chapman), Cambridge University Press, Cambridge.

Hewitt, G.M. (1988) Hybrid zones – natural laboratories for evolutionary studies. *Trends in Ecology and Evolution* **3**, 138–147.

Hull, D.L. (1976) Are species really individuals? *Systematics and Zoology* **25**, 174–191.

Hunt, P.F. (1986) The nomenclature and registration of orchid hybrids at specific and generic levels. In *Infraspecific Classification of Wild and Cultivated Plants*, (ed. B.T. Styles), The Systematics Association, special volume, Clarendon Press, Oxford.

Hutchinson, G.E. (1957) Concluding remarks. *Cold Spring Harbour Symposium on Quantitative Biology* **22**, 415–427.

Hutchinson, G.E. (1959) Homage to Santa Rosalia, or why there are so many kinds of animals? *American Naturalist* **93**, 145–159.

Jansen, R.K. and Palmer, J.D. (1988) Phylogenetic implications of chloroplast DNA restriction site variation in the Mutisieae (Asteraceae). *American Journal of Botany* **75**, 753–766.

Janzen, D.H. (1977) What are dandelions and aphids? *American Naturalist* **111**, 586–589.

Janzen, D.H. (1979) Reply. *American Naturalist* **114**, 156–157.

Janzen, D.H. (1979) How to be a fig. *Annual Review of Ecology and Systematics* **10**, 13–51.

Jeffrey, C. (1989) *Biological Nomenclature* (3rd edn), Edward Arnold, London.

Knobloch, I.W. (1972) Intergeneric hybridization in flowering plants. *Taxon* **21**, 97–103.

Larson, A. (1989) The relationship between speciation and morphological evolution. In *Speciation and its Consequences* (eds D. Otte and J.A. Endler), Sinauer Associates, Sunderland, Massachusetts.

Lewis, K.B. and John, B. (1968) The chromosomal basis of sex determination. *International Review of Cytology* **23**, 277–379.

Linnaeus, C. (1753) *Species Plantarum*, Laur Salvius, Holmiae.

Linnaeus, C. (1758) *Systema Naturae*, Laur Salvius, Holmiae.

Lupton, F.G.H. (1987) *Wheat Breeding: Its Scientific Basis*, Chapman & Hall, London.

MacNair, M.R. and Cumbes, Q. (1987) Evidence that arsenic tolerance in *Holcus lanatus* L. is caused by an altered phosphate uptake system. *The New Phytologist* **107**, 387–394.

Majerus, M., Amos, W. and Hurst, G. (1996) *Evolution. The Four Billion Year War*, Longman, Harlow.

Mallet, J. (1995) A species definition for the modern synthesis. *Trends in Ecology and Evolution* **10**, 294–299.

Margulis, L. and Schwartz, K.V. (1982) *Five Kingdoms*, W.H. Freeman, San Francisco.

Mayr, E. (1942) *Systematics and the Origin of Species from the Viewpoint of a Zoologist*, Columbia University Press, New York.

Minnelli, A. (1993) *Biological Systematics: The State of the Art*, Chapman & Hall, London.

Moritz, C., Dowling, T.E. and Brown, W.M. (1987) Evolution of animal mitochondrial DNA: relevance for population biology and systematics.

Annual Review of Ecology and Systematics **18**, 269–292.

Nylander, W. (1866) Circa novum in studio lichenum criterium chimicum. *Flora* **49**, 98.

Peakall, R., Beattie, A.J. and James, S.J. (1987) Pseudocopulation of an orchid by male ants: a test of two hypotheses accounting for the rarity of ant pollination. *Oecologia* **78**, 522–524.

Pianka, E.R. (1976) Competition and niche theory. In *Theoretical Ecology Principles and Applications* (ed. R.M. May), Blackwell, Oxford.

Pianka, E.R. (1988) *Evolutionary Ecology* (4th edn), Harper and Row, New York.

Poulton, E.B. (1904) Presidential address. *Proceedings of the Entomological Society of London*, pp. 73–116.

Richards, A.J. (1986) *Plant Breeding Systems*, Allen and Unwin, London.

Sarich, V.M., Schmid, C.W. and Marks, J. (1989) DNA hybridisation as a guide to phylogenies. A critical analysis. *Cladistics* **5**, 3–32.

Schilewen, U.K., Tautz, D. and Paabo, S. (1994) Sympatric speciation by monophyly of crater lake cichlids. *Science* **368**, 629–632.

Schluter, D. (1994) Experimental evidence that competition promotes divergence in adaptive radiation. *Science* **266**, 798–801.

Schoener, T.W. (1989) The ecological niche. In *Theoretical Ecology Principles and Applications* (ed. R.M. May), Blackwell, Oxford.

Smith, R.L. (1992) *Elements of Ecology* (3rd edn), Harper Collins, New York.

Stace, C.A. (1989) *Plant Taxonomy and Biosystematics* (2nd edn), Edward Arnold, London.

Sterck, A.A., Groenhart, M.C. and Mooren, J.F.A. (1983) Aspects of the ecology of some microspecies of *Taraxacum* in the Netherlands. *Acta Bot. Neerlandica* **32**, 385–415.

Traverse, A. (1988) Plant evolution dances to a different beat. Plant and animal evolutionary mechanisms compared. *Historical Biology* **1**, 277–301.

Uzzell, T. and Berger, L. (1975) Electrophoretic phenotypes of *Rana ridibunda*, *R. lessonae*, and their hybridogenetic associate, *Rana esculenta*. *Proceedings of the Academy of Natural Sciences Philadelphia* **127**, 13–23.

Vandermeer, J.H. (1972) Niche theory. *Annual Review of Ecology and Systematics* **3**, 107–132.

Wagner, D.F., Furnier, G.R., Saghai-Maroof, M.A. *et al.* (1987) Chloroplast DNA polymorphism in lodgepole and jack pines and their hybrids. *Proceedings of the National Academy of Science of the United States of America* **84**, 2097–2100.

Wagner, W.H. Jr (1970) Biosystematics and evolutionary noise. *Taxon* **19**, 146–151.

Walters, S.M. (1986) The name of the rose: a review of ideas on the European bias in angiosperm classification. *New Phytologist* **104**, 527–540.

Wayne, R.K. and Jenks, S.M. (1991) Mitochondrial DNA analysis implying extensive hybridzation of the endangered red wolf *Canis rufus*. *Science* **351**, 565–568

Weibes, J.T. (1979) Co-evolution of figs and their insect pollinators. *Annual Review of Ecology and Systematics* **10**, 1–12.

White, M.J.D. (1978) *Modes of Speciation*, Freeman, San Francisco.

Whittaker, R.H. and Margulis, L. (1978) Protist classification and the kingdoms of organisms. *BioSystematics* **10**, 3–18.

Young, J.P.W. (1988) The construction of protein and nucleic acid homologies. In *Prospects in Systematics* (ed. D.L. Hawksworth), Clarendon Press, Oxford.

Zuckerkandl, E. (1987) On the molecular evolutionary clock. *Journal of Molecular Evolution* **26**, 34–46.

3 POPULATIONS

Ashley, M.V., Melnick, D.J. and Western, D. (1990) Conservation genetics of the Black Rhinoceros (*Diceros bicornis*). 1. Evidence from the mitochondrial DNA of three populations. *Conservation Biology* **4**, 71–77.

Caughley, G., Dublin, H.T. and Parker, I. (1990) Projected decline of the African elephant. *Biological Conservation* **54**, 157–164.

Cohn, J.P. (1990) Elephants: remarkable and endangered. *BioScience* **40**, 10–14.

Cushing, D.H. (1981) *Fisheries Biology*, University of Wisconsin Press, Madison, Wisconsin.

Dinerstein, E. and McCracken, G.F. (1990) Endangered greater one-horned rhinoceros carry high levels of genetic variation. *Conservation Biology* **4**, 417–422.

Douglas-Hamilton, I. (1987) African elephants: population trends and their causes. *Oryx* **21**, 11–24.

Dublin, H.T., Sinclair, A.R.E. and McGladem, J. (1990) Elephants and fire as causes of multiple stable states in the Serengeti-Mara woodlands. *Journal of Animal Ecology* **59**, 147–164.

Haynes, G. (1991) *Mammoths, Mastodonts and Elephants: Biology, Behaviour and the Fossil Record*, Cambridge University Press, Cambridge.

Holmes, R. (1994) Biologists sort the lessons of fisheries collapse. *Science* **264**, 1252–1253.

Jedrzejewska, B., Okarma, H., Jedrzejewska, W. and Milkowski, L. (1994) Effects of exploitation and protection on forest structure, ungulate density and wolf predation in Bialowieza Primeval Forest, Poland. *Journal of Applied Ecology* **31**, 664–676.

4 INTERACTIONS

Addicot, J.F. (1986) Variation in the costs and benefits of mutualism: the interaction between yuccas and yucca moths. *Oecologia* **70**, 486–494.

Beattie, A.J., Turnbull, C., Knox, R.B. and Williams, E.G. (1984) Ant inhibition of pollen function: a possible reason why ant pollination is rare. *American Journal of Botany* **71**, 421–426.

Belsky, A.J. (1986) Does herbivory benefit plants? A review of the evidence. *American Naturalist* **127**, 870–892.

Blum, M.S., Rivier, L. and Plowman, T. (1981) Fate of cocaine in the lymantriid *Elonia noyesii* a predator of *Erythroxylum coca*. *Phytochemistry* **20**, 2499–2500.

Bond, W. and Slingsby, P. (1984) Collapse of an ant–plant mutualism: the Argentinian ant (*Iridomyrmex humilis*) and myrmecochorous proteaceae. *Ecology* **65**, 1031–1037.

Bronowski, J. (1973) *The Ascent of Man*, BBC Publications, London.

Case, T.J., Bolger, D.T. and Richman, A.D. (1992) Reptilian extinctions: the last ten thousand years. In *Conservation Biology* (eds P.L. Fiedler and S.K. Jain), Chapman & Hall, New York.

Cox, F.E.G. (ed.) (1982) *Modern Parasitology*, Blackwell, Oxford.

Crombie, A.C. (1946) Further experiments on insect competition. *Proceedings of the Royal Society of London, Series B* **133**, 76–109.

Dawkins, R. and Krebs, J.R. (1979) Arms race between and within species. *Proceedings of the Royal Society of London, Series B* **205**, 489–511.

Eisener, T. and Meinwald, J. (1966) Defensive secretions of arthropods. *Science* **153**, 1341–1350.

Eisener, T., Alsop, D., Hicks, K. and Meinwald, J. (1978) Defensive secretions of millipedes. In *Arthropod Venoms (Handbook of Experimental Pharmacology, 48)* (ed. S. Bettini), Springer, Berlin.

Feeny, P.P. (1970) Seasonal changes in oak leaf tannins and nutrients as a cause of spring feeding by winter moth caterpillars. *Ecology* **51**, 565–581.

Grime, J.P. (1979) *Plant Strategies and Vegetation Processes*, John Wiley, Chichester.

Grime, J.P., Hodgeson, J. and Hunt, R. (1988) *Comparative Plant Ecology: a Functional Approach to Common British Species*, Unwin Hyman, London.

Harborne, J.B. (1982) *Introduction to Ecological Biochemistry* (2nd edn), Academic Press, London.

Hassell, M.P. and Anderson, R.M. (1989) Predator–prey and host–pathogen interactions. In *Ecological Concepts* (ed. J.M. Cherrett), Blackwell, Oxford.

Janzen, D.H. (1966) Coevolution of mutualism between ants and acacias in Central America. *Evolution* **20**, 249–275.

Jones, D.A. (1973) Co-evolution and cyanogenesis In *Taxonomy and Ecology* (ed. V.H. Heywood), Academic Press, New York.

Krebs, J.R. and Davies, N.B. (1987) *An Introduction to Behavioural Ecology*, Blackwell, Oxford.

LaPage, G. (1963) *Animals Parasitic in Man*, Dover, New York.

Law, R. and Watkinson, A.R. (1989) Competition. In *Ecological Concepts* (ed. J.M. Cherrett), Blackwell, Oxford.

Monro, J. (1967) The exploitation and conservation of resources by populations of insects. *Journal of Animal Ecology* **36**, 531–547.

Muller, C.H. (1966) The role of chemical inhibition (allelopathy) in vegetational composition. *Bulletin of the Torrey Botanical Club* **93**, 332–351.

Owen, D.F. (1980) *Camouflage and Mimicry*, Oxford University Press, Oxford.

Payne, C.C. (1977) The ecology of brood parasitism in birds. *Annual Review of Ecology and Systematics* **8**, 1–28.

Peterson, S.C., Johnson, N.D. and LeGuyader, J.L. (1987) Defensive regurgitation of allelochemicals derived from cyanogenesis by eastern tent caterpillars. *Ecology* **68**, 1268–1272.

Pierkarski, G. (1962) *Medical Parasitology*, Bayer, Leverkusen.

Pimental, D. and Stone, F.A. (1968) Evolution and population ecology of parasite–host systems. *Canadian Entomologist* **100**, 655–662.

Potts, G.R. and Aebischer, N.J. (1989) Control of population size in birds: the grey partridge as a case study. In *Towards a More Exact Ecology* (eds P.J. Grubb and J.B. Whittaker), Blackwell, Oxford.

Pundir, Y.P.S. (1981) A note on the biological control of *Scurrula cordifolia* (Wall.) G. Don. by another mistletoe in the Sivalik Hills (India) *Weed Research* **21**, 233–234.

Rosenthal, G.A. and Janzen, D.H. (1979) (eds) *Herbivores: their Interaction with Secondary Plant Metabolites*, Academic Press, New York.

Silvertown, J.W. (1980) The evolutionary ecology of mast seeding in trees. *Biological Journal of the Linnean Society of London* **14**, 235–250.

Southwood, T.R.E. (1977) Habitat, the template for ecological strategies. *Journal of Animal Ecology* **46**, 337–365.

Southwood, T.R.E. (1988) Tactics, strategies and templates. *Oikos* **52**, 3–18.

Tilman, D., Mattson, M. and Langer, S. (1981) Competition and nutrient kinetics along a temperature gradient: an experimental test of a mechanistic approach to niche theory. *Limnology and Oceanography* **26**, 1020–1033.

White, G.G. (1981) Current status of prickly pear control by *Cactoblastis cactorum* in Queensland. In *Proceedings of the Fifth International Symposium for the Biological Control of Weeds*, Brisbane, 1980.

Wigglesworth, V.B. (1964) *The Life of Insects*, Weidenfeld & Nicholson, London.

5 SYSTEMS

Ammerman, A. and Cavalli-Sforza, L.L. (1971) Measuring the rate of spread of early farming in Europe. *Man* **6**, 674–688.

Andrews, S.M., Johnson, M.S. and Cooke, J.A. (1989) Distribution of trace element pollutants in a contaminated grassland ecosystem established on metalliferous fluorspar tailings. 1: Lead. *Environmental Pollution* **59**, 73–85.

Andrews, S.M., Johnson, M.S. and Cooke J.A. (1989) Distribution of trace element pollutants in a contaminated grassland ecosystem established on metalliferous fluorspar tailings. 2: Zinc. *Environmental Pollution* **59**, 241–252.

Balch, C.C. and Reid, J.T. (1976) The efficiency of conversion of feed energy and protein into animal products. In *Food Production and Nutrient Cycles* (eds A.N. Duckham, J.G.W. Jones and E.H. Roberts), North Holland Publishing Company, Amsterdam.

Begon, M., Harper, J.L. and Townsend, C.R. (1990) *Ecology: Individuals, Populations, and Communities* (2nd edn), Blackwell, Oxford.

Breznak, J.A. (1975) Symbiotic relationships between termites and their intestinal biota. In *Symbiosis* (eds D.H. Jennings and D.L. Lee), Symposium 29, Society for Experimental Biology, Cambridge University Press, Cambridge.

Briand, F. (1983) Environmental control of food web structure. *Ecology* **64**, 253–263.

Briand, F. and Cohen, J.E. (1984) Community food webs have scale-invariant structure. *Nature* **307**, 264–266.

Cohen, J.E. and Briand F (1984) Trophic links of community food webs. *Proceedings of the National Academy of Sciences of the United States of America* **81**, 4105–4109.

Colinvaux, P. (1980) *Why Big Fierce Animals are Rare*, Penguin Books, Harmondsworth, Middlesex.

Cooper, J.P. (ed.) (1975) *Photosynthesis and Productivity in Different Environments*, Cambridge University Press, Cambridge.

Cousins, S. (1987) The decline of the trophic level. *Trends in Ecology and Evolution* **2**, 312–316.

Duckham, A.N. (1976) Environmental constraints. In *Food Production and Nutrient Cycles* (eds A.N. Duckham, J.G.W. Jones and E.H. Roberts), North Holland Publishing Company, Amsterdam.

Evans, L.T. (1993) *Crop Evolution, Adaptation and Yield*, Cambridge University Press, Cambridge.

Fitter, A.H. and Hay, R.K.M. (1987) *Environmental Physiology of Plants* (2nd edn), Academic Press, London.

Golley, F.B. (1960) Energy dynamics of an old-field community. *Ecological Monographs* **30**, 187–200.

Goudie, A. (1993) *The Human Impact on the Natural Environment* (4th edn), Blackwell, Oxford.

Hay, R.K.M. and Walker, A.J. (1989) *An Introduction to the Physiology of Crop Yield*, Longman, Harlow.

Heal, O.W. and Maclean, S.F. (1975) Comparative productivity in ecosystems – secondary productivity. In *Unifying Concepts in Ecology* (eds W.H. Van Dobben and R.H. Lowe-McConnell), Dr W Junk, The Hague.

Humphreys, W.F. (1979) Production and respiration in animal populations. *Journal of Animal Ecology* **48**, 427–474.

Hungate, R.E. (1975) The rumen microbiological ecosystem. *Annual Review of Ecology and Systematics* **6**, 39–66.

Jones, H.G. (1983) *Plants and Microclimate*, Cambridge University Press, Cambridge.

Lawton, J.H. (1989) Food webs. In *Ecological Concepts* (ed. J.M. Cherrett), Blackwell, Oxford.

Morgan, R.P.C. (1995) *Soil Erosion and Conservation* (2nd edn), Longman, Harlow.

Nature Conservancy Council (1984) *Nature Conservation in Great Britain*, Nature Conservancy Council, Peterborough.

Odum, E.P. (1989) *Ecology and Our Endangered Life-support System*, Sinauer Associates, Sunderland, Massachusetts.

Odum, H.T. (1957) Trophic structure and productivity of Silver Springs, Florida. *Ecological Monographs* **27**, 55–112.

Osmond, C.B., Winter, K. and Zeigler, H. (1981) Functional significance of different pathways of CO_2 fixation in photosynthesis. In *Encyclopedia of Plant Physiology*, New Series (ed. A. Pirson and M.H. Zimmermann), Springer Verlag, Berlin.

Phillips, J. (1966) *Ecological Energetics*, Edward Arnold, London.

Pimm, S.L. (1982) *Food Webs*, Chapman & Hall, London.

Pimm, S.L. and Kitching, R.L. (1987) The determinants of food chain length. *Oikos* **50**, 302–307.

Pimm, S.L., Lawton, J.H. and Cohen, J.E. (1991) Food web patterns and their consequences. *Nature* **350**, 669–674.

Price, P.W. (1975) *Insect Ecology*, Wiley, New York.

Salisbury, F.B. and Ross, C. (1985) *Plant Physiology* (3rd edn), Wadsworth, Belmont, California.

Schultz, E.D. (1970) Der CO_2-Gaswechsel de Buche (*Fagus sylvatica* L.) in Abhangigkeit von den Klimafaktoren in Feiland. *Flora Jena* **159**, 177–232.

Schultz, E.D., Fuchs, M. and Fuchs, M.I. (1977) Spatial distribution of photosynthetic capacity and performance in a mountain spruce forest in northern Germany. I. Biomass distribution and daily CO_2 uptake in different crown layers. *Oecologia* **29**, 43–61.

Schultz, E.D., Fuchs, M. and Fuchs, M.I. (1977b) Spatial distribution of photosynthetic capacity and performance in a mountain spruce forest in northern Germany. III. The significance of the evergreen habit. *Oecologia* **30**, 239–248.

Simmons, I.G. (1989) *The Changing Face of the Earth: Culture and Environment* (4th edn), Blackwell, Oxford.

Slesser, M. (1975) Energy requirements of agriculture. In *Food, Agriculture and the Environment* (eds J. Lenihan and W.W. Fletcher), Blackie, Glasgow and London.

Smith, D.F. and Hill, D.M. (1975) Natural agricultural ecosystems. *Journal of Environmental Quality* **4**, 143–145.

Spedding, C.R.W. (1975) *Biology and Agricultural Systems*, Academic Press, London.

Tilman, D. (1982) *Resource Competition and Community Structure*, Princeton University Press, Princeton.

Tivy, J. (1975) Environmental impact of cultivation. In *Food, Agriculture and the Environment* (eds J. Lenihan and W.W. Fletcher), Blackie, Glasgow and London.

Tivy, J. (1990) *Agricultural Ecology*, Longman, Harlow.

Varley, G.C. (1970) The concept of energy flow applied to a woodland community. In *Animal Populations in Relation to their Food Resource* (ed. A. Watson), Blackwell, Oxford.

Tivy, J. and O'Hare, G. (1981) *Human Impact on the Ecosystem*, Oliver & Boyd, Glasgow.

Usher, M.B. and Thompson, D.B.A. (eds) (1988) *Ecological Change in the Uplands*, Blackwell, Oxford.

Whittaker, R.H. (1975) *Communities and Ecosystems* (2nd edn.), Macmillan, New York.

Whittaker R.H. and Likens, G.E. (1973) The primary production of the biosphere. *Human Ecology* **1**, 299–369.

Woodward, F.I. (1987) *Climate and Plant Distribution*, Cambridge University Press, Cambridge.

Zscheile, F.P. and Comar, C.L. (1941) Influence of preparative procedure on the purity of chlorophyll components as shown by absorption spectra. *Botanical Gazette* **102**, 463–481.

6 BALANCES

Baler, A., Brooks, R. and Reeves, R. (1988) Growing for gold ... and for copper ... and zinc. *New Scientist* **117**, 44–48.

Barry, R.G. and Chorley, R.J. (1970) *Atmosphere, Weather and Climate*, Holt, Rinehart and Winston, New York.

Bolin, B. and Cook, R.B. (eds) (1983) *The Major Biogeochemical Cycles and their Interactions*, John Wiley, Chichester.

Bradshaw, A.D. (1983) The reconstruction of ecosystems. *Journal of Applied Ecology* **20**, 1–17.

Bradshaw, A.D. (1984) Ecological principles and land reclamation practice. *Landscape Planning* **11**, 35–48.

Bradshaw, A.D. (1987) Restoration: an acid test for ecology. In *Restoration Ecology: A Synthetic Approach to Ecological Research* (eds W.R. Jordan, M.E. Gilpin and J.D. Aber), Cambridge University Press, Cambridge.

Bradshaw, A.D. (1989) Management problems arising from successional processes. In *Biological Habitat Reconstruction* (ed. G.P. Buckley), Belhaven Press, London.

Bradshaw, A.D. (1993) Understanding the fundamentals of succession. In *Primary Succession on Land* (eds J. Miles and D.H. Walton), Blackwell, Oxford.

Bradshaw, A.D. and Chadwick, M.J. (1980) *The Restoration of Land: the Ecology and Reclamation of Derelict and Degraded Land*, Blackwell, Oxford.

Bradshaw, A.D., Humphreys, R.N., Johnson, M.S. and Roberts, R.D. (1978) The restoration of vegetation on derelict land produced by industrial activity. In *The Breakdown and Restoration of Ecosystems* (eds M.W. Holdgate and M.J. Woodward), Plenum, New York.

Brown, C.M., McDonald-Brown, D.S. and Meers, J.L. (1974) Physiological aspects of inorganic nitrogen metabolism. *Advances in Microbial Physiology* **11**, 1–52.

Buckley, G.P. (ed.) (1989) *Biological Habitat Reconstruction*, Belhaven Press, London.

Bunce, R.G.H. and Jenkins, N.R. (1989) Land potential for habitat reconstruction in Britain. In *Biological Habitat Reconstruction* (ed. G.P. Buckley), Belhaven Press, London.

Department of the Environment (1994) *The Reclamation of Metalliferous Mining Sites*, Her Majesty's Stationary Office, London.

Down, G.S. and Morton, A.J. (1989) A case study of whole woodland transplanting. In *Biological Habitat Reconstruction* (ed. G.P. Buckley), Belhaven Press, London.

Earle, S. (1992) Assessing the damage one year later. *National Geographical* **179**, 122–134.

Edmondson, W.T. (1979) Lake Washington and the predictability of limnological events. *Archiv fur Hydrobiologie, Beiheft* **13**, 234–241.

Focht, D.D. and Verstraete, W. (1977) Biochemical ecology of nitrification and denitrification. *Advances in Microbial Ecology* **1**, 135–214.

Freedman, B. (1995) *Environmental Ecology: the Ecological Effects of Pollution, Disturbance, and Other Stresses* (2nd edn), Academic Press, San Diego.

Helliwell, D.R. (1989) Soil transfer as a means of moving grassland and marshland vegetation. In *Biological Habitat Reconstruction* (ed. G.P. Buckley), Belhaven Press, London.

Jordan, C.F. and Kline, J.R. (1972) Mineral cycling: some basic concepts and their application in a tropical rain forest. *Annual Review of Ecology and Systematics* **3**, 33–49.

Jordan, W.R., Gilpin, M.E. and Aber, J.D. (eds) (1987) *Restoration Ecology: A Synthetic Approach to Ecological Research*, Cambridge University Press, Cambridge.

Lee, I.W.Y. (1985) A review of vegetative slope stabilisation. *The Journal of the Hong Kong Institution of Engineers* (July) 9–22.

Lee, J. (1988) Acid rain. *Biological Sciences Review* **1**, 15–18.

Lehman, J.T. (1986) Control of eutrophication in Lake Washington. In *Ecological Knowledge and Environmental Problem-solving, Concepts and Case Studies*, National Academy Press, Washington DC.

McNaughton, S.J. (1988) Mineral nutrition and spatial concentrations of African ungulates. *Nature* **334**, 343–345.

Marrs, R.H. and Bradshaw, A.D. (1993) Primary succession on man-made wastes: the importance of resource acquisition. In *Primary Succession on Land* (eds J. Miles and D.H. Walton), Blackwell, Oxford.

Miller, R.M. (1987) Mycorrhizae and succession. In *Restoration Ecology: A Synthetic Approach to Ecological Research* (eds W.R. Jordan, M.E. Gilpin and J.D. Aber), Cambridge University Press, Cambridge.

Moss, B. (1988) *Ecology of Fresh Waters: Man and Medium* (2nd edn), Blackwell, Oxford.

Nature Conservancy Council (1990) *On Course Conservation: Managing Golf's Natural Heritage*, Nature Conservancy Council, Peterborough.

Odum, E.P. (1989) *Ecology and Our Endangered Life-support System*, Sinauer Associates, Sunderland, Massachusetts.

Piatt, J.F. and Lensink, C.J. (1989) *Exxon Valdez* oil spill. *Nature* **342**, 865–866.

Ricklefs, R.E. (1990) *Ecology* (3rd edn), Freeman, New York.

Ritchie, W. and O'Sullivan, M. (eds) (1994) *The environmental impact of the wreck of the Braer*, The Scottish Office, Edinburgh.

Schindler, D.W. (1977) Evolution of phosphorus limitation in lakes. *Science* **195**, 260–262.

Sprent, J.I. (1983) *The Biology of Nitrogen-fixing Organisms*, McGraw Hill, New York.

Van Donk, G. and Gulati, R.D. (1991) Ecological management of aquatic ecosystems: a complementary technique to reduce eutrophication-related perturbations. In *Terrestrial and Aquatic Ecosystems: Perturbation and Recovery* (ed. O. Ravera), Ellis Horwood, Chichester.

Ward, D.M., Atlas, R.M., Boehm, P.D. and Calder, J.A. (1980) Microbial degradation and chemical evolution from the Amoco spill. *Ambio* **9**, 277–283.

7 COMMUNITIES

Archibold, O.W. (1995) *Ecology of World Vegetation*, Chapman & Hall, London.

Begon. M., Harper, J.L. and Townsend, C.R. (1990) *Ecology: Individuals, Populations, and Communities* (2nd edn), Blackwell, Oxford.

Bottema, S., Entjes-Nieborg, G. and Van Zeist, W. (eds) (1990) *Man's Role in the Shaping of the Eastern Mediterranean Landscape*, Balkema, Rotterdam.

Boucher, C. and Moll, E.J. (1981) South African mediterranean shrub-lands. In *Mediterranean-type Shrublands* (eds F. Di Castri *et al.*), Elsevier, Amsterdam. pp 233–248.

Braun-Blanquet, J. (1932) *Plant Sociology: the Study of Plant Communities* (trans. G.D. Fuller and H.S. Conrad), McGraw Hill, New York.

Clements, F.E. (1936) Nature and structure of the climax. *Journal of Ecology* **24**, 252–284.

Cody, M.L. and Mooney, H.A. (1978) Convergence versus nonconvergence in Mediterranean-climate ecosystems. *Annual Review of Ecology and Systematics* **9**, 265–351.

Connell, J.H. (1978) Diversity in tropical rainforests and coral reefs. *Science* **199**, 1302–1310.

Connell, J.H. and Slatyer, R.O. (1977) Mechanisms of succession in natural communities and their role in community stability and organisation. *American Naturalist* **111**, 1119–1144.

Connell, J.H., Noble, I.R. and Slatyer, R.O. (1987) On the mechanisms of producing successional change. *Oikos* **50**, 136–137.

Diamond, J. and Case, T.J. (eds) (1986) *Community Ecology*, Harper and Row, New York.

Di Castri, F. (1981) Mediterranean-type shrublands of the world. In *Mediterranean-type Shrublands* (eds F. Di Castri *et al.*), Elsevier, Amsterdam, pp. 1–52.

Di Castri, F. and Mooney, H.A. (eds) (1973) *Mediterranean Type Ecosystems: Origin and Structure*, Springer, New York.

Di Castri, F. and Vitali-Di Castri, V. (1981) Soil fauna of mediterranean-climate regions. In *Mediterranean-type Shrublands* (eds F. Di Castri *et al.*), Elsevier, Amsterdam.

Di Castri, F., Goodall, D.W. and Specht, R.L. (1981) *Ecosystems of the World 11. Mediterranean-type Shrublands*, Elsevier, Amsterdam.

Gleason, H.A. (1926) The individualistic concept of the plant association. *Bulletin of the Torrey Botanical Club* **53**, 7–26.

Greuter, W. (1995) Extinctions in mediterranean areas. In *Extinction Rates* (eds J.H. Lawton and R.M. May), Oxford University Press, Oxford.

Grey, A.J., Crawley, M.J. and Edwards, P.J. (eds) (1987) *Colonization, Succession and Stability*, Blackwell, Oxford.

Hanes, T.L. (1981) California chaparral. In *Mediterranean-type Shrublands* (eds F. Di Castri *et al.*), Elsevier, Amsterdam, pp. 1339–1373.

Harper, J.L. (1977) *The Population Biology of Plants*, Academic Press, London.

Horn, H.S. (1976) Succession. In *Theoretical Ecology: Principles and Applications* (ed. R.M. May), Blackwell, Oxford.

Janzen, D.H. (1986) Keystone plant resources in the tropical forest. In *Conservation Biology: the Science of Scarcity and Diversity* (ed. M.E. Soulé), Sinauer Associates, Sunderland, Massachusetts.

Krebs, C.J. (1985) *Ecology, the Experimental Analysis of Distribution and Abundance* (3rd edn), Harper and Row, New York.

Le Houréau, H.N. (1981) Impact of man and his animals on Mediterranean vegetation. In *Mediterranean-type Shrublands* (eds F. Di Castri *et al.*), Elsevier, Amsterdam.

MacArthur, R.H. (1972) *Geographical Ecology: Patterns in the Distribution of Species*, Harper and Row, New York.

MacArthur, R.H. and Wilson, E.O. (1967) *The Theory of Island Biogeography*, Princeton University Press, Princeton.

Naveh, Z. (1990) Ancient man's impact on the Mediterranean landscape in Israel – ecological and evolutionary perspectives. In *Man's Role in the Shaping of the Eastern Mediterranean Landscape* (eds S. Bottema *et al.*), Balkema, Rotterdam.

Paine, R.T. (1969) A note on trophic complexity and community stability. *American Naturalist* **100**, 65–75.

Quezel, P. (1981) Floristic composition and phyto-sociological structure of sclerophyllous matorral around the Mediterranean. In *Mediterranean-type Shrublands* (eds F. Di Castri *et al.*), Elsevier, Amsterdam, pp. 107–122.

Rodwell, J.S. (ed.) (1989) *British Plant Communities*, Vol. 1, Woodlands and scrub, Cambridge University Press, Cambridge.

Rodwell, J.S. (ed.) (1991) *British Plant Communities, Vol. 2, Mires and Heaths*, Cambridge University Press, Cambridge.

Rodwell, J.S. (ed.) (1992) *British Plant Communities, Vol. 3, Grasslands and Montane Communities*, Cambridge University Press, Cambridge.

Rundel, P.W. (1981) The matorral zone of central Chile. In *Mediterranean-type Shrublands* (eds F. Di Castri *et al.*), Elsevier, Amsterdam, pp. 175–201.

Specht, R.L. (1981) Mallee ecosystems in Southern Australia. In *Mediterranean-type Shrublands* (eds F. Di Castri *et al.*), Elsevier, Amsterdam, pp. 203–231.

Terborgh, J. (1976) Island biogeography and conservation: strategy and limitations. *Science* **193**, 1029–1030.

Tomaselli, R. (1981) Main physiognomic types and geographic distribution of shrub systems related to mediterranean climates. In *Mediterranean-type Shrublands* (eds F. Di Castri *et al.*), Elsevier, Amsterdam.

Trabaud, L. (1981) Man and fire: impacts on Mediterranean vegetation. In *Mediterranean-type Shrublands* (eds F. Di Castri *et al.*), Elsevier, Amsterdam.

Van Andel, T.H. and Zanger, E. (1990) Landscape stability and destabilisation in the prehistory of Greece. In *Man's Role in the Shaping of the Eastern Mediterranean Landscape* (eds S. Bottema *et al.*), Balkema, Rotterdam.

8 SCALES

Adams, J.M., Faure, H., Faure-Denard, L. *et al.* (1991) Increases in terrestrial carbon storage from the last glacial maximum to the present. *Nature* **348**, 711–714.

Aspinall, R. and Matthews, K. (1994) Climate change impact on distribution and abundance of wildlife species: an analytical approach using GIS. *Environmental Pollution* **83**, 217–223.

Baker, J.T. and Allen, L.H. (1994) Assessment of the impact of rising carbon dioxide and other potential climate changes on vegetation. *Environmental Pollution* **83**, 223–235.

Bekkering, T.D. (1992) Using tropical forests to fix atmospheric carbon: the potential in theory and practice. *Ambio* **21**, 414–419.

Bierregaard, R.O., Lovejoy, T.E., Kapos, V. *et al.* (1992) The biological dynamics of tropical rain forest fragments. *BioScience* **42**, 859–866.

Clayton, K. (1995) The threat of global warming. In *Environmental Science for Environmental Management* (ed. T. O'Riordan), Longman, Harlow.

Forman, R.T.T. (1995) *Land Mosaics: the Ecology of Landscapes and Regions*, Cambridge University Press, Cambridge.

Houghton, J.T., Jenkins, G.J. and Ephraums, J.J. (eds) (1990) *Climate Change: the IPCC Scientific Assesment*, Cambridge University Press, Cambridge.

Innes, J.L. (1994) Climatic sensitivity of temperate forests. *Environmental Pollution* **83**, 237–243.

Leemans, R. and Zuidema, G. (1995) Evaluating changes in land cover and their importance for global change. *Trends in Ecology and Evolution* **10**, 76–81.

Lloyd, J. and Farquhar, G.D. (1996) The CO_2 dependence of photosynthesis, plant growth responses to elevated atmospheric CO_2 concentrations and their interaction with soil nutrient status. 1. General principles and forest ecosystems. *Functional Ecology* **10**, 4 –32.

Lloyd, D. and Jenkinson, D.S. (1995) The exchange of trace gases between land and atmosphere. *Trends in Ecology and Evolution* **10**, 2–4.

Mitchell, J.F.B., Johns, T.C., Gregory, J.M. and Tett, S.F.B. (1995) Climate response to increasing levels of greenhouse gases and sulphate aersols. *Nature* **376**, 501–504.

Norby, R.J., Gunderson, C.A., Wullschleger, S.D. *et al.* (1992) Productivity and compensatory responses of yellow-poplar trees in elevated CO_2. *Nature* **357**, 322–324.

Pastor, J. and Post, W.M. (1988) Response of northern forests to CO_2-induced climate change. *Nature* **334**, 55–58.

Sedjo, R.A. (1992) Temperate forests ecosystems in the global carbon cyle. *Ambio* **21**, 274–277.

Thompson, R.D. (1992) The changing atmosphere and its impact on Planet Earth. In *Environmental Issues in the 1990s* (eds A.W. Mannion and S.R. Bowlby), John Wiley, Chichester.

Turner, M.G., Romme, W.H., Gardner, R.H. *et al.* (1993) A revised concept of landscape equilibrium: disturbance and stability on scaled landscapes. *Landscape Ecology* **8**, 213–227.

Whittaker, R.H. (1975) *Communities and Ecosystems* (2nd edn), MacMillan, New York.

Wigley, T.M.L. and Raper, S.C.B. (1992) Implications for climate and sea level of revised IPCC emission scenarios. *Nature* **357**, 293–300.

Wigley, T.M.L., Richels, R. and Edmonds, J.A. (1996) Economic and environmental choices in the stabilization of atmospheric CO_2 concentrations. *Nature* **379**, 240–243.

9 CHECKS

Angel, M.V. (1994) Spatial distribution of marine organisms: patterns and processes. In *Large-scale Ecology and Conservation Biology* (eds P.J. Edwards, R.M. May and N.R. Webb), Blackwell, Oxford.

Auspurger, C.K. (1983) Offspring recruitment around tropical trees: changes in cohort distance with time. *Oikos* **40**, 189–196.

Biswas, M.R. (1994) Agriculture and environment: a review, 1972–1992. *Ambio* **23**, 192–197.

Briand, F. and Cohen, J.C. (1987) Environmental correlates of food chain length. *Science* **238**, 956–960.

Chapin, F.S., Schulze, E.-D. and Mooney, H.A. (1992) Biodiversity and ecosystem processes. *Trends in Ecology and Evolution* **7**, 107–108.

Clark, D.A. and Clark, D.B. (1984) Spacing dynamics of a tropical rain forest tree: evaluation of the Janzen–Connell model. *American Naturalist* **124**, 769–788.

Clarke, A. (1992) Is there a latitudinal diversity cline in the sea? *Trends in Ecology and Evolution* **7**, 286–287.

Collins, S.L. (1995) The measurement of stability in grasslands. *Trends in Ecology and Evolution* **10**, 95–96.

Coope, G.R. (1995) Insect faunas in ice age environments: why so little extinction? In *Extinction Rates* (eds J.H. Lawton and R.M. May), Oxford University Press, Oxford.

Crosson, P.R. and Rosenberg, N.J. (1989) Strategies for agriculture. *Scientific American* **261**, 128–135.

Currie, D.J. (1991) Energy and large-scale patterns of animal- and plant-species richness. *American Naturalist* **137**, 27–49.

Dennett, D.C. (1991) *Consciousness Explained*, Penguin, Harmondsworth.

Hannah, L., Lohse, D., Hutchinson, C. *et al.* (1994) A preliminary inventory of human disturbance of world ecosystems. *Ambio* **23**, 246–250.

Huston, M. (1993) Biological diversity, soils and economics. *Science* **262**, 1676–1679.

Jablonski, D. (1995) Extinctions in the fossil record. In *Extinction Rates* (eds J.H. Lawton and R.M. May), Oxford University Press, Oxford.

Jackson, J.B.C. (1991) Adaptation and diversity of reef corals, *BioScience* **41**, 475–482.

Jackson, J.B.C. (1995) Constancy and change of life in the sea. In *Extinction Rates* (eds J.H. Lawton and R.M. May), Oxford University Press, Oxford.

Jordan, C.F. (1995) *Conservation*, John Wiley, New York.

Kendall, H.W. and Pimentel, D. (1994) Constraints on the expansion of the global food supply. *Ambio* **23**, 198–205.

Keyfitz, N. (1989) The growing human population. *Scientific American* **261**, 71.

Labandeira, C.C. and Sepkoski, J.J. (1993) Insect diversity in the fossil record. *Science* **261**, 310–315.

Lawler, S.P. (1993) Species richness, species composition and population dynamics of protists in experimental microcosms. *Journal of Animal Ecology* **62**, 711–719.

Lubchenco, J., Olson, A.M., Brubaker, L.B. *et al.* (1991) The Sustainable Biosphere Initiative: an ecological research agenda. *Ecology* **72**, 371–412.

Mace, G.M. (1995) Classification of threatened species and its role in conservation. In *Extinction Rates* (eds J.H. Lawton and R.M. May), Oxford University Press, Oxford.

May, R.M. (1992) Bottoms up for the oceans. *Nature* **357**, 278–279.

May, R.M. (1992) How many species inhabit the Earth? *Scientific American* **261**, 18–24.

McNaughton, S.J. (1988) Diversity and stability. *Nature* **333**, 204–205.

Moore, J.C. and Hunt, H.W. (1988) Resource compartmentation and the stability of real ecosystems. *Nature* 333, 261–263.

Myers, N. (1993) Questions of mass extinction. *Biodiversity and Conservation* **2**, 2–17.

Naeem, S., Thompson, L.J., Lawler, S.P. *et al.* (1994) Declining biodiversity can alter the performance of ecosystems. *Nature* **368**, 734–737.

Norton, B.G. and Ulanowicz, R.E. (1992) Scale and biodiversity policy: hierarchical approach. *Ambio* **21**, 244–249.

Pinker, S. (1994) *The Language Instinct*, Morrow, New York.

Raymo, M.E. and Ruddiman, W.F. (1992) Tectonic forcing of the late Cenozoic climate. *Nature* **359**, 117–122.

Rex, M.J., Stuart, C.T., Hessler, R.R. *et al.* (1993) Global-scale latitudinal patterns of species diversity in the deep sea benthos. *Nature* **365**, 636–639.

Ricklefs, R.E. (1990) *Ecology* (3rd edn), Freeman, New York.

Solow, A.R. (1993) Measuring biological diversity. *Environmental Science and Technology* **27**, 25–26.

Steele, J.H. (1991) Marine functional diversity. *BioScience* **41**, 470–474.

Stiles, D. (1994) Tribals and trade: a strategy for cultural and ecological survival. *Ambio* **23**, 106–111.

Stocking, M. (1995) Soil erosion and land degradation. In *Environmental Science for Environmental Management* (ed. T. O'Riordan), Longman, Harlow.

Tilman, D. and Downing, J.A. (1994) Biodiversity and stability in grasslands. *Nature* **367**, 3633–3635.

Tuckwell, H.C. and Koziol, J.A. (1992) World population. *Nature* **359**, 200.

Vincent, A. and Clarke, A. (1995) Diversity in the marine environment. *Trends in Ecology and Evolution* **10**, 55–56.

Waltho, N. and Kolasa, J. (1994) Organization of instabilities in multispecies systems, a test of hierarchy theory. *Proceedings of the National Academy of Sciences, USA* **91**, 1682–1685.

Index

Page numbers appearing in *italics* refer to figures, those appearing in **bold** refer to tables.

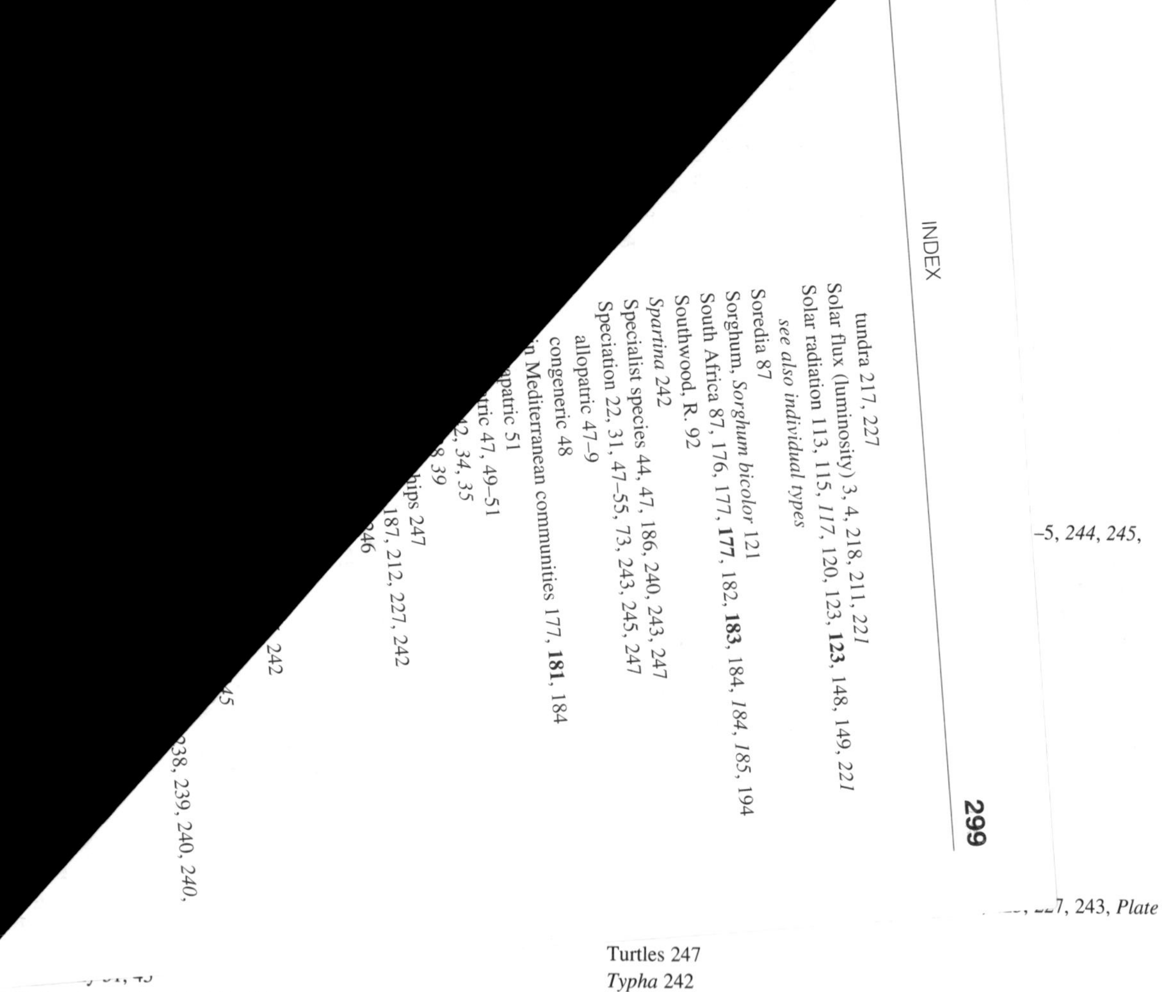